Also in the Variorum Collected Studies Series:

GEOFFREY V. SCAMMELL
Ships, Oceans and Empire: Studies in European Maritime and Colonial History, 1400–1750

RICHARD W. UNGER
Ships and Shipping in the North Sea and Atlantic, 1400–1800

PIETER EMMER
The Dutch in the Atlantic Economy, 1580–1880

JOHN GUY
Politics, Law and Counsel in Tudor and Early Stuart England

MARION LEATHERS KUNTZ
Venice, Myth and Utopian Thought in the 16th Century: Bodin, Postel and the Virgin of Venice

JOSEPH F. FREEDMAN
Philosophy and the Arts in Central Europe, 1500–1700: Teaching and Texts at Schools and Universities

BODO NISCHAN
Lutherans and Calvinists in the Age of Confessionalism

JOHN M. HEADLEY
Church, Empire and World: The Quest for Universal Order, 1520–1640

DAVID B. QUINN
European Approaches to North America, 1450–1640

JOHN W. O'MALLEY
Religious Culture in the 16th Century: Preaching, Rhetoric, Spirituality and Reform

SCOTT H. HENDRIX
Tradition and Authority in the Reformation

THOMAS F. MAYER
Cardinal Pole in European Context: A *via media* in the Reformation

VARIORUM COLLECTED STUDIES SERIES

Geography, Cartography and Nautical Science in the Renaissance

W.G.L. Randles

Geography, Cartography and Nautical Science in the Renaissance

The Impact of the Great Discoveries

Ashgate
VARIORUM

Aldershot · Burlington USA · Singapore · Sydney

Published in the Variorum Collected Studies Series by

Ashgate Publishing Limited
Gower House, Croft Road,
Aldershot, Hampshire GU11 3HR
Great Britain

Ashgate Publishing Company
131 Main Street
Burlington, Vermont 05401–5600
USA

Ashgate website: http://www.ashgate.com

ISBN 0–86078–836–9

British Library Cataloguing-in-Publication Data
Randles, W.G.L.
Geography, Cartography and Nautical Science in the Renaissance: The Impact of the Great Discoveries.
(Variorum Collected Studies Series: CS689).
1. Discoveries in geography–History. 2. Navigation–History. 3. Cartography–History. 4. Nautical charts–History.
I. Title.
910. 9' 4' 09024

US Library of Congress Card Number: 00–102911

The paper used in this publication meets the minimum requirements of the American National Standard for Information Sciences – Permanence of Paper for Printed Library Materials, ANSI Z39.48–1984. ∞ ™

Printed by St Edmundsbury Press, Bury St Edmunds, Suffolk

VARIORUM COLLECTED STUDIES SERIES CS689

CONTENTS

This volume contains xii + 354 pages

PUBLISHER'S NOTE

The articles in this volume, as in all others in the Variorum Collected Studies Series, have not been given a new, continuous pagination. In order to avoid confusion, and to facilitate their use where these same studies have been referred to elsewhere, the original pagination has been maintained wherever possible.

Each article has been given a Roman number in order of appearance, as listed in the Contents. This number is repeated on each page and is quoted in the index entries.

PREFACE

The studies brought together in this collection, written intermittently over some forty years, examine broadly a varied series of themes in the history of the Maritime Discoveries in the early modern period. Their ordering is intended to show a certain historical development: (1) The image of the world before the Discoveries; (2) the preparation and carrying out of the Discoveries; (3) the scientific consequences; (4) the more general cultural consequences.

Chapter I reviews classical hypotheses of the distribution of land and water over the surface of the globe before the Discoveries and describes the emergence of new geographical theories to which they gave rise. The study is a substantially re-worked version of an earlier short work: *De la terre plate au globe terrestre. une mutation épistémologique rapide, 1480–1520*, (Paris: Librairie Armand Colin, 1980)[1]. New classical texts were added to re-enforce the basic argument and some less probative vernacular ones eliminated.

With the exception of chapters XV and XIX, written much earlier in the 1950s and 1960s, many of these articles emerged out of papers researched on the encouragement of the late Professor Luís de Albuquerque for the numerous international conferences on the history of nautical science that he organized in the 1980s and 1990s at Lisbon, Sagres, Madeira and Porto Santo. May these articles stand as a tribute to his tireless energy and communicative enthusiasm.

Chapter II surveys the Atlantic in European culture through the Middle Ages to the Discoveries, tracing the changing image of it from that of an infinite unknown debarred space to one surveyed and measured in charts and texts as a result of systematic oceanic exploration. First published in Italian, it appears here in the English original together with added references and reproductions of charts and maps.

Chapter III re-appraises the alleged rôle of Prince Henry's nautical school at Sagres in the beginnings of Portuguese maritime expansion. Chap-

[1] Italian translation: Sansoni Editore, Florence, 1986; Spanish translation: Fondo de Cultura Económica, Mexico City, 1990; Portuguese translation: Gradiva, Lisbon, 1990.

ter IV, while not pretending to any fundamental originality, reviews for English-speaking readers the present state of Portuguese work on the history of the development of nautical astronomy in Portugal and notably of the contribution of Luís de Albuquerque; emphasized are the origins of the two main phases of nautical astronomy, first, for the northern hemisphere recourse to observations of the height of the Pole Star to establish the position of a ship, then, later for the southern hemisphere, the obtaining of latitude positions calculated from observations of the meridian altitude of the sun using tables giving the sun's declination throughout the year. Chapter V analyses the impact in Portugal, for the planning of the Discoveries, of the recovery in Italy of Ptolemy's *Geography* at the beginning of the 15th century. Chapter VI, deals with the problems of navigating the high latitudes and distant shores of the Atlantic with charts designed for navigation in the narrow waters Mediterranean.

Chapers VII and VIII trace out the navigational and cartographical problems encountered by the Portuguese navigators Diogo Cão and Bartolomeu Dias in their endeavours to round the African continent at the end of the 15th century. Although poorly documented on account of the state's policy of secrecy, the explorations of both navigators played a vastly more determinative rôle than Vasco da Gama's later more widely publicized voyage.

Chapter IX examines Columbus' plan to reach India by sailing across the Atlantic and the nature of the scientific arguments that he could not succeed, as marshalled by Portuguese and Spanish cosmographers; Chapter X analyses his ideas concerning the cartography of the Atlantic at the moment of his departure; Chapter XI reviews the varied assessments by Columbus' contemporaries of the accounts of his voyages.

Chapter XII returns to the problem discussed in chapter I regarding the clash, with the contradictory experience of the Discoveries, of traditional Aristotelian scholastic science relating to the distribution of land and water over the globe. New texts are presented showing early awareness of the problem and how it led to the emergence of the new specifically non-Aristotelian concept of the terraqueous globe.

Chapter XIII recounts the persistent endeavours by Portuguese and Spaniards to resolve the difficulties of calculating longitude in the period before the invention of the chronometer. The problem of calculating longitude accurately is shown to have hindered settling the demarcation of the Portuguese and Spanish spheres of sovereignty in the Pacific. The document *Apontamentos pera el Rey de Portugal que fez Lopo Homem...*, referred to in note 30, is printed in full in Portuguese as an appendix to an earlier version of the paper published in the *Boletim da Biblioteca da Universidade de Coimbra*, Vol. 39 (1984), pp. 143–160. The same paper, together with the appendix, is reprinted as Vol. 179 of the *Serie Separatas* published by the Centro de Estudos de História e Cartografia Antiga, (Coimbra, 1985), 21 pages.

Chapter XIV discusses the difficulties of navigating with a plane chart across the surface of a spherical globe. The solution of introducing corrections for the loxodromic curve, which the Portuguese mathematician Pedro Nunes had discovered in mid-16th century, is shown to have appeared too early to be applicable in practice by navigators through lack of instruments of sufficient precision.

Chapters XV and XVI deal with aspects of the geographical, scientific and cultural consequences of the Discoveries in European thought.

Chapter XVII, admittedly somewhat unrelated to the title of the book, deals with an important and little studied aspect of the Discoveries, the technological advances of the Portuguese in the use of artillery on ships and to defend land fortifications in the Indian Ocean in the early period of the Discoveries. It also examines Asian attempts to copy Portuguese technology, leading to an ensuing arms race.

Chapter XVIII analyses attempts in the 16th century Europe to classify non-European societies in terms of European political and social categories inherited from Antiquity.

Finally, chapter XIX traces out the origin of the distortion in the cartography of the interior of the African continent apparent on European printed maps of the 16th and 17th centuries; the entire empty inner space as profiled on Portuguese nautical charts being arbitrarily filled by European cartographers with Ethiopian towns and rivers copied from sketch maps of Ethiopia made by visiting European monks of the 15th century. The paper is a condensed version of an original longer study, later published in full with 17 reproductions of maps in *Stvdia*, No. 2 (Lisbon, 1958), pp. 103–163.

W.G.L. RANDLES

Mézin, France
January 2000

ACKNOWLEDGEMENTS

Acknowledgement for permission to reproduce the articles in this volume is gratefully made to the following: Walter de Gruyter, G.m.b.H and Co., Berlin, publisher of *The Classical Tradition and the Americas*, eds Wolfgand Haase and Meyer Reinhold (Berlin/New York, 1994) (I); Professor Guglielmo Cavallo, editor of *Cristoforo Colombo e l'apertura degli spazi*, Istituto Poligrafico e Zecca dello Stato, Libreria dello Stato (Rome, 1992), for permission to reprint the original English text (II); Dr Catherine Delano-Smith, editor of *Imago Mundi* (III, VI, IX and XIX); the editor of the *Journal of Navigation* (IV and XIV); Drs Francisco Contente Domingues and Luís Filipe Baretto of the University of Lisbon, co-editors of *A Abertura do Mundo. Estudios de História dos Descobrimentos Europeus* (Lisbon, 1986) (V): Dr Inácio Guerreiro, Vice-President of the *Instituto de Investigação Científica Tropical* in Lisbon, responsible for the publication of the series with green covers known as the *fascículos verdes* and for the proceedings of the *VIII Reunião Internacional de História da Náutica e da Hidrografia*, Viana do Castelo, Portugal, 19–24 September 1994 (Cascais, 1998) (VIII and XVII); Professors Luís António de Oliviera Ramos and José Marques of the University of Oporto, editors of the proceedings of the *Congresso Internacional Bartolomeu Dias e a sua época*, Oporto, 1989 (VIII); Dr Alberto Vieira, Director of the *Centro de Estudos de História do Atlântico* in Funchal, Madeira, editor of the *Actas do II Colóquio Internacional de História da Madeira* (X); Professor Domenico Ruocco of the *Dipartimento di Scienze dell'Antichità e del Medioevo* of the University of Genoa, editor of the *Atti conclusivi del XXVI Congresso Geografico Italiano, Genova, 4–9 Maggio 1992 (Genoa, 1993)* (XI); the Commissioner General of the *Comissão Nacional para as Comemorações dos Descobrimentos Portugueses* and the editor of the review *Mare Liberum* (XII and XVIII); Professor Michael Duffy, editor of the *Mariner's Mirror* (XIII); Madame Luísa Braz de Oliveira, Secretary of the *Bureau d'édition* of the *Série Actes de Colloque* of the *Centre Culturel Calouste Gulbenkian*, Paris (XVI).

I

Classical Models of World Geography and Their Transformation Following the Discovery of America

Contents

Grateful thanks are expressed to Professor Léon Bourdon of the Sorbonne who, with great patience, helped me over many years with his vast competence in elucidating the meaning of many of the Latin texts here quoted. I take however full responsibility for the English translations.

I. Introduction

In the late fifteenth century, on the eve of COLUMBUS' first voyage, four rival and greatly differing conceptions of the shape and ordering of the landmasses and their relation to the seas were alive and current in Europe, all of them inherited through the classical tradition.

The first goes back to Homer: The inhabited earth is represented as a single flat disc (the *oikoumene* surrounded by the Ocean).

The second is that of a spherical earth in which the relation of water to land was based on Aristotle's doctrine of the concentric ordering of the elements.

The third, of four small *oikoumenes* placed symmetrically on a spherical globe otherwise covered by water, can be traced to Crates of Mallos, a Greek who flourished c. 150 B. C.

The fourth, in which the oceans are separate lakes lying in hollows in a spherical earth, appears in Ptolemy's *Geography*, the text of which had only become known in Europe in the early fifteenth century, when a manuscript of it was brought to Florence from Constantinople in 1406.

These rival theories had, during the preparation of the great ocean voyages of discovery, varying and contrary influences on those who organized them. The outcome and the experience of the voyages were to reorder and transform the theories to produce the geographical image of the globe in the modern period.

In order to understand the context in which these geographical theories appeared and how they claimed greater or lesser attention among scholars in successive periods, we must go back and retrace the history of their emergence from the earliest knowledge we have of them among the Greeks.

II. Early Antiquity. Homer, Pythagoras and Aristotle

For Homer (VIIIth century B. C.) the earth was a plane circular surface entirely surrounded by the Ocean which flowed round it like a river.[1]

Geminos (c. 70 B. C.) tells us that,

> "Homer and the poets of old, almost without exception, imagined the earth as a flat surface extending to the sky; the Ocean lay all around as a circle acting as the horizon ..."[2]

The idea of the earth as a flat disc is found among the Ionians (VIth – Vth century B. C.) and such was the conception held by Anaximander and Hecataeus (Vth century B. C.)[3]

This early Greek representation of the inhabited earth, founded on the immediate evidence of the senses, was to persist long after the subsequent appearance of the notion of a spherical earth based on hypothetico-deductive reasoning from the observation of celestial phenomena.

What we shall call the Homeric representation of the earth, referred to by the Greeks as the *oikoumene*, was taken up by the Romans who translated the Greek term by *orbis terrarum*.[4]

The division of the *oikoumene* or *orbis terrarum*, a single landmass, into three "parts" (not "continents"): Asia, Europe and Africa is attested among the Greeks by Dicaearchus (fl. c. 326 – 296 B. C.), Eratosthenes (c. 275 – 194)[5] and Geminos (c. 70 B. C.)[6] and among the Romans by Sallust (86 – 35 B. C.)[7] and by Pomponius Mela (fl. 40 – 44 A. D.).[8] Hyginus (1st century A. D.) writes:

[1] T. H. MARTIN, "Mémoire sur la cosmographie grecque à l'époque d'Homère et d'Hésiode", in: *Mémoires de l'Institut National de France, Académie des Inscriptions et Belles Lettres*, Tome 28, Paris, 1874, pp. 212 – 235. DICKS notes, that "... the earth (according to Homer) the shape of which is unspecified, is apparently surrounded by the Ocean river, although this is nowhere stated in so many words by Homer". Cf. D. R. DICKS, *Early Greek Astronomy*, London, 1970, p. 29. The reference to Homer is *Iliad*, XVIII, 483 – 489, Loeb edition, Greek text and English translation by A. T. MURRAY, London, 1925, pp. 322 – 24.

[2] Geminos, *Introduction aux Phénomènes*, text and French translation by G. AUJAC, Paris, 1975, Bk. XVI, 27 (p. 81).

[3] Cf. F. GISINGER, art. "Oikumene" in: PAULY's *Realencyclopädie der classischen Altertumswissenschaft*, Vol. XVII, Pt. 2 (1937), cols. 2128 – 29.

[4] J. VOGT, *Orbis Romanus. Zur Terminologie des römischen Imperialismus, Philosophie und Geschichte*, Vol. XXII, Tübingen, 1929, p. 6 (= IDEM, *Vom Reichsgedanken der Römer*, Leipzig, 1942, p. 171 – 72 = IDEM, *Orbis. Ausgewählte Schriften zur Geschichte des Altertums*, ed. by F. TAEGER and K. CHRIST, Freiburg – Basel – Wien, 1960, p. 152).

[5] F. GISINGER, *art. cit.*, col. 2170.

[6] Geminos, *op. cit.*, *loc. cit.*

[7] Sallust, *Bellum Jugurthinum*, ed. and translated by ALFRED ERNOUT, Paris, 1958, chap. 17, p. 151.

[8] Pomponius Mela, *De Chorographia libri tres*, ed. GUNNAR RANSTRAND, *Studia Graeca et Latina Gothoburgensia*, Vol. XXVIII, Göteborg, 1971, Chap. I (p. 4).

"... the whole region lying between the Arctic (circle) and the tropic (of Cancer) is divided into three parts: Europe, Asia and Africa".[9]

The Christian Middle Ages were to perpetuate the early Greek concept of a flat inhabited earth (*oikoumene/orbis terrarum*; see § 4) while simultaneously accepting, within the context of astronomy, the notion of the earth as a sphere.

Pythagoras (fl. 530 B. C.) was probably the first to postulate the sphericity of the earth.[10] Three criteria seem to have led him to this opinion: (a) the circular shadow of the earth during eclipses, (b) the hypothesis of a spherical heaven based on the observation of the movement of stars, (c) mathematico-aesthetical considerations grounded in the belief that the sphere was the most beautiful of the solid geometrical figures. It was probably the last criterion which carried the most weight.[11]

Aristotle (384 – 322 B. C.) took over from the Pythagoreans their obsession with the perfect sphericity of the earth and he made it a key feature of his physics. He added a further principle which was that weight determined the earth's sphericity, all the latter's parts ordering themselves as closely as possible to the centre of the universe with which the centre of earth coincided.

For Aristotle, the earth was only one of the four elements, the others being the spheres of the water, the air and the fire all arranged in concentric fashion[12] according to the order of their decreasing densities, around the universe's centre.

For each element to be in its proper place undergoing no violence (and violence according to Aristotle cannot endure), not only must the centre of each sphere coincide with the centre of the universe, but the outer surface of the sphere of each element must also be tightly lodged[13] within the inner surface of the element coming immediately after it in accordance with the hierarchy of their respective densities. The application of this rigorous principle produced the result that the earth could not be said to be in its proper place unless surrounded and entirely covered by water.[14] The existence of dry land was thereby rendered impossible.[15]

9 Hyginus, *L'Astronomie*, Latin text and French translation by André Le Bœuffle, Paris, 1983, p. 12.

10 Cf. Sir Thomas Heath, *Aristarchus of Samos*, Oxford, 1913, p. 48. Heidel only finds this concept for the first time in Plato, but we prefer Heath's opinion. Cf. W. A. Heidel, *The Frame of the Ancient Greek Maps. With a discussion of the sphericity of the earth*, *American Geographical Society Research Series* No 20, New York, 1937, pp. 81 – 84.

11 Sir Thomas Heath, *op cit.*, *loc. cit.*

12 Aristotle, *De Caelo* (*On the Heavens*), Bk. II, Chap. 4, 287a English translation by W. K. C. Guthrie, Loeb edition, London, 1971; *Physics*, Bk.IV, Chap. 5, 212b, English translation by P. H. Wicksteed and F. M. Cornford, Loeb edition, London, 1970; *Meteorologica*, Bk. II, Chap. 2, 354b, English translation by H. D. P. Lee, Loeb edition, London, 1978.

13 The notion of a vacuum is unthinkable in Aristotle's system.

14 Aristotle, *De Caelo*, Bk. IV, Chap. 3, 310b.

15 Pierre Duhem, *Le système du monde*, Vol. 9, Paris, 1958, pp. 88 – 91.

As DUHEM noted, none of Aristotle's Greek commentators attempted to resolve the contradiction between the Stagirite's implacable logic and observed experience.[16] It remained for the European Middle Ages to attack the problem, and then only in the 13th century.

In one of his works, the *Meteorologica*, Aristotle expounded a theory of the relation between land and water entirely different from that of his concentric ordering of the elements. Following Parmenides (c. 515 – c. 450 B.C.), he divided the earth into zones:[17] two habitable temperate zones between the tropics and the Arctic circle, two uninhabitable frigid zones between the Arctic and Antarctic circles and the poles and an uninhabitable torrid zone lying between the two tropics.

The habitable regions Aristotle described as "drum shaped". Of the southern one he said nothing, but of the northern one, corresponding to the *oikoumene*, he admitted that between India and the Pillars of Hercules (i.e. eastwards from the Far East to Gibraltar), the Ocean severed it, preventing it from forming a continuous belt around the globe.[18] In a passage in another work, *On the Heavens*, Aristotle suggested that this gap was very small and that the Atlantic Ocean was probably quite narrow:

> "... those who imagine that the region around the Pillars of Hercules joins on to the regions of India, [...] are not, it would seem, suggesting anything utterly incredible".[19]

In the Middle Ages, with one exception, Roger Bacon (c. 1219 – 1292), the great majority of scholars who commented on Aristotle's works ignored the contradiction between the above passage and Aristotle's theory of the concentric ordering of the elements, preferring to direct all their efforts to interpreting the latter, in doing which they were led, as we shall see, to postulating an immensely wide Atlantic ocean. In spite of their preponderant position, especially in the universities, it was Bacon's insistence on a narrow Atlantic, copied by Pierre d'Ailly (1350 – 1420) in his *Imago Mundi*, that came to COLUMBUS' knowledge and provided him with the intellectual authority for pressing his project.

Aristotle's Greek commentators were to misinterpret the Stagirite's doctrine of the transmutation of the elements and passed on to the Middle Ages a notion that came to be regarded as authentically Aristotelian. While Aristotle had stated that the volumes of the elements were inversely proportional to their densities, he had submitted as a hypothetical example that a given quantity of water could, for instance, produce as much as ten times that volume of air. Olympiodorus (6th century A.D.) went further than the other

[16] IDEM, op. cit., p. 91.

[17] On Parmenides cf. KARLHANS ABEL, art. "Zone" in: PAULY's *Realencyclopädie*, Supplement Vol. XIV (1974), col. 1000.

[18] Aristotle, *Meteorologica*, Bk. II, Chap. 5, 362b.

[19] Aristotle, *De Caelo*, II, 14, 298a.

commentators and, taking Aristotle's hypothetical case literally, wrote of a continuous proportionality of the volumes of the elements.

> "There is a rule", he wrote, "that the totality is to the totality as the parts are to another part. We see that a *kotylos* [Greek measure] of water is changed into ten *kotyloi* of air ..."[20]

This erroneous doctrine, leading to the belief that the volume of the Ocean was ten times that of the earth was, as we shall see, to constitute a hindrance to projects of transatlantic navigation and was later, on the evidence of the Discoveries, to be proved groundless by COPERNICUS.

III. The Cratesian theory

The Greeks did however seek to resolve the contradiction between the Homeric representation of the inhabited earth as a flat disc and the Pythagorean one of a spherical earth, about the inhabited nature of whose surface the Pythagoreans had nothing to say.

Crates of Mallos (fl. c. 150 B. C.) is usually credited with being the first to postulate the existence of four small landmasses situated symmetrically on a sphere over the greater part of which flowed two wide ocean rivers at right angles to one another.[21] GISINGER claims that the Cratesian theory can be traced much farther back, to Eudoxus of Cnidus (c. 408 – 355 B. C.).[22]

In the Roman period the Cratesian theory was taken up and made well-known by Cicero in his *Dream of Scipio*[23] and by the Neoplatonist Macrobius (*Commentary on the Dream of Scipio*) in the late IVth or early Vth century and by Martianus Capella (*The Marriage of Philology and Mercury*) before the middle of the Vth century A. D.[24]

Cicero imagines Scipio in his dream hearing his grandfather speaking:

> "You see, Scipio, that the inhabited portions of the earth are widely separated and narrow, and that vast spaces lie between these inhabited spots (*maculae*) as we might call them; the earth's inhabitants are so cut

[20] Olympiodorus, *In Meteora Aristotelis*, Venice, 1551, f. 5 (r.) = Ed. by G. STÜVE, *Commentaria in Aristotelem Graeca*, Vol. XII, Part 2, Berlin, 1900, p. 18, 5 – 7. Cf. also P. DUHEM, *op. cit.*, Vol. 9, pp. 91 – 96.

[21] HANS JOACHIM METTE, *Sphairopoiia. Untersuchungen zur Kosmologie des Krates von Pergamon*, Munich, 1936, and GISINGER, *art. cit.*, cols. 2142 – 2144.

[22] F. GISINGER, *art. cit.*, cols. 2132 – 2143.

[23] English translation in: W. H. STAHL (ed.), *Macrobius' Commentary on the Dream of Scipio*, New York, 1952, pp. 74 – 75.

[24] Martianus Capella, *The Marriage of Philology and Mercury*, in: W. H. STAHL, *Martianus Capella and the Seven Liberal Arts*, New York, 1977, Vol. II, § 603, pp. 225 – 26; Macrobius, *Commentary on the Dream of Scipio*, trans. and notes by W. H. STAHL, New York, 1952, pp. 200 – 201, and pp. 214 – 215.

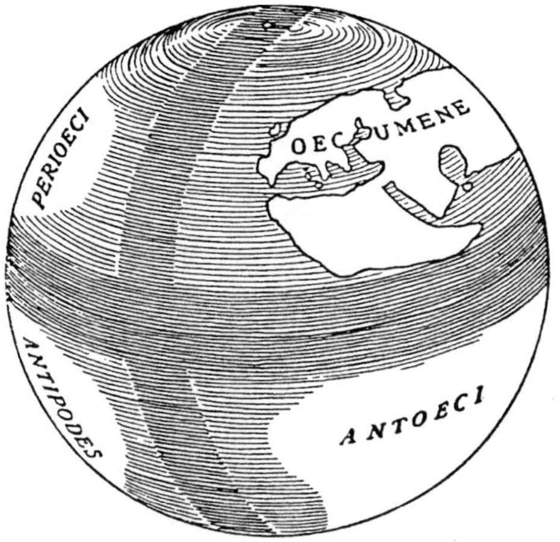

Fig. 1. Diagram illustrating the Cratesian theory of the four land masses situated symmetrically on the surface of the globe. From E. L. STEVENSON

> off that there can be no communication among the different groups; moreover, some nations stand obliquely, some transversely to you, and some stand directly opposite you ..."[25]

Macrobius glossed Cicero's words with greater precision. The nations referred to by Cicero

> "... must," he wrote, "be divided as follows: those who are separated from us by the torrid zone, whom the Greek call *antoikoi*, the Antoeci; next those who live on the underside of the southern hemisphere, the Antipodes, separated from the Antoeci by the southern frigid zone; those who are separated from their Antoeci, that is, the inhabitants of the underside of our zone, by their torrid zone; they are in turn separated from us by the north frigid zone."[26]

The latter group, which Macrobius does not name, Martianus Capella calls the Antichthones. Cleomedes (2nd century A. D.) calls them the Perioeci, but his work was only known in the West in a Latin translation published in 1497.[27] From then on the term was frequently used in the Renaissance, as we shall see.

[25] Cicero, *The Dream of Scipio*, in: *op. cit.*, p. 74.

[26] Macrobius, *op. cit.*, p. 206, ed. J. WILLIS (Leipzig, Teubner, 1963), *Comm.*, Bk. II, chap. 5,33.

[27] Cleomedes, *De Contemplatione excelsorum orbium disputatio*, Brescia, 1497. Cf. modern French translation by RICHARD GOULET, *Cléomède, Théorie élémentaire (De motu circulari corporum caelestium)*, Latin text and French translation, Paris, 1980, p. 95.

Macrobius saw the four islands as divided by two streams of the Ocean, one flowing along the equator and the other over the poles.

> "Separating us from the people of the southern hemisphere, Ocean flows along the whole extent of the equator; again, as its streams branch out at the extremities of both regions, it forms two islands on the upper face of the earth and two on the underside."[28]

Martianus Capella described the Cratesian theory thus:

> "You must know that the earth is divided into two distinct parts or hemispheres: there is an upper hemisphere which we inhabit and which Oceanus encircles; and there is a lower hemisphere [...] Now in as much as the five zones encompass both parts, upper and lower, they actually form ten regions. The one inhabited by us extends northwards. Another, tending southward, is believed to be inhabited by men called antecians (*antoikoi*). Correspondingly, there are two habitable regions in the lower hemisphere. Those diametrically opposite us are called antipodes and those who are diametrically opposite our antecians are called antichthones".[29]

Martianus Capella refers at the same time to the Greek theory of the five climatic zones into which the globe was divided.

> "The two zones that border upon the north and south poles have reason to be deserted, with enormous chill and cold, and severe snowfalls; the belt in the middle is scorched by fires and stifling heat, and burns all living things that draw near. The two belts between, tempered by the exhalations of life-giving breezes, are conducive to human and animal habitation."[30]

The Cratesian theory, as a geographic model of the arrangement of land and water on the globe, enjoyed a wide currency in the early Middle Ages, especially in the 12th century. From the beginning of the 13th century it was progressively abandoned in the face of the rising enthusiasm for Aristotle, whose texts were beginning to enter the West for the first time.

Eclipsed during the 13th and 14th centuries, the Cratesian theory was revived in the Renaissance in the 15th and, as we shall see, scholars turned to it as a model to transmit an understanding of how the newly discovered lands of the Discoveries were situated over the globe.

IV. The term orbis

Both Cicero and Macrobius call the four Cratesian islands "spots" (*maculae*). Only one of them corresponds to the inhabited portion of the earth known to the Romans as the *orbis terrarum*.

28 Macrobius, *op. cit.*, pp. 214 – 215.
29 Martianus Capella, *op. cit.*, pp. 225 – 26.
30 Idem, p. 225.

The Romans remained oddly ambiguous in their use of the term *orbis*, sometimes giving it the meaning of a two-dimensional flat disc and sometimes using it in an astronomical context with the meaning of a three-dimensional sphere. As J. VOGT has noted, in Latin no new word was coined by the Romans to designate the earth as a sphere.[31] The ambiguity in the meaning of the word *orbis* was to persist throughout the Middle Ages even into the Renaissance.

A striking example of this ambiguity appears in a passage from Lucius Ampelius' *Liber Memorialis* written between the 2nd and 4th centuries A. D.:

> *Orbis terrarum qui sub caelo est quattuor regionibus incolitur. Una pars eius est in qua nos habitamus; altera huic contraria, quam qui incolunt vocantur antichthones; quarum inferiores duae ex contrario harum sitae, quas qui incolunt vocantur antipodes. Orbis terrarum quem nos colimus in tres partes dividitur, totidemque nomina: Asia, quae est inter Tanain et Nilum; Libya, quae est inter Nilum et Gaditanum sinum; Europa, quae est inter fretum et Tanain.*[32]
>
> ("The *orbis terrarum* beneath the heaven is made up of four regions. One of them is that in which we live; another is opposite it and its inhabitants are called antichthones, while the latter two, placed below and opposite the former, are inhabited by those who are called antipodes. The *orbis terrarum* which we inhabit is divided into three parts respectively named Asia situated between the Tanais [= Don] and the Nile, Libya [= Africa] between the Nile and the Gulf of Gades [= Cadiz] and Europe lying between the straits [of Gades = Gibraltar] and the Tanais").

The term *orbis terrarum* here first designates the spherical earth on which lie the four Cratesian landmasses; then, in the very same paragraph, it is seen to refer to the *oikoumene* divided into three parts. Any doubt about the ambiguity is removed by observing the contradiction between a fourfold division and a threefold one.

The Elder Pliny (23 – 79 A. D.) well illustrates in his *Historia Naturalis* the confusion over the term *orbis*. At one point he uses it to describe a spherical earth: *Orbem certe dicimus terrae, globumque verticibus includi fatemur*[33] ("We speak with certainty of the *orbis* of the earth and we agree that the globe is contained within the poles"). At another point he writes of the earth in a Homeric context: *Terrarum orbis universus in tres dividitur partes, Europam Asiam Africam*[34] ("The *orbis terrarum* is divided into three parts Europe, Asia and Africa").

[31] J. VOGT, *Orbis Romanus*, Tübingen, 1929, p. 7.

[32] Lucius Ampelius, *Liber Memorialis*, edited by ERWIN ASSMANN, Leipzig, Teubner, 1935, Chap. 6,1 (p. 9).

[33] Pliny, *Historia Naturalis*, Bk. II, Chap. 64, Latin text and English translation by H. RACKHAM, Loeb edition, Vol. I, London, 1972, p. 294.

[34] IDEM, Bk. III, Chap. 1, Loeb ed. p. 4.

Other Roman authors sometimes hold to one meaning and sometimes to the other. While Seneca in his *Quaestiones Naturales* uses *orbis* in the sense of a sphere: ... *quod nisi esset, non diceremus orbem terrarum pilam esse*[35] ("If it were not thus, we would not say that the *orbis* is a ball"), Cicero makes *orbis* the equivalent of an island: ... *quasi magnam quandam insulam* [...] *quam nos orbem terrae vocamus* ...[36] ("Almost a sort of large island [...] which we call the *orbis terrae*"). Macrobius tries to clarify the Cratesian nature of Cicero's words by saying that "... Cicero [...] did not say, 'The whole earth is a small island', but rather 'The whole portion that you inhabit is a small island'".[37]

Orbis was sometimes used to designate islands or landmasses separated, or alleged to be separated, from the *oikoumene* by sea. Pliny wrote that the island of Taprobane [Ceylon/Sri Lanka] was believed to be an *alterum orbem*,[38] and Pomponius Mela reported that Hipparchus had said that it was either a large island or the beginning of a second *orbis* (*orbis alterius*)[39] Britain was considered to be an *alterum orbem*[40] and Pliny spoke of Scandinavia as an *alterum orbem terrarum*.[41] Ample precedent thus existed for later calling the American continent a *Novus Orbis*.

Yet a further example of the ambiguity of the term *orbis* appears in a passage from Hyginus (1st century A.D.):

> *Terra mundi media regione conlocata omnibus partibus aequali dissidens intervallo centrum obtinet sphaerae* [...] *Oceanus autem* (e) *regione circumductionis sphaerae profusus prope totius orbis adluit fines*.[42]
>
> ("The earth, placed in the middle of the universe, at an equal distance from all its parts, lies at its centre [...] The Ocean, spread out over the entire surface of the sphere, bathes the confines of nearly all the *orbis*".)

The first sentence describes a spherical earth; in the second, *orbis* is obviously used in a Homeric context.

35 Seneca, translated by THOMAS H. CORCORAN, Loeb edition, London, 1971, Bk. IV, Chap. II,2.

36 Cicero, *De Natura Deorum*, translated by H. RACKHAM, Loeb edition, London, 1972, Bk. II, 165 (p. 280).

37 Macrobius, *op. cit.*, p. 215.

38 Pliny, *Hist. Nat.* Bk. VI, Chap. 24, Loeb ed. translated by H. RACKHAM, Vol. II, Cambridge, Mass. – London, 1947, p. 398.

39 Pomponius Mela, *op. cit.*, Chap. III (p. 59).

40 Velleius Paterculus, *Compendium of Roman History*, Latin text and English translation by FREDERICK W. SHIPLEY, Loeb edition, London, 1955, Bk. II, Chap. 46 (p. 150) and Lucius Annaeus Florus, *Epitome of Roman History*, Latin text and English translation by EDWARD SEYMOUR FORSTER, Loeb edition, London, 1947, Bk. I, Chap. 46 (p. 204).

41 Pliny, *op. cit.*, Bk. IV, Chap. 196.

42 Hyginus, *L'Astronomie*, Latin text and French translation by ANDRÉ LE BŒUFFLE, Paris, 1983, p. 12.

V. *Pomponius Mela and Isidore of Seville*

In the *De Chorographia* of Pomponius Mela (composed between 41 and 44 A. D.), largely a systematic geography of the known world, the Cratesian theory is relegated to a paragraph at the beginning, where it is roughly described (pl. I):

> "The earth (*terra*) is surrounded on all sides by the sea, which penetrates it from east to west and divides it into two parts or hemispheres in which five zones can be distinguished. While the middle one suffers torrid heat, the ones at the ends undergo extreme cold. The two others are uninhabitable and have the same seasons but at opposite times. Those who have their feet opposite us (antipodes) inhabit one hemisphere and we inhabit the other. But as the former are unknown to us because of the extreme heat in the zone which divides us from them, I shall say no more on the subject."[43]

The rest of Pomponius Mela's work deals with the world as known to the Romans; the Cratesian theory had for him little interest.

Isidore of Seville (c. 570 – 636 A. D.) perpetuated into the Middle Ages, when he was widely read, the ambiguity of the Romans regarding the term *orbis*, and if anything he was to lend support to a Homeric interpretation of the term. Isidore shows the influence of both Hyginus and Pomponius Mela.

In his *De Rerum Natura* (composed between 612 and 615 A. D.) Isidore repeated to the letter Hyginus' definition given above,[44] but in a passage in another of his works, the *Etymologies*, he writes quite unequivocally of a flat *orbis*.

> "The *orbis*" he wrote, "is called thus because of its rotundity and because it resembles a wheel and that is why a little wheel is called an *orbiculus*. The all encompassing Ocean surrounds its confines on all sides. The *orbis* is divided into three parts; one of its parts is Asia, another is Europe and the third is called Africa."[45]

Elsewhere, repeating the same cursory recognition of the Cratesian theory given by Pomponius Mela, Isidore writes that beyond the three parts of the *orbis* lies a fourth part across the Ocean to the south, which because of the heat of the sun is unknown to us and within whose limits are said to live the

[43] Pomponius Mela, *De Chorographia*, *cit.*, pp. 3 – 4.

[44] Isidore of Seville, *De Rerum Natura*, ed. and French translation by Jacques Fontaine, Bordeaux, 1960, Chap. 48, p. 326.

[45] Isidore of Seville, *Etymologiarum libri XX*, Bk. XIV, Chap. 2, in: Migne, *Pat. Lat.*, Tome LXXXII, cols. 495 – 96 (= *Isidori Hispalensis Episcopi Etymologiarum Sive Originum Libri XX*, ed. W. M. Lindsay, Vol. II, London, 1911).

antipodes of fable,[46] in whose real existence, he adds, with due respect for the Church for he was a Christian, there is no reason to believe (*nulla ratione credendum est*).[47]

It is not until the 13th century that we see systematic attempts made to reduce the ambiguity inherited from the Romans on the subject of geography, the principal motivation being the clash between the speculations of antiquity and the authority of Biblical myth.

VI. Ptolemy: The Almagest *and the* Geography

A writer of great importance in Late Antiquity, whose work appears largely unrelated to that of his Roman contemporaries, is the Alexandrine Greek Claudius Ptolemy (c. 90 – 168 A. D.). Ptolemy's major works are two in number. His treatise on astronomy, which came to be known in the West by its Arabic name the *Almagest*, was brought to Europe in the 12th century and made known through Gerard of Cremona's Latin translation of 1175.[48] The other work, his *Geography*, only reached Europe in the Renaissance to be translated into Latin in Florence in 1406 by Manuel Chrysoloras and Jacopo d'Angiolo della Scarperia.[49]

In his *Geography*, Ptolemy's representation of the relation of the landmasses to the seas was exactly the reverse of Aristotle's. Where Aristotle had insisted that the earth lay within the water, Ptolemy envisaged the oceans as lakes separated from one another, lying in hollows over the earth surface. Ptolemy looked at geography and the relation between land and sea strictly from an astronomer's point of view and he appeared totally indifferent to Aristotle's concern to integrate them into his theory of the concentric ordering of the elements in which the latter's densities were in inverse proportion to their volumes. Among the scholars of Late Antiquity, whether Greek or Roman, one finds no trace of any debate on how the conflicting approaches of these two authorities were received or interpreted.

On the relation between land and sea, Ptolemy states in his *Geography* in clear and precise terms that "... the continuous surface of the earth and the seas forms one sphere".[50] Such an exact definition does not appear in the *Almagest*. There he first gives a proof of the sphericity of the earth deduced from astronomical phenomena and then, following it, a different one for the sphericity of the water.

[46] Idem, Bk. XIV, Chap. 5, col. 512.

[47] Idem, Bk. IX, Chap. 2, col. 341.

[48] Ptolemy, *Almagest*, translated and annotated by G. J. Toomer, London, 1984, p. 3.

[49] Vespasiano da Bisticci, (1421 – 1498), *Vite di Uomini illustri del secolo XV*, 3rd edition, Vol. III, Bologna, 1893, pp. 9 – 10.

[50] Ptolemy, *Traité de Géographie*, Greek text and French translation by the Abbé Halma, Paris, 1828, Bk. I, Chap. 2, p. 11.

"That the earth, too, taken as a whole, is sensibly spherical can best be grasped from the following considerations. We can see again that the sun, moon and other stars do not rise and set simultaneously for everyone on earth, but do so earlier for those more to the east, later for those towards the west. For we find that the phenomena at eclipses, especially lunar eclipses, which take place at the same time [for all observers] are nevertheless not recorded as occurring at the same hour (that is at an equal distance from noon) by all observers. Rather, the hour recorded by the more easterly observers is always later than that recorded by the more westerly. We find that the differences in the hour are proportional to the distances between the places [of observation]. Hence one can reasonably conclude that the earth's surface is spherical, because its evenly curving surface (for so it is when considered as a whole) cuts off [the heavenly bodies] for each set of observers in turn in a regular fashion. [...] In fact, the farther we travel toward the north, the more of the southern stars disappear and the more of the northern stars appear. Hence it is clear that here too the curvature of the earth cuts off [the heavenly bodies] in a regular fashion in a north-south direction, and proves the sphericity [of the earth] in all directions. There is the further consideration that if we sail towards mountains or elevated places from and to any direction whatever, they are observed to increase gradually in size as if rising up from the sea itself in which they had previously been submerged: this is due to the curvature of the surface of the water."[51]

The apparent disjoining ("There is the further consideration ...") of the proof given in the *Almagest* for the sphericity of the earth from that of the water was sufficient to mask for Medieval scholars the conflict between Ptolemy's representation of the relation between these two elements and Aristotle's doctrine of their concentricity. Only in the Renaissance, when the clear definition given in the *Geography* became known, did the conflict reveal itself in all its acuity.

In antiquity, the Greek geographer Strabo (64/3 B.C. – 21 A.D. or later) had, in his *Geography*, given the same definition of the relation between the earth and the seas as Ptolemy had:

"Now let us take as hypothesis that the earth together with the sea is sphere-shaped and the surface of the earth is one and the same with that of the high seas."[52]

But Strabo remained unknown to the Middle Ages and his *Geography* was only rediscovered and brought to Italy in 1423 and published c. 1469.[53]

[51] Ptolemy, *Almagest*, translated and annotated by G. J. TOOMER, London, 1984, pp. 40–41.

[52] Strabo, *Geography*, II,5,5, translation by HORACE LEONARD JONES, Loeb edition, London, 1969, Vol. I, p. 43.

[53] Cf. A. DILLER, *The textual tradition of Strabo's Geography*, Amsterdam, 1975, pp. 102 and 132.

A pale reflection of the definition given in Ptolemy's *Geography* was transmitted to the Middle Ages by the Arabic astronomer Alfraganus (al-Farghani) (died after 861). His work was translated into Latin in 1135 by John of Spain and became well known in the Medieval period. In it he wrote:

> *Convenerunt quoque sapientes quod terra cum universis partibus suis tam terrestribus quam marinis sit similis spere.*[54]
>
> ("The experts agree that the earth together with all its parts, those made up of the earth as well as those made up of the seas, is like a sphere").

There is nothing here of Ptolemy's rigorous definition that the earth and the seas form "one continuous surface". Although widely read throughout the Middle Ages, Alfraganus had, on this point, no braking influence on the growing deference for Aristotle as an authority after the 12th century.[55]

VII. Christian Antiquity: Lactantius and St Augustine

In the early fourth century A. D. the Christian writer Lactantius took vigorous issue with the Aristotelian and pagan doctrine of a spherical earth, to the centre of which all heavy bodies were attracted. For Lactantius there could only be one absolute "up" or "down" and bodies on the "underside" of the earth would fall into the sky. Refusing to entertain the existence of men living on the opposite side of the globe, the antipodes, he wrote:

> "Is there anyone so senseless as to believe that there are men whose footsteps are higher than their heads? or that things that are with us in a recumbent position, with them hang in an inverted direction? that crops and trees grow downwards? that the rains, and snow and hail fall upwards to the earth? [...] But if you inquire from those who defend those marvellous fictions, why all things do not fall into that lower part of the heaven, they reply that such is the nature of things, that heavy bodies are borne to the middle as we see spokes in a wheel; [...] I am at a loss what to say respecting those who, when they have once erred, consistently persevere in their folly, and defend one vain thing by another ..."[56]

54 Al-Farghani, *Differentie scientie astrorum*, edited by Francis J. Carmody, Berkeley, California, 1943, p. 6. Another Latin translation was made by Gerard of Cremona. Cf. the modern edition by Romeo Campani: Alfraganus, *Il libro dell'aggregazione delle stelle*, Città di Castello, 1910.

55 The only writer in the early Medieval West to quote Alfragan seems to have been Alexander Neckam (1157 – 1217). Alexander Neckam, *De Naturis rerum*, edited by Thomas Wright, London, 1863, p. 159 (... *cum tamen Alfraganus dicat unam esse sphaeram aquarum et terrae*, "... since however Alfraganus says that there is but one sphere of earth and water").

56 Lactantius, *Divine Institutes*, Bk. III, Chap. 24, in W. Fletcher, *The Works of Lactantius*, Edinburgh, 1871, pp. 196 – 7. Cf. G. Moretti, "The Other World and the 'Antipodes'.

This Father of the Church has always been held up by modern writers as an absurd figure, the original "flat earther". It is true that in the Middle Ages he is scarcely mentioned, but in the Renaissance he was rediscovered, commented on and respected perhaps more for his Ciceronian style than for the originality of his thought. His *Divine Institutes* was printed fifteen times in the 15th century.[57] After the discovery of America, he was to become throughout Europe in the 16th century a constant target of ridicule.

Unlike Lactantius, St Augustine (354–430), in his *City of God*, did not hold that the earth was flat, but he did refuse to accept the existence of antipodean men (pl. II).

> "But in regard to the story of the antipodes, that is, that there are men on the other side of the earth, where the sun rises when it sets for us, who plant their footprints opposite ours, there is no logical ground for believing this. Its authors do not claim that they have learned it from any historical evidence, but offer it as a sort of logical hypothesis. Their theory is that the earth hangs suspended within the heavenly sphere, so that the lowest and middle points of the world are one and the same. From this they conjecture that the other half of the earth, which lies beneath our portion cannot lack human occupants. They fail to observe that even if the world is held to be global or rounded in shape, or if some process of reasoning should prove this to be the case, it would still not necessarily follow that land on the opposite side is not covered by masses of water. Furthermore, even if the land there be exposed [= uncovered by water], we must not jump to the conclusion that it has human inhabitants. For there is absolutely no falsehood in Scripture, which gains credence for its account of past events by the fact that its prophecies are fulfilled. And the idea is too absurd to mention that some men might have sailed from our part of the earth to the other and have arrived there by crossing the boundless tracts of the ocean, so that the human race be established there also by descent from one first man."[58]

Saint Augustine agrees with Aristotle that the earth is spherical and that bodies fall towards its centre ("the lowest and middle parts of the world are one and the same"). He seems however to entertain the possible existence of two rival models, one a version of the Cratesian theory in which the other landmasses would be uninhabited ("we should not jump to the conclusion

The Myth of Unknown Countries between Antiquity and the Renaissance," below in this volume (*CTA* I 1), p. 252–53; J. Romm, "New World and "*novos orbes*": Seneca in the Renaissance Debate over Ancient Knowledge of America, below, p. 108.

[57] Johannes Quasten, *Patrology*, Vol. II, Utrecht/Antwerp, 1953, p. 394 (= Idem, *Initiation aux Pères de l'Eglise*, translation by J. Laporte, Paris, 1958, pp. 464–65).

[58] Saint Augustine, *The City of God against the pagans*, Latin text and English translation by Eva Matthews Sanford & William McAllen Green, Loeb edition, London, 1965, Bk. XVI, Chap. 9, pp. 49–51.

that it [the land on the opposite side] has human inhabitants") and an alternative model in which the parts of the globe opposite our *oikoumene* "might be covered by masses of water". Saint Augustine here appears to be alluding to a notion that, as we shall see, was to become common in the Middle Ages (and which can already be glimpsed in Manilius and Pliny) of the earth immersed in a sphere of water out of which, as an "island", emerges our *oikoumene*.

His main concern is however to refute the possibility of inhabited Cratesian landmasses, which he does invoking the absence of historical evidence and insisting on the impossibility, because of "the boundless tracts of the ocean", for the descendants of Adam to have colonized them.

The *City of God*, unlike the *Divine Institutes*, was well known in the Middle Ages, but this did not prevent certain followers of the Neoplatonists from accepting inhabited Cratesian landmasses.

VIII. The Middle Ages (12th century): William of Conches, Manegold of Lautenbach, and Geoffroy of St Victor

In the early Middle Ages, while certain writers expounded the Cratesian theory in complete indifference to its incompatibility with Biblical tradition, others showed their Christian sensitivity to the point of openly attacking the Neoplatonists and Macrobius in particular. The question still turned, not on the existence of the four Cratesian landmasses, but on whether they were inhabited by members of the human race. If they were inhabited, not only would the unity of mankind be impossible to sustain, but furthermore how would Christ's Apostles have brought the Gospel to such men, since the doctrine of the *divisio apostolorum* held that they had brought it to the whole of mankind[59], while communication between the landmasses was impossible because of the vastness of the Ocean? Thus was Greek speculation on the geography of the globe submitted to the constraints of Christian myth.

Among those who remained indifferent to the problem were the French scholars, Lambert of St Omer (c. 1120)[60] who simply repeated Macrobius, and William of Conches (1080 – 1145) who, after describing the five zones of Parmenides, wrote as follows:

> "Therefore the part of the habitable zone in which we are, is divided into two. [...] ... the Ocean rivers (*refluxiones*) which surround the shores of the earth as far as the horizon, divide it into two regions [one to the east and one to the west]. We inhabit the upper part and our antipodes the lower part. However none of us can go to them and none of them can

[59] Cf. Hrabanus Maurus (776 – 856), *De Vita Beatae Mariae Magdalenae*, in: MIGNE, *Pat. Lat.*, Vol. CXII, col. 1491.

[60] Lambert of St Omer, *Liber floridus*, ed. by A. DEROLEZ, Gand, 1968, f. 224 (r.) (p. 447).

> come to us [...] Likewise, the other habitable region is divided into two parts. The upper part is inhabited by our *antoikoi* and the lower part by their antipodes. Thus in the two habitable regions there are four kinds of populations."[61]

Apart from explaining how each of the populations had their days and nights and seasons at different times, William did not elaborate any further. "And since", he concluded, "we have spoken enough of the three inhabited regions and their inhabitants, let us begin by speaking of our own region."[62]

Manegold of Lautenbach (c. 1045 – 1103/19) in his *Opusculum contra Wolfelmum* felt that he had sound reason to attack this complacent acceptance of Macrobius.

> "Supposing for a moment that there are four spaces inhabited by men who, from the nature of things, have absolutely no possibility of crossing from one space to another, tell me, I pray, how it can be true that the Holy Apostolic Church can declare as reasonable that our Saviour [...] came for the salvation of the whole of humankind, if three races of men of whom the above mentioned Macrobius tells us that, given the climate of the zones and the earth, they can live beyond this habitable part in which we live, be excluded from this salvation and have not been able to receive the news of such a great hope of obtaining it."[63]

Geoffroy of St Victor (12th century), in his *Microcosmus*, was another to emphasize the clash between Biblical orthodoxy and pagan philosophy arising from the Cratesian theory.

> "... many Christian philosophers have strayed from the path of the true faith and under the influence of this opinion [Crates] have imagined that there are as many kinds of men as there are landmasses (*arida*) in the world. But who does not see how far this idea lies from the orthodoxy of the faith. For it follows that not all men descend from a single man [Adam] and that [eternal] life was not restored to the world through the merit of a sole individual [Christ]."[64]

The twelfth century Renaissance was to see the recovery in the West of new sources of Greek culture, among them Aristotle and Ptolemy. The obvious intellectual superiority of the latter over the Neoplatonists tended to force the

61 William of Conches, *De philosophia mundi*, in: MIGNE, *Pat. Lat.* Tome CLXXII, col. 85.

62 Idem, col. 86.

63 Manegold, *Opusculum contra Wolfelmum*, in: MIGNE, *Pat. Lat.*, Tome CLV, cols. 154 – 55. On Manegold also see G. MORETTI "The Other World and the 'Antipodes.' The Myth of Unknown Countries between Antiquity and the Renaissance," below in this volume (*CTA* I 1), p. 265 – 6.

64 Geoffroy of St Victor, *Microcosmus*, ed. P. DELHAYE, *Mémoires et Travaux des Facultés Catholiques de Lille*, fasc. LVI, Lille, 1951, pp. 67 – 68 (manuscript of the second half of the 13th century).

Cratesian theory into the background and, by a curious paradox, the theological problem it had raised was, as we shall see, eliminated by the protagonists of Aristotelian thought themselves.

IX. The Middle Ages (13th century): Abélard, Sacrobosco, Michael Scot, and Robert Grosseteste

Aristotle, as we have seen, had left his successors with the difficult problem of reconciling his theory of the concentric ordering of the elements with the very evident reality that the earth was not entirely covered by water.

The earliest Medieval attempt to show how the earth as a sphere (as opposed to a flat *oikoumene*), was not covered by water may not be directly a response to Aristotle, but simply repeated from an intermediate classical author who had sought in rough terms to make the Stagirite seem more plausible. The twelfth century theologian Pierre Abélard (1079 – 1142) gives the following description of the earth and its relation to the seas:

> *Quasi enim aliquis globus ita in aqua constituatur, ut una pars eius superemineat, ita ille globus terrae in aquis insedit, ut ex una parte eum mare contingeret ...*[65]
>
> ("Just as when a globe is plunged in water in such a way that a part of it emerges, so the globe of the earth lies (*insedit*) in the waters so that one part of it is in contact with the sea").

The source of Abélard may have been Manilius or Pliny or both, although it should be noted that neither of the latter uses the word *globus*: Manilius (1st century A. D.) writes:

> *Ipsa natat tellus pelagi lustrata corona*
> *cingentis medium liquidis amplexibus orbem.*[66]
>
> ("The earth itself is afloat (*natat*) encircled by the crown of the ocean which clasps the *orbis* in its midst in an embrace of water").[67]

Pliny's definition is very similar:

> *Pars nostra terrarum, de qua memoro, ambienti, ut dictum est, oceano velut innatans.*[68]

[65] Pierre Abélard, *Expositio in Hexameron*, in: MIGNE, *Pat. Lat.*, Tome CLXXVIII, col. 748 AB. (I am grateful to Madame DANIÈLLE LECOQ for this reference.)

[66] Manilius, *Astronomica*, 4, 595 – 96, Latin text and English translation by G. P. GOOLD, Loeb edition, London, 1974, p. 268.

[67] Author's translation.

[68] Pliny, *Historia Naturalis*, Bk. II, Chap. 112, Latin text and English translation by H. RACKHAM, Loeb ed. London, 1972, pp. 366 – 67.

> ("Our own portion of the earth which is my subject, swims as it were in the ocean by which, as we have said, it is surrounded").

It should be emphasized that the words *insedit* used by Abélard, *natat* by Manilius and *innatans* used by Pliny can scarcely be construed to imply "floating" in any Archimedian sense. Archimedes' work on floating bodies, which was translated into Latin by William of Moerbeke in 1269,[69] does not seem to have had any influence on Medieval thinkers concerned with the problem of the spheres of the earth and the water.

In the early 13th century there appeared a short treatise on physics and astronomy by an obscure Englishman, John of Holywood (or in Latin form, Johannes de Sacrobosco). His *Sphaera* (c. 1220), which became a handbook of cosmography in the universities, drew on Ptolemy for astronomy and on Aristotle for its physics. In it, Sacrobosco supplied an Aristotelian explanation to solve the problem of the earth being covered by water, as Aristotle had implied.

> "The machine of the universe is divided into two, the ethereal and the elementary region. The elementary region, existing subject to continual alteration, is divided into four. For there is earth, placed, as it were, in the middle of all, about which is water, about water air, about air fire, which is pure and not turbid there and reaches to the sphere of the moon as Aristotle says in his book of *Meteorology* [...] Three [of the elements] in turn surround the earth on all sides spherically (*orbiculariter*) except in so far as the dryness (*siccitas*) of the land stays (*obsistit*) the wetness (*humori*) of the water to protect the life of animate beings."[70]

Sacrobosco is the first Medieval writer to use the very Aristotelian qualities of dryness and wetness to bring the Stagirite's doctrine into line with observed reality. This ploy of scholastic writers of using Aristotelian arguments to clarify problems in Aristotle's thought is not unusual, although it frequently led rather to a displacement of the problem than to its resolution.

A contemporary of Sacrobosco, Michael Scot (1175? – 1234?),[71] about whom we know very little, wrote a commentary on Sacrobosco's *Sphaera* (c. 1230). In it he shows that he had quite fully grasped Aristotle's theory of the elements and of their places and its clash with observed experience.

69 Cf. Edward Grant (ed.), *A Source Book in Medieval Science*, Cambridge, Mass., Harvard University Press, 1974, p. 40.

70 Johannes de Sacrobosco, *Sphaera*, in: Lynn Thorndike, *The Sphere of Sacrobosco and its Commentators*, Chicago, University of Chicago Press, 1949, p. 119. On Sacrobosco cf. the article by John F. Daly S. J., in: *Dictionary of Scientific Biography*, Vol. XII (1975), pp. 60 – 63.

71 Michael Scot, *Super auctore spere cum questionibus*, in: Lynn Thorndike, *op. cit.*, pp. 248 – 342. Cf. also the article on Scot by Lorenzo Minio – Paluello in: *Dictionary of Scientific Biography*, Vol. IX, (1974), pp. 361 – 65.

> "If it is agreed that the whole earth is not covered by water, but that parts of the earth remain uncovered by water, one therefore asks why the water is contained by the air and the air by the fire, since according to Aristotle an element is the place (*locus*) of another element, that is to say that the [outer] surface of [the sphere of] one element is the place (*locus*) of [the sphere of] another element, just as the [outer] surface of [the sphere of] the air is the place (*locus*) of [the inner surface of the sphere of] the fire."[72]

To explain how the earth emerged from the water Michael Scot invoked, like Sacrobosco (but in different terms), the Aristotelian qualities of wetness and dryness.

> "No dry body", he said, "is limited by a limit other than itself and it is limited only by itself. The earth is a dry body, therefore it is not limited by a limit other than itself such as water or air and thus it does not have a spherical shape [...] A liquid [on the other hand] is limited by a limit other than itself and therefore it has a spherical shape."[73]

To explain what had brought about the existence of dry land, Michael Scot adduced a higher final cause, the perfection of the universe, having priority over the principle of the concentric ordering of the elements. Without it, warm blooded living creatures and plants would have no dry land on which they could survive. To provide for them, a certain part of the earth lay uncovered by the waters.[74]

> "[Aristotle]", declared Michael Scot, "says that [all] round the earth lies the water. [But] this [opinion] seems to be unfounded, since it rather appears that the earth lies above the water, like an island in the sea."[75]

In saying that "... the earth lies above the water like an island", Michael Scot was careful to distinguish between the quite appreciable emergence of the whole earth out of the water and the minimal outcropping of mountains whose importance was insignificant in interrupting the exact sphericity of the earth's surface.[76] Pliny had long before declared that the elevation of mountains was insignificant in relation to the total size of the earth.[77]

Although he had absorbed the principles of Aristotle's thought and undertaken a critique of it, in saying that "parts [not a part] of the earth remain uncovered by the water" and that instead of the earth being surrounded by water, "the earth lies above the water like an island", Michael Scot had still not broken with the Cratesian theory. Such a simultaneous adhesion to two

72 Michael Scot, in: *op. cit.*, p. 296.
73 Idem, pp. 295–96.
74 Idem, *loc. cit.*
75 Michael Scot, in: *op. cit.*, p. 274.
76 Idem, p. 296.
77 Pliny, *Historia Naturalis*, Bk. II, Chap. 64.

contradictory theories, is not uncommon among the scholastics. His implicit insinuation that the "earth island" was but one of several is confirmed by a passage where he examines the question of whether the zone situated between the tropic of Capricorn and the Antarctic circle is temperate and habitable. Arguing like Manegold and Geoffrey of St Victor from a Christian point of view, Michael Scot asks why, if it is temperate, our zone is habitable and the other not.

> "Nature in its natural function (*Natura naturans*) does nothing in vain and we are, as it were, the final purpose of all things. But since none of those [who live there] can come to us and none of us can go to them because of the torrid zone which lies between the two [temperate] zones, it follows that it is neither temperate nor habitable. If it were habitable and temperate, there would there be men who were mortal or immortal. If they were immortal they would not be composed of the four elements, and so they would not undergo generation and corruption, since they would not be composed of contraries, something which is not to be examined here. Thus if there were there men who are immortal, they would not be descended from Adam because all those who are descended from Adam are mortal because of the sin of our first parent. And if there were there men who are mortal, it would be contrary to the faith, since the reason why God became incarnate and suffered for us in our habitable zone would be the same for them as for us. But this is impossible because Christ would have been born twice and would have suffered twice, which is false. [...] It must be said that, in truth, this zone is not habitable by mortals and the reasons which prove this must be accepted because it would be contrary to the faith if there were there men who were mortal."[78]

That Michael Scot still thought in terms of the Cratesian theory is further shown when he says that

> "We see that the Ocean or [= and] Amphitrides [*sic* for Amphitrites] surrounds the whole earth".[79]

Of the two "rivers" that encircle the earth in the Cratesian schema, one called Ocean flows along the equator; the other called Amphitrides flows at right angles to the first through the poles.

A contemporary of Michael Scot, Robert Grosseteste (c. 1168 – 1253), also wrote a *Sphaera* in which he expounded the Cratesian theory of the Ocean rivers:

> "Let there be a great circle going round the body of the earth from one pole to another and another great circle going round the body of the earth

[78] Michael Scot, in: *op. cit.*, p. 321.
[79] Idem, p. 274.

> along the equator. Corresponding to these two circles, two seas circle the earth. And the sea which circles the earth through the poles is called Amphitrides and the other sea is called Ocean. These two seas divide the earth into four parts of which only one is inhabited."[80]

Robert Grosseteste says nothing about the four islands of the Cratesian theory, but, in a passage elsewhere in his work, he seems closer to Ptolemy's view of the relation between earth and water and indifferent to Aristotle's concentric ordering of the elements; yet his explanation of the existence of dry land has nothing to do with the definition in Ptolemy's *Geography*, or with Alfraganus' pale reflection of it.

> "Nevertheless for living beings who are accustomed to live on land to have a place to live and to obtain shelter, the water withdrew into the cavities of the earth whose dry surface [then] appeared. And the earth together with the waters contained in it are nothing more than the sphere of the earth."[81]

The sources of Grosseteste's idea that "the water withdrew into the cavities of the earth" can be traced to the Greek writers of Late Antiquity, John Philoponus and Olympiodorus,[82] but one cannot conclude that he had read them himself directly.

X. The Middle Ages (13th century): Albertus Magnus, Roger Bacon, Campanus de Novara, and Thomas Bradwardine

Two writers of the 13th century, with very different minds, one a Dominican, Albertus Magnus (c. 1193 – 1280), the other a Franciscan, Roger Bacon (c. 1219 – 1292), showed that they were fully conversant with the issues discussed so far. Both have been claimed as appearing as "moderns", that is in advance of their time, but this judgement is illusory for their sources were largely the same ones as those of their predecessors and contemporaries.

Albertus Magnus in his *De Natura loci* (c. 1250) asserts that the torrid zone is difficult, but not impossible, to cross, and that the southern hemisphere is not only inhabitable, but is inhabited.[83] His only evidence was drawn from

80 Robert Grosseteste, *De Sphaera*, in: L. Bauer, *Die Philosophischen Werke des Robert Grosseteste*, in: *Beiträge zur Geschichte der Philosophie des Mittelalters*, Vol. IX, Münster, 1912, p. 24.

81 Idem, p. 12.

82 Cf. P. Duhem, *Le système du monde*, Vol. IX, Paris, 1958, pp. 96 – 97.

83 Albertus Magnus, *De Natura loci*, in: *Opera omnia*, Tome V, Pars II, Münster, 1980, pp. 12 – 14.

Homer's remark that there are two sorts of Aethiopians: those who live under the summer tropic and those who live under the winter one, whose feet are opposite ours,[84] and from Lucan's description of the Arabs who came to Rome and saw the sun's shadow fall in the opposite direction to that in which it fell where they came from.[85]

Albertus is fully aware of Aristotle's theory of the elements. Almost all the philosophers and mathematicians are agreed, he wrote, that the earth is not everywhere inhabitable, because the sphere of the water is greater than the sphere of the earth. Since the water does not cover the earth in the northern hemisphere, it seems to them that it must cover it in the southern one. Against this view, Albertus quotes Albumasar and his followers, who claim, on astronomical grounds, that the southern hemisphere is exactly the same as ours.[86]

If the water is said to occupy more space than the earth as an element, it is said to do so because it has less matter than earth and because it is less dense and not because it does so in fact (*in effectu*), but because it is said that the proportion of water to earth is such, (*sed quod ita est proportione aquae ad terram*). That the water is greater [in quantity] than the earth in fact (*in effectu*), is not certain, for if this were so, the water would not be limited to the shorelines, but it would inundate the whole earth. Albertus finally concludes that the lower hemisphere is divided up just as ours is, with regions that are habitable and uninhabitable on account of cold and heat.[87]

Roger Bacon's arguments in his *Opus Majus* (1264) in favour of an inhabitable southern hemisphere were similar to Albertus' but they were not the same ones.

Bacon first quotes Ptolemy's *De Dispositione Sphaerae*[88] that a sixth part of the earth is inhabitable and that the rest is covered by water; he then quotes Ptolemy's *Almagest* that only the quarter which we inhabit is inhabitable. Aristotle, however, he notes, declares that more than a quarter is inhabitable, for Aristotle says that there is only a short sea distance between the [western] limit of Spain and the beginning [i.e. eastern] limit of India.[89] Seneca, also, writes Bacon, asserts that this distance can be navigated in a few days with

[84] Homer, *Odyssey*, I, 22–24, translation by A. T. MURRAY, Loeb edition, London, 1974 (Vol. I, pp. 4–5).

[85] Lucan, *The Civil War*, III, 247–48, translation by J. D. DUFF, Loeb edition, London, 1969 (p. 132).

[86] On Albumasar, cf. RICHARD LEMAY, *Abu Ma'shar and Latin Aristotelianism in the twelfth century*, Beirut, 1962.

[87] Albertus Magnus, *op. cit.*, pp. 20–21.

[88] This work seems to have disappeared for it is not mentioned among Ptolemy's extant works. There is however a possibility that it might have been a re-worked introduction to the *Almagest*.

[89] Aristotle, *De Caelo*, Bk. II, Chap. 14.

a favorable wind.[90] The prophet Esdras wrote that six parts of the earth were dry land and that the seventh was covered by water.[91]

Bacon, besides quoting the example of the two races of Aethiopians cited by Albertus, also quoted Pliny's reports of a port in India called Pathalis where the shadow of the sun's rays fell to the south and his description of Ceylon where the sun's rays also fell to the south.[92]

Bacon's conclusion therefore formed an argument claiming that more than a quarter of the earth was inhabited, yet he never committed himself to specifying how much. He did admit however that, although the region beyond the tropic of Capricorn had excellent qualities for being inhabitable, no one had ever reached us from there nor had anyone from among us ever gone there.[93]

Nowhere in the *Opus Majus* does Bacon take up Aristotle's theory of the elements and the problem it posed for the Medieval scholastics. In the IVth Book of his Commentary on Aristotle's *Physics*, Bacon does however deal with the problem and asks what is the place of the earth (*Queritur postea de loco terrae*) and whether the water is the place of the earth or whether the place of the earth is the centre of the universe (*queritur utrum aqua sit locus terrae vel centrum*). In this question he was asking whether the place (*locus*) of the earth was the same as the centre of the universe or was the concave surface of the sphere of the water surrounding it. His answer was that the place (*locus*) of the earth was not the centre of the universe, as this centre was quite indivisible and the earth was wholly divisible. The place of the earth was the inner surface of the sphere of the water.[94] Bacon had quite clearly perceived Aristotle's demonstration that for the earth to be in its proper place, it must be entirely lodged and surrounded by the sphere of the water, but, as Duhem noted,[95] he failed to explain the emergence of dry land, nor did he apparently see any contradiction on this point with what he had written in the *Opus Majus*.

Apart from Pierre d'Ailly's copying two hundred years later of Bacon's words, the arguments of both Bacon and Albertus Magnus in favour of a greatly extended *oikoumene* and of an inhabited southern hemisphere were largely ignored by their successors and their attempts to construct a critical synthesis of Ptolemaic astronomy and Aristotelian physics were not taken up

[90] Seneca, *Naturales Quaestiones*, Bk. I, Pref. 13, English translation by T. H. Corcoran, Vol. I, Cambridge, Mass. – London, 1991, p. 10.

[91] Esdras, Bk. II, Chap. 6, v. 42.

[92] Pliny, *Historia Naturalis*, Bk. VI, Chap. 22.

[93] Roger Bacon, *Opus Majus*, ed. John Henry Bridges, London, 1897, facsimile ed. Minerva, Frankfurt/Main, 1964, Vol. I, pp. 290 – 309.

[94] Roger Bacon, *Quaestiones super libros octo physicorum Aristotelis*, in: *Opera hactenus inedita*, ed. F. M. Delorme and Robert Steele, Oxford, 1935, fascicule XIII, pp. 215 – 216.

[95] P. Duhem, *Le Système du Monde*, Paris, 1958, Vol. IX, p. 112.

and built on. On the whole, the discourses of the physicists and the astronomers tended to be kept in separate channels, even when dealt with by a single writer.

A good example of this is that of Campanus de Novara (1205 – 1296) who, in his *Tractatus de Sphaera*, written after 1268, shows, unlike the two previous writers studied, no awareness of the incompatibility between the physics of Aristotle and the astronomy of Ptolemy.

After giving in Chapter IIII of his *Tractatus* a short clear account of Aristotle's theory of the concentric ordering of the elements, which he declares is their final shape and order (and by "final" he may be suggesting some sort of eschatological perfection), Campanus then asks in Chapter V why the sphere of the water does not continuously envelop the earth. His explanation is that of Michael Scot, whom he had obviously read: a higher final cause providing living creatures with dry land to live on. He specifically relates this higher final cause to the action of Divine Providence, when, according to Genesis, the Creator pronounced the words: "Let the waters beneath the heavens be gathered together in one place and let the dry land appear". The earth, said Campanus, had risen up like an island, interrupting the sphere of the water and itself losing its exact spherical shape. Drawing on Aristotelian concepts of dryness and wetness to which both Sacrobosco and Michael Scot had had recourse, he adds that whereas the water, because of its wetness, cannot lose its spherical shape and can only be limited by a limit other than itself, the earth because of its dryness and compactness can limit itself by itself. Water cannot lose its spherical shape, but earth can. "The whole of the dry land", Campanus concluded, "is like a very large island elevated up in the air above the surface of the water".[96]

In this mix of providentialism and of Aristotelian qualities, there appears no longer any mention of the Cratesian theory. Further on, in Chapter XIX, Campanus proceeds to give a careful repetition of Ptolemy's proofs demonstrating the sphericity of the earth and of the water, which the Alexandrine had given in the opening pages of the *Almagest*. In this repetition there is one important difference which separates Campanus from the *Almagest*. Campanus gives Ptolemy's proofs for the sphericity of the earth as being valid for the earth *together* with the water. The behaviour of celestial phenomena illustrating the sphericity of the earth in both an east-west direction and a north-south direction is exactly identical for observers, says Campanus, whether they be on land or on sea or on islands or on board ship.[97] Such a definition, equivalent to that of Ptolemy in his *Geography*, which Campanus could not have known, is obviously incompatible with the definition already given in his Chapter V of "the earth as a large island elevated up in the air above the surface of the water". Campanus seemed not to be aware of the clash.

96 Campanus de Novara, *Tractatus de Sphaera*, Florence, 1518, f. 153 (r.).
97 Campanus de Novara, *op. cit.*, f. 154 (v.).

The first to see the mathematical incompatibility between the physicists and the astronomers, behind whom stood the figures of Aristotle and Ptolemy, was an English scholar Thomas Bradwardine (c. 1290 – 1349) of Merton College Oxford.

Bradwardine had noted, copied in Alfraganus' treatise, Ptolemy's assertion in the *Almagest*, that the distance between the earth and the moon was 33⅓ times the radius of the earth. This doctrine, for the Aristotelians, was equivalent to the radius of the four elements: earth, water, air and fire taken together, but certain Aristotelians (not Aristotle) had been, in the wake of his Greek commentators, claiming, as we have seen, that the volumes of the elements were in the proportion of 10 : 1. Their total volume, as contained by the inner surface of the sphere of the moon, would thus be 1 + 10 + 100 + 1000 = 1111. The application of Euclid's proposition 18 in Book XII of his *Elements*, that "Spheres are to one another in the triplicate [= cube] ratio of their respective diameters"[98] showed that it was obvious that Ptolemy's distance of 33⅓ earth radii separating the moon from the earth gives a volume of the space occupied by the elements beneath the moon of 33.3^3 times that of the volume of the earth (the ratio of the diameters being the same as that of the radii). This figure was vastly different from that proposed by the Aristotelians.

Bradwardine in his *Tractatus de proportionibus* (1328)[99] proposed a solution to this problem. He retained the assertion of the commentators of Aristotle that the elements maintained a constant proportion in the relations to one another of their volumes, but he eschewed any attempt to advance any exact figure for it. He also retained the figure of the astronomers of the distance between the earth and the moon. Into this distance he attempted to fit the four elements spaced proportionally between each other. Expressed in modern mathematical terms this can be formulated as $x^3 + x^2 + x + 1 = K^3$, where K is the distance between the moon and the centre of the earth. Since Bradwardine did not have a mathematics capable of resolving this problem in algebraic terms, he proceeded by approximation, asserting that the proportions were greater than 32,768 : 1024 : 32 : 1 and smaller than 35,937 : 1089 : 33 : 1. In the Middle Ages he was acclaimed as a great mathematician.[100]

Very few scholars in later centuries took up Bradwardine's demonstration; the greater number contented themselves with repeating in the context of physics, the ratio of 10 : 1 for the elements and, in the context of astronomy, the Ptolemaic figure of 33.3 earth radii for the distance between the earth and the moon, often giving the two figures in the same work without seeming to realize that they clashed.

[98] Euclid, *The Thirteen Books of Euclid's Elements*, translation by Sir Thomas Heath, Dover Books, New York, 1956, pp. 434 – 436.

[99] Thomas Bradwardine, *Tractatus de proportionibus*, edited and translated by H. Lamar Crosby, Jr., Madison, University of Wisconsin Press, 1961, pp. 133 – 141.

[100] P. Duhem, *Le Système du Monde*, Vol. IX, Paris, 1958, pp. 164 – 5.

XI. The Middle Ages (14th century): John Buridan and Albert of Saxony: Aristotle interpreted with the aid of Archimedes

The University of Paris produced in the 14th century a remarkable group of physicists, whose principal contribution was in seeking explanations in terms of natural philosophy, rather than in having recourse to Divine Providence, to clarify the difficulties in Aristotle's doctrine of the elements.

The most original of these physicists was John Buridan (c. 1300 – 1358).[101] Before expounding his own doctrine, he argued first against the Cratesian theory.

> "It remains now to speak of the places which are uninhabitable on account of the water. On this there are three important opinions. Some say that only one quarter, or nearly a quarter, is inhabitable and others say that there are inhabitable places in all the quarters of the earth. And of this latter opinion we must speak first.
>
> Those in favour of it say that both the earth and the water are concentric with the universe, that is to say that the centre of the universe is the same as the centre of both the former two. However they say that in each quarter of the earth there are many tracts of land uncovered by the waters because of the many humps (*gibbositates*) emerging almost like the elevations of mountains above the water. And they say that many other parts of the earth are covered by the water because of the depressions like valleys in between the said elevations. And they say that this is so in each of the quarters of the earth and the proof is that from one very large landmass uncovered by the water, we cross a very large and extended sea and we come to another very large landmass uncovered by the water and it is probable that this is the case all round the earth.
>
> But against this view there are two important objections. The first is that all the seas which can be crossed by men and all the habitable lands which can be found are contained in this quarter of the earth which we inhabit. And some have labored [in navigation] to reach by sea the other quarters and they have never been able to reach any habitable land. And that is why it is said that Hercules placed pillars at the limits of our quarter as a sign that beyond them there was no habitable land, nor sea that was navigable.
>
> The second objection, which should have been made in the first place, is even graver, because if the world is eternal, [in Aristotle's view] this opinion is incapable of explaining how the elevations of the earth have been

[101] On Buridan cf. the article by E. A. Moody, in: *Dictionary of Scientific Biography*, Vol. II (1970), pp. 603 – 608.

able, from the beginning of time, to escape destruction, since innumerable particles of earth have been continually carried down from them in rivers to the depths of the seas. Now from infinite time back, these depths would have been filled up and the elevations of the earth eroded away, something which should not be mentioned to those who want to keep the world perpetually in a favorable condition for living creatures and plants, such as it now is."[102]

Should we be surprised to see this university scholar rely on the myth of Hercules and his pillars to refute the Cratesian theory, when navigation beyond the straits of Gibraltar had for long been regularly practised by Mediterranean peoples and a Portuguese expedition led by the Genoese Lanzarotto Malocello had in 1336 reached the Canary islands probably shortly before Buridan wrote?[103] As will be continually evident further on, the information furnished by theoretically oriented scholars and practical mariners was frequently out of phase, and lack of communication between them, except in rare cases, was a constant feature in our context.

Buridan's argument against the Cratesian theory, that in the course of time, soil erosion would have worn away the elevations of the land emerging out of the water, is basic to his own theory of the relation between water and earth.

He next expounds the providentialist theory that, for living beings and plants to survive, God had originally placed the earth in the centre of the universe, but had made the centre of the water eccentric to the centre of the universe. Thus a part of the earth, almost a quarter in surface area, had emerged above the water. Against this theory, Buridan sees two objections, one that the eccentricity of the water is not compatible with the circular movement of celestial bodies, nor with the physical nature of the elements of water and earth, and secondly that the emergent elevations of the earth would over time be eroded away. He therefore proposes a third solution which seems to him most probable and which "perpetually saves all the phenomena."[104] Both the earth and the water, he declares, are concentric with the universe (and on this point he appears to contradict himself in relation to what he says later on), but how then will the elevations of the earth be preserved through time? For the survival of living beings and animals, a part of the earth, a quarter, lies uncovered by the water and emerges above it and it remains and will, he insists, remain "naturally uncovered" (sic!) in spite of the concentricity (!) and even if we overlook the mountains.

"And such is this theory (*imaginatio*) that the earth, in the part that is not covered by the water, is altered by the air and the heat of the sun and a

102 Jean Buridan, *Quaestiones super libris quattuor de caelo et mundo*, Lib. II, Quaestio 7, edited by E. A. Moody, Cambridge, Mass., 1942, pp. 157–8.

103 Charles Verlinden, "Lanzarotto Malocello et la découverte portugaise des Canaries", in: *Revue belge de philologie et d'histoire*, Tome XXXVI (1958), pp. 1173–1209.

104 Jean Buridan, *op. cit.*, p. 159.

large quantity of air is mixed with it and thus this part of the earth becomes less dense and lighter and it has many pores full of air or subtle bodies. But the part of the earth which is covered by the water is not altered by the air and the sun and it remains denser and heavier. And thus if the earth were divided through the middle of its magnitude, one part would be much heavier than the other and that part in which the earth is uncovered by the water would be much lighter. And thus it appears that the centre of gravity of the earth is different from its centre of magnitude, for the centre of gravity is there where there is as much weight on one side as on the other and this is not in the middle of its magnitude, as has been said. Furthermore, if the earth, because of its heaviness tends towards the centre of the universe, it is because its centre of gravity coincides with the centre of the universe, and not with its centre of magnitude. It is because of this that the earth is elevated above the water on one side and on the other it is totally submerged."[105]

Buridan's introduction into Aristotelian discourse of Archimedes' (c. 287–212 B. C.) concept of a centre of gravity of a body, as a means of overcoming the difficulty in accommodating Aristotle's doctrine of the elements with observed experience, was inspired from his reading of Alexander of Aphrodisias (fl. 193–217 A. D.) as repeated by Simplicius (1st half of the 6th century A. D.) in the latter's commentary on Aristotle's *De Caelo*.[106] But Buridan in his mechanical solution to the problem,[107] went beyond Alexander of Aphrodisias; his innovatory thought was followed by his contemporaries and successors into the 15th and early 16th century, when the maritime voyages of the Portuguese, revealing the great extent of land in the southern hemisphere, were then, as we shall see, to prove his theory untenable. By placing three quarters of the earth's surface under water, it eliminated until then the theological difficulties posed by the Cratesian model, as well as those raised by the existence of inhabited antipodes.

Though Buridan did not explicitly relate the part of the earth emerging out of the water to the *orbis terrarum* of antiquity, this was in fact done by his contemporary Albert of Saxony (c. 1316–1390).

In his commentary on Aristotle's *De Caelo*, Albert first refutes the Cratesian theory as Buridan had done, using as proof of its untenability the myth of Hercules and his pillars.[108] He then, after making the earth emerge out of the water by distinguishing its centre of gravity from its centre of magnitude, writes as follows:

[105] Idem, *loc. cit.*

[106] Cf. P. Duhem, *Le Système du Monde*, Vol. IX, pp. 80–81.

[107] Idem, p. 201.

[108] Albert of Saxony, *Quaestiones in libros de caelo et mundo*, Lib. II, Quaestio XXVI, ed. Paris, 1518, f. cxix (r.)–(v.).

> "One must imagine a line running from east to west along the equator, and, parallel to it, a second line running along the Arctic circle. Between them one imagines a third line parallel to the other two. Furthermore one imagines that all that lies from the equator [southward] passing through the south pole and [back and round through] the north pole as far as the Arctic circle is covered by water and that what remains, which is no more than a quarter of the earth's [surface] and yet not entirely that, is uncovered by water. Further one imagines that, from the middle of the line placed over the [part of the] earth which is not covered by the water [= along the equator] a line is drawn perpendicularly toward the line which is parallel to it and placed along the Arctic circle. With these lines thus imagined, one sees clearly in a diagram how this part of the earth that is not covered by water is divided into four parts, of which two are oriented to the west and two to the east and between those which are oriented to the west, that which is closest to the equator is called Africa and that which is closest to the north is called Europe. Of the two others oriented to the east, that which is turned towards the equator is called Asia major and that which is turned toward the north is called Asia minor."[109]

Thus did Albert equate the emerged part of the earth in Buridan's model with the *orbis terrarum* of antiquity, traditionally divided into three parts: Europe, Asia and Africa. There is here proof that physics and cartography were linked in the minds of Medieval scholars. The emerged part of the earth, shaped like an upturned saucer, can still be seen to correspond, to all intents and purposes, with the Homeric model of a flat earth.

In another of his works, a commentary on Aristotle's *Physics*, Albert of Saxony modified Buridan's use of the concept of the centre of gravity of a body. Instead of making the centre of gravity of the earth alone coincide with the centre of magnitude of the water and with the centre of the universe, he conceived of the earth and the water as making up what he called an "aggregate" having a single centre of gravity (*unam totalem gravitatem*) which coincided with the centre of the universe and with the centre of the magnitude of the water.[110] The object of Albert's modification was perhaps to assure that both elements together had a common centre both of gravity and of joint magnitude in relation to celestial bodies. His innovation was preferred to Buridan's version by later scholars, and in particular by Pierre d'Ailly, as we shall see.

Albert's innovation of treating the two Aristotelian elements as an "aggregate" and of giving them a single centre of gravity was scarcely compatible with the principles of the Stagirite's thought and only goes to show the freedom with which Medieval scholars interpreted and built on Aristotle's doctrine.

109 Albert of Saxony, *op. cit.*, Lib. II Quaest. XXV, f. cxix (v.).

110 Albert of Saxony, *In octo libros physicorum*, Lib. IV, Quaest. V, Paris, 1518, f. xlv (v.). Cf. also P. Duhem, *Le système du monde*, Vol. IX, pp. 212–213.

Neither of these two Parisian scholars appear in their works to make any allusion to the difficulties in accommodating the pseudo-Aristotelian tradition of a continuous proportionality in the volumes relating the elements to one another, with the planetary distances in the Ptolemaic tradition, such as Bradwardine had tried to find a solution for. Their concerns remained solely in the domain of non-mathematical physics.

XII. The Renaissance (15th century): *Ptolemy*, GIOVANNI FONTANA, AENEAS SYLVIUS PICCOLOMINI, *Pierre d'Ailly*, GASPARINO BORRO, ANTONIO DE NEBRIJA, JACOB PEREZ DE VALENCIA, *and* JEROME MÜNZER

The salient innovation of the Renaissance period was the introduction into Western Europe of Ptolemy's *Geography*. On the initiative of the Florentine patron of letters Palla Strozzi (1370 – 1462) a Greek manuscript of the *Geography* was obtained in Constantinople in 1400 and translated into Latin in Florence in 1406 by Manuel Chrysoloras (died 1415) and Jacopo d'Angiolo della Scarperia (died after 1410). Francesco di Lapacino and Domenico di Lionardo Boninsegni re-drew the maps and translated the Greek place-names into Latin ones (pl. III).[111]

No work on geography was so to revolutionize the subject in the period of the Renaissance. Ptolemy's principal contribution was to mathematicize the subject, something which the Middle Ages had been unable to do. He proposed three different projections of the curved surface of the earth on a flat sheet of paper; the first known as the conical projection (in reality pseudo-conical) in which the meridians are straight lines and the parallels curved ones and the second later known as the homoeotheric projection in which both meridians and parallels are curved. The third, known simply as Ptolemy's "third projection", was never taken up by cartographers of the Renaissance since it was too complicated.[112]

Ptolemy's projections involved the use of coordinates of latitude and longitude spaced at regular intervals so as to form a grid covering the whole of the globe. A necessary precondition for the reception of the postulate of this grid was agreement that the land and the sea form together one continuous curved surface and on this point the clash with the Aristotelians was inevitable.

[111] VESPASIANO DA BISTICCI (1421 – 1498), *Vite di Uomini illustri del secolo XV*, 3rd edition, Vol. III, Bologna, 1893, pp. 9 – 10 and pp. 216 – 217 and pp. 232 – 235.

[112] On Ptolemy's projections cf. HANS V. MŽIK & F. HOPFNER, *Des Klaudios Ptolemaios Einführung in die darstellende Erdkunde*, übertragen und mit Erläuterungen versehen, in: *Klotho*, No. 5, Vienna, 1938, pp. 1 – 109; O. NEUGEBAUER, *A History of Ancient Mathematical Astronomy*, Part II, Springer Verlag, Berlin, Heidelberg and New York, 1975, pp. 879 – 890; JOHANNES KEUNING, "The History of geographical map projections until 1600", in: *Imago Mundi*, Vol. XII (1955), pp. 1 – 24.

Ptolemy's grid had its exact reflection in the heavens and through the adoption of a convention, or module, relating angular distance in degrees to units of terrestrial distance, it made it possible to calculate, from observations of the positions of celestial bodies, distances separating places on land or the positions of ships at sea. Ptolemy, in applying his method, was only able to do so partially, as many of his coordinates were obtained from the estimated distances of itineraries.

Although the Medieval West knew of the concept of climates, which corresponded to an irregular succession of parallels of latitude, it was Ptolemy's *Geography* which first introduced into Europe the concept of regularly spaced parallels and of converging meridians also spaced at regular intervals (Ptolemaic maps space meridians at intervals of 5 degrees). It was this latter concept which was crucial in Europe's final understanding of the surface of the *oikoumene* as curved,[113] and which made it possible to show places in high latitudes as lying closer together than ones in lower latitudes when separated by the same number of degrees.

Ptolemy's *oikoumene*, only slightly larger in extent than the Homeric one postulated by Albert of Saxony (it accorded a greater importance to the surface area of Asia) in no way covered the whole of the globe. Its east-west extent was of 180° and its north-south limits ran from 63° N to 16⁵⁄₁₂° S. Since Ptolemy saw the oceans as lying in separate basins, the rest of his globe, it was implied, was land surface and not sea. None of his maps show the whole globe in spherical form. The first to extend the Ptolemaic grid to cover the complete sphere were the humanists of the German Renaissance and this can be seen in a world map accompanying the so-called "German Ptolemy" c. 1490.[114]

Ptolemy's *Geography* circulated in Europe in manuscript copies before being printed. His mathematical principles of cartography were first taken up in the monasteries of Southern Germany and attempts were made, from about 1427 on, to apply them to areas of central Europe.[115] Only later, in about

[113] This point is missed by JÓSEF BABICZ and HERIBERT M. NOBIS, "Die Mathematisch-Geographischen und Kartographischen Ideen von Albertus Magnus und ihre Stelle in der Geschichte der Geographie", in: *Die Kölner Universität im Mittelalter, Miscellanea Mediaevalia*, Bd. 20, Berlin, New York, Walter de Gruyter, 1989, pp. 98–103. The concept of converging meridians is also absent in the description given by Roger Bacon of his map (lost). DAVID WOODWARD in his paper presented to the XIIth International Congress of the History of Cartography, Paris, 1987 gives a reconstruction of Bacon's map (reproduced by BABICZ and NOBIS), showing regularly spaced convergent meridians, whereas in fact Bacon only speaks in his text of "arcs of great circles passing through the poles and through the longitudes of cities as given along the equator". The longitudes of the cities cited by Bacon are obviously not spaced at regular intervals and as such cannot constitute a grid, which is a pre-condition for any mathematical cartography.

[114] ERWIN ROSENTHAL, "The German Ptolemy and its world map", in: *Bulletin of the New York Public Library*, Vol. 48, No. 2 (1944), pp. 135–148.

[115] Cf. DANA B. DURAND, *The Vienna Klosterneuburg Map Corpus in the fifteenth century*, Leiden, 1952.

1460 – 61, is a copy of the maps (it is not certain if the text accompanied them) known to have reached the maritime circles of Portugal,[116] where its influence before the end of the 15th century is difficult to assess.

The *Geography* was first printed at Vicenza in 1475, but without the maps, then at Bologna in 1477 with the maps. Other editions followed: Rome, 1478, Florence, 1482, Ulm, 1482 and 1486 and Rome, 1490.[117]

For most of the 15th century, the incompatibility between the principles of Ptolemy's *Geography* and those of Aristotelian physics was scarcely noticed by the majority of scholars, who often expounded both side by side in their works. Giovanni Fontana (c. 1395 – c. 1455) writing about the middle of the century in a book only published in 1544 under the name of Pompilius Azalus, after repeating Buridan's theory of the surface of the earth emerging from the water on account of the earth's centre of gravity being lower than its centre of magnitude[118] then proceeded to show that he was quite sufficiently conversant with Ptolemy's grid of mathematical coordinates and of his principle of convergent meridians to note that a parallel passing over the Indian Ocean and the regions of India contained a greater number of miles in a degree than a parallel passing along the tropic of Cancer and a still greater number of miles than a parallel lying along the Arctic circle. The farther north a region of the habitable earth lay, the less was its east-west extent.[119] Such an understanding of Ptolemy's principles of geography was necessarily incompatible with Buridan's doctrine, but Fontana was not able to see this.

Aeneas Sylvius Piccolomini (1405 – 1464), who became Pope Pius II from 1458, wrote a work published in 1477 that was later read and attentively annotated by Columbus, though it is uncertain whether he had read it before his first voyage. In this *Historia rerum ubique gestarum*, Piccolomini expounded almost every model in geography that we have examined up till now, without appearing to see any logical incompatibility among them. He began first by noting the scholastics' interpretation of Aristotle that heavy

116 Virginia Rau, "Bartolomeo di Iacopo di Ser Vanni, mercador-banqueiro florentino « estante » em Lisboa nos meados do século XV", in: *Do Tempo e da História*, Vol. IV, Lisbon, 1971, p. 113.

117 Cf. W. H. Stahl, "Ptolemy's Geography, a select bibliography", in: *Bulletin of the New York Public Library*, Vol. 55, No 9 (1955), pp. 419 – 432; 484 – 495; 554 – 564; 604 – 614. On the transmission and adaptation of Ptolemy's *Geography* before and after 1492 see now M. Dilke and O. A. W. Dilke, "The Adjustment of Ptolemaic Atlases to feature the New World," below in this volume (*CTA* I 1), pp. 117 – 134.

118 Pompilius Azalus [= Giovanni Fontana], *Liber de omnibus rebus naturalibus*, Venice, 1544, f. 11 (v.). That the work is by Giovanni Fontana is argued by Alexander Birkenmajer, „Zur Lebensgeschichte und wissenschaftlichen Tätigkeit von Giovanni Fontana (1395? – 1455?)", in: *Isis*, Vol. XVII (1932), pp. 34 – 53 and Lynn Thorndike, *A History of Magic and Experimental Science during the first thirteen centuries of our era*, Vol. IV, New York, 1934, Chap. 45, pp. 150 – 163.

119 [Giovanni Fontana], f. 94 (r.).

bodies are attracted to the centre of the earth, the greater part of which lay submerged under water.[120] He then evoked the Cratesian theory:

> "There are some who believe that there are four tracts of dry land that the great Ocean, which surrounds them, separates from one another by means of two great rivers, one of which runs under the zodiac [= along the equator?], while the other, running through the poles from right to left for those whose viewpoint lies in the east, joins the first beyond the pillars of Hercules. It is thus that four portions of the whole earth emerge like enormous islands so that benefiting from a temperate climate they provide habitats for mortal beings."[121]

After mentioning the theory of the five zones, which he correctly attributes to Parmenides, PICCOLOMINI again repeats the scholastic modification of Aristotle's theory of the elements that

> "only a part of the earth remains emerged [out of the water] between the Arctic circle and the equator and that the Divine Will has decided this for the sake of men, an opinion worthy of approval by a Christian".
>
> "Some", he continues, "make the shape of the earth spherical [by this he no doubt meant the appearance of the spherical earth as an upturned saucer as it emerged out of the water], others make it oblong and with them Ptolemy agrees and this opinion appears to be the most likely. Homer tells us that the whole of the earth which we inhabit is bathed by the Ocean which makes it nothing less than an island, for from wherever mortals may go to reach the limits of the earth, they find the sea, which we call Ocean and in whichever direction they proceed, they find that it constitutes the limits of our *oikoumene* (*nostrae habitabilis*). In all the four cardinal points [of the *oikoumene*] Ptolemy surrounds the region inhabited by man by an unknown land, though in most places the former is limited by the Ocean".[122]

At no point in his generous eclecticism, does PICCOLOMINI make a critical choice between the conflicting theories he reviews, not even to choosing between Homer and Ptolemy.

Cardinal Pierre d'Ailly (1350 – 1420) wrote two major works on cosmography, one a series of "Quaestiones" on the *Sphaera* of Sacrobosco and the other his celebrated *Imago Mundi*, from which COLUMBUS is alleged to have drawn inspiration for his voyages.

In his *Quaestiones* on Sacrobosco, first published in 1498, d'Ailly repeated the modification of Buridan's model, which Albert of Saxony had expounded

[120] AENEAS SYLVIUS PICCOLOMINI, *Historia rerum ubique gestarum*, Venice, 1477, Cap. I, sign. a. i. (v.) – sign. a. ii. (r.).

[121] IDEM, *loc. cit.*

[122] IDEM, *loc. cit.*

in his commentary on Aristotle's *Physics*. The earth has, within it, three distinct centres, the first its centre of magnitude, the second its centre of gravity and the third the centre of the universe. It is the centre of gravity of the aggregate formed by the sphere of the earth and the sphere of the water together, which coincides with the centre of the universe.[123]

In his *Imago Mundi*, first printed in Louvain c. 1483, d'Ailly only makes a passing allusion to the problem of the spheres of the water and of the earth, merely giving a very succinct statement of the definition given in his *Quaestiones*. He accompanied it however with a diagram clearly showing the three centres within the earth mentioned in the latter work.[124]

Further on in the *Imago Mundi*, d'Ailly tries to make an estimation of the "quantity of inhabitable earth". Quoting abundantly from Roger Bacon's *Opus Majus*, d'Ailly brings forward Aristotle's and Seneca's contention that the distance between Spain and India (across the Atlantic) was very small. The extent of water, it could thus be deduced, would not be so great as to cover three quarters of the earth. Quoting, after Bacon, the prophet Esdras that six parts of the earth were inhabited while a seventh lay covered by water, he concluded that the inhabitable earth, though reduced to less than a quarter of its size by Ptolemy and his disciples, extended, in fact, beyond such limits. Then, without realizing the contradiction, he proceeded without acknowledgement to quote from the *Sphere* of Nicole Oresme (1320/25 – 1382).[125] The inhabitable earth, said d'Ailly, quoting Oresme, was not round like a circle as Aristotle (Oresme meant the Aristotelians) would have it, but had the form of a quarter of the surface of a sphere of which the two extreme limits (i. e. the north and the south) were somewhat shortened, that is those limits which were uninhabitable because of heat or cold.[126] According to Oresme's definition, the inhabitable earth would thus have an east – west extension of half the circumference of the earth, or 180 degrees. Yet d'Ailly appears not to realize that this definition is quite incompatible with the statements of the authorities he had quoted about there being a small distance between Spain and India, or with the definition of the relation between the spheres of the earth and the water he had himself given in his *Quaestiones*. The inconsistancy of his thought and his plagiarism are flagrant.

In Venice in 1490 a Servite monk, Gasparino Borro (died 1498), a teacher of philosophy and theology, well known for his knowledge of astronomy and

[123] Pierre d'Ailly, „*Quaestiones*", in: Pedro Ciruelo, *Uberrimum Sphere mundi comentum intersertis etiam questionibus domini Petri de Aliaco*, Paris, 1498, sign. e. ii. (v.). Cf. also the French translation in P. Duhem, *Le système du monde*, Vol. IX, Paris, 1958, pp. 232 – 33.

[124] Pierre d'Ailly, *Imago Mundi*, ed. E. Buron, Tome I, Paris, 1930, p. 135.

[125] Nicole Oresme, *Sphere*, ed. Paris, 1508, Chap. xxx, sign. g. ii (v.).

[126] Pierre d'Ailly, *op. cit.*, pp. 207 – 215.

mathematics, published a commentary on Sacrobosco's *Sphaera.*[127] In it he discusses the problem of the existence of the antipodes, comparing the Cratesian view of Macrobius with the strict anti-Aristotelian Biblical orthodoxy of Lactantius. Surprisingly, BORRO declares that he prefers the opinion of Lactantius from a philosophical as well as from a theological point of view. His argument on philosophical grounds proves to be none other than Buridan's interpretation of Aristotle's doctrine of the elements in the light of Archimedes' principle of the centre of gravity of a body.

> "The centre of gravity of the universe is not the same as the centre of the earth because the earth has two centres; one is its centre of gravity which is the same as the centre of the universe and the other is its centre of magnitude which is not the same as the centre of the universe. The earth's two centres are different from each other because the earth does not have uniform weight and that is why a part of the earth is not covered by water and is closer to the heaven, while the other part is more distant from it so that the water flows toward the centre of the universe. But the centre of magnitude of the earth is different from the centre of the universe and on account of this only one part of the earth, that which is closest to the heaven, is not covered by the water. From the theological point of view it must be said that there are no Antipodes and the reason is that such men would have to be descended from Adam or from some other man. If they were descended from Adam, how would they have reached these regions which are inaccessible? And if they were descended from another man, either he would have been in a state of original sin or else he would not be. If he were not, then a state of innocence would be found there, which is something which cannot be entertained. And if he were in a state of original sin, how then would he have been cleansed from it? Not by Christ whom these beings never knew, nor by the Apostles. Therefore another law and another true faith would have been given them, which is all absurd."[128]

BORRO's thought recalls closely that of Michael Scot in the early 13th century; and his use of Buridan's theory to defend the doctrine of Lactantius, however paradoxical it may seem, is proof that until the great voyages of Discovery at the close of the 15th century, consolidated Renaissance geographical theory had advanced little beyond that of two centuries before.

Just at the turn of the 15th century, the justly celebrated Spanish grammarian ANTONIO DE NEBRIJA (1444–1522) wrote a little book on geography entitled

127 GASPARINO BORRO, *Commentum electum et pretiosum super tractatum sphaerae mundi*, Venice, n.p.n.d. (c. 1490). On BORRO and the date attributed to his work see the article by B. RECCHILONGO in: *Dizionario Biografico degli Italiani*, Vol. 13 (Rome, 1971), pp. 22–23.

128 GASPARINO BORRO, *op. cit.*, sign. c. iii. (r.) – (v.).

In Cosmographiae libros introductorium, which was published at Salamanca sometime between 1497 and 1505. On the first page of his first chapter he describes simultaneously in the same breath and without noticing that they clash, the two contradictory theories, that of the scholastics' interpretation of Aristotle's theory of the elements and Ptolemy's definition in his *Geography* of the relation between land and water.

Showing that he is thoroughly familiar with the work of the Alexandrine, he first gives Ptolemy's definition.

> "To begin with, we should consider that which is easily proved by the physicists and the mathematicians that the surface of the earth and the surface of the water together (*simul*) have a spherical shape and that they have in common the same centre as the centre of the universe."[128a]

Then in the same paragraph NEBRIJA expounds the Medieval doctrine of the relation between earth and water.

> "The surface of the earth is, over its greater part, covered by sea. And if we are certain that in the northern hemisphere which is known to us through land travel and sea voyages, half of it, that is a quarter of the whole surface is covered by water, the quarter which remains, in which lies the *orbis terrarum* known to us, emerges above the water like an island."[129]

The first outright recognition of the clash between the physics of the Aristotelian scholastics and the mathematical geography of Ptolemy derived from astronomy, appears in the work of another Spaniard, JACOB PEREZ DE VALENCIA (c. 1408 – 1490). In his Commentary on the Psalms published in Valencia in 1484, he wrote:

> "Some consider that all the seas are interconnected and that the Ocean is much larger than the whole of the earth and that it surrounds it on all sides and that the earth is like a light ball or like an apple in a basin full of water of which only the top appears above the water. And they say that God made it so from the beginning. For, since the whole earth was then covered with water, God separated the waters and this part of the earth [= the *oikoumene*] remained uncovered as dry land for men to live on [...]. But this position is manifestly irrational [...] And thus it must be said that Ptolemy's position and his description concerning the arrangement of land and water is much more rational, more in conformity with Holy Scripture and closer to truth, as is obvious to the senses and as has been

128a ANTONIO DE NEBRIJA, *In Cosmographiae libros introductorium*, Salamanca, n.p.n.d. (between 1497 and 1504 according to FRANCISCO RICO, "El nuevo mundo de Nebrija y Colón", in: VÍCTOR GARCÍA DE LA CONCHA, *Nebrija y la introducción del Renascimiento en España*, Salamanca, 1983, pp. 157 – 185), sign. a.i(r).

129 IDEM, *loc. cit.*

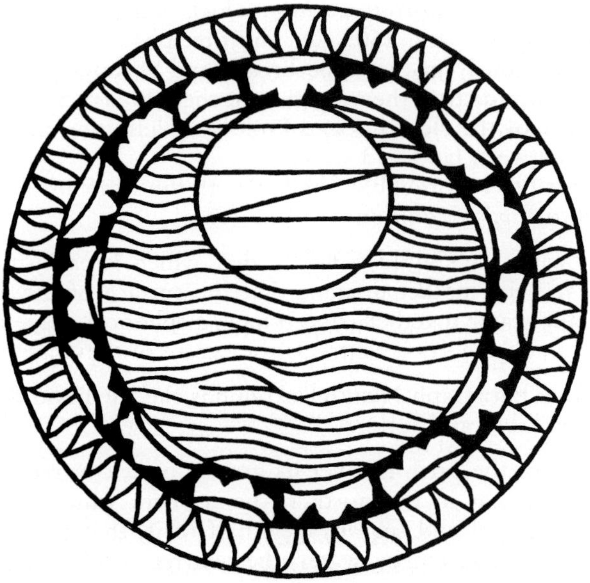

Fig. 5. Engraving showing the sphere of the earth just emerging from the sphere of the water, both being surrounded by the air and the fire, in Johannes de Sacrobosco, *Sphaera*, Venice, 1485 and 1490

> proven by the experience of a good number of people who have travelled and sailed across different parts of the world."[130]

In this clear choice in favour of Ptolemy against the scholastics PEREZ invokes the experience reported by travellers and navigators, but he does not elaborate or give examples.

Another writer who made a clear choice in favour of Ptolemy, was a German of Nuremberg, JEROME MÜNZER (or HIERONYMUS MONETARIUS), who wrote a letter dated 14 July 1493 to the King of Portugal encouraging him to pursue maritime exploration with the aim of reaching China by sailing across the Atlantic. Enjoining him not to believe in the scholastic view, he wrote:

> "And do not allow yourself to be influenced by Alfraganus and others without experience who say that only a quarter of the earth lies above the level of the sea and that three quarters of the earth is submerged beneath it."[131]

[130] JACOB PEREZ DE VALENCIA [*Comentum noviter edditum* [*sic*] ... *in Psalmos*], Valencia, 1484, sign. ii.ii (v.) – ii.iii (r.). There was a second edition published in Paris in 1518.

[131] JEROME MÜNZER, Letter to the King of Portugal, 14 July 1493. Truncated Latin text and Portuguese integral text together with Portuguese translation of the whole followed

The reference to Alfraganus is aberrant, since the latter had no connection with the pseudo-Aristotelian doctrine.

MÜNZER then shows that he prefers Ptolemy's view of the oceans as lakes lying in the hollows of the earth, to the scholastic view of the sphere of the earth to a great extent enveloped by water.

> "There is no doubt that the earth is not spread out under the water; on the contrary the sea is immersed [within the earth]."
>
> ("*Nam seja duvida que a terra nam esta alagada sob o mar, mas pelo contrario o mar esta immerso.*"[132]

If PEREZ' views appeared early enough to influence the climate of opinion in which COLUMBUS prepared his first voyage, MÜNZER's letter is dated after the Genoese's departure. Both however bear witness to the growing recognition of the clash between the two rival views and to the existence of writers who were prepared to take a stand.

XIII. CHRISTOPHER COLUMBUS' *"India" project in the context of the clash between the pseudo-Aristotelian physicists and the Ptolemaic geographers*

The inspiration of COLUMBUS' project[133] was drawn, we believe, ultimately from two sources: the first, Aristotle's statement in the *De Caelo* that a small sea distance separated the coasts of Spain from those of Asia,[134] a statement that was repeated by Roger Bacon and after him by Pierre d'Ailly, and secondly Ptolemy's definition of land and water occupying the same continuous surface on the surface of the globe.[135]

It is difficult to determine how and when COLUMBUS reached these two convictions, since the question remains the subject of a long-standing polemic. It has been asserted that the books of his private library which are known to us and which he abundantly annotated, either had little influence on him[136]

by complete French translation in: HENRY VIGNAUD, *Histoire critique de la grande entreprise de Christophe Colomb*, Tome II, Paris, 1911, pp. 620–625.

132 IDEM, *loc. cit.*

133 Although HENRI VIGNAUD (*Le vrai Christophe Colomb et la légende*, Paris, 1921) for years defended the theory that COLUMBUS never sought to reach Asia, but only to discover new lands in the Atlantic, a document discovered in the 1920's clearly indicates that the Catholic Kings understood the aim of his first voyage as being an attempt to reach Asia. Cf. "Carta Comendaticia" 17/4/1492, written by the Catholic Kings, in ANTONIO RUMEU DE ARMAS, *Nueva luz sobre las capitulaciones de Santa Fé de 1492, consertadas entre los reyes catolicos y Cristobal Colón*, Madrid, 1985, p. 94.

134 Cf. supra p. 9.

135 Cf. supra p. 16.

136 GEORGE E. NUNN, "The *Imago Mundi* and Columbus", in: *American Historical Review*, Vol. XL (1935), pp. 646–661.

or were only read by him after his first voyage.[137] It has been shown that the famous TOSCANELLI letter[138] in which the Florentine cosmographer, applying the principles of Marinus of Tyre's map projection,[139] had proposed that a voyage to Asia could be made in a shorter time than one round Africa, was only copied into one of COLUMBUS' books between 1498 and 1504, thus long after his first voyage.[140]

If COLUMBUS' own writings are of doubtful value in any reconstruction of his convictions regarding the make-up of the globe before his first voyage, there yet remain certain passages in the histories of BARTOLOMÉ DE LAS CASAS and of his son FERNANDO COLUMBUS, both probably based, as regards our subject, on an unknown common source.[141]

As is well known, COLUMBUS submitted his project of sailing to Asia across the Atlantic, first to the Portuguese King D. João II and then, after its rejection, on two successive occasions to the Catholic Kings of Spain. In each case the project was examined by a board of cosmographers designated by royal authority. These cosmographers were all hostile to COLUMBUS and his project.

The arguments of the Spanish cosmographers, as far as can be judged from the fragmentary nature of the descriptions that have reached us, were all based on the scholastic view of the relations between earth and water, and in particular, on the doctrine of the converted Jew, Paul de Burgos (c. 1350–1435),[142] whose thought was much influenced by his reading of Aristotle.

Paul de Burgos' work appeared as "Additiones" (1429) to the *Postillae* on the Bible of Nicolas de Lyra (1270–1349) (first edition Rome in 1461–72). Because both texts were printed together, Paul de Burgos' "Additiones" were often attributed to Nicolas de Lyra.

Instead of invoking Buridan's use of Archimedes' principle of the centre of gravity of a body to make the sphere of the earth emerge out of the water, Paul de Burgos relied simply on Divine intervention. On the first Day of the

137 JUAN GIL, *Mitos y Utopías del Descubrimiento* I. *Colón y su tiempo*, Madrid, 1989, pp. 123–126.

138 Cf. N. SUMIEN, *La correspondance du savant florentin Paolo dal Pozzo Toscanelli*, Paris, 1927 (original with French translation).

139 HERMANN WAGNER, "Die Rekonstruktion der Toscanelli-Karte v. J. 1474 und die Pseudo-Facsimilia des Behaim-Globus v. J. 1492" (= *Vorstudien zur Geschichte der Kartographie* III), in: *Nachrichten der K. Gesellschaft der Wissenschaften zu Göttingen*, Phil.-Hist. Kl. (1894) pp. 238–242.

140 ILARIA L. CARACI, "La cultura di Colombo," in: *Atti del IV Convegno Internazionale di Studi Colombiani*, Genoa, 1987, Vol. II, pp. 226–28; EADEM, *Colombo vero e falso*, Genoa, 1989, pp. 153–177.

141 On the complex problem of these two histories and their sources, cf. ILARIA CARACI, *Colombo vero e falso, passim.*

142 On Paul de Burgos, or by his other name Pablo de Santa Maria, cf. MANUEL MARTÍNEZ AÑIBARRO Y RIVES, *Intento de un diccionario bibliografico de autores de la provincia de Burgos*, Madrid, 1889–90, pp. 469–89 and LUCIANO SERRANO, *Los conversos don Pablo de Santa Maria y don Alonso de Cartagena*, Madrid, 1942.

Fig. 6. Engraving of the spheres of the earth and the water before and after the *congregatio aquae* on the Third Day of the Creation, according to Paul de Burgos, *Additiones* (written in 1429), in Nicolai de Lyra, *Postillae Nicolai de Lyra super totam Bibliam cum additionibus*, Nuremberg, 1481

Creation of the world, as described in the Book of Genesis, the Aristotelian elements, declared Paul de Burgos, were all arranged concentrically and the earth was entirely covered by the waters. On the third Day of the Creation, God ordained that the waters be gathered together in one place, so that dry land might appear. In so doing he shifted the sphere of the water so that its centre no longer coincided with that of the earth and with that of the other elements; a part of the surface of the earth was thus able to emerge above the water.[143]

It was this theory of Paul de Burgos that, as we shall see later confirmed, the Spanish cosmographers had in mind at their meeting at Salamanca in 1486 – 87. In the words of Fernando Columbus, they declared that:

> "... of this lower sphere of water and of earth, there was only inhabited a "crown" (*corona*) or "little belt" (*picciola cinta*) which lay upon the

[143] Paulus Burgensis, "Additiones", in: Nicolas de Lyra, *Postillae Nicolai de Lyra super totam Bibliam cum additionibus Pauli Burgensis* ..., Nuremberg, 1481, Cap. I (no pagination).

water in our hemisphere and that all the rest was water and that one could only navigate along the coasts and shorelines. And that when the experts were agreed that it was possible to reach [by land] the limits of the East [i. e. when once the east-west extension of the *oikoumene* was known] they would then [and only then] agree that it was possible to go [by sea] from the [western] limit of Spain to the farther limit of the West [i. e. the western limit of the Atlantic]."[144]

LAS CASAS repeated FERNANDO COLUMBUS', or their common source's, words in much the same vein, saying that the cosmographers were convinced that:

"... of this lower sphere of water and earth, no more than a very small part lay uncovered, since all the rest was covered by water and as a result one could only navigate along the shorelines and coasts as the Portuguese were doing along the Guinea coast [i. e., the west coast of Africa]."[145]

In this criticism they implied, very reasonably, that COLUMBUS' project could only be entertained when the east-west extension of the Eurasian landmass was known, as only then would it be possible to know the exact extent of the Atlantic. Since Paul de Burgos' representation made it clear that only a small extent of the *oikoumene* emerged above the water, the Atlantic was likely to be very great in extent and impossible to cross.

At the second meeting of the Spanish cosmographers at Santa Fé in 1491 to re-examine COLUMBUS' project, an Italian, ALEXANDER GERALDINI (1455–1525), was present. He later left a report of the proceedings, written between 1520 and 1524, but published only in 1631. The cosmographers, said GERALDINI, claimed that

"Nicolas de Lyra [i. e. Paul de Burgos] had stated that the whole of man's habitat (*totam Terrae humanae compaginem* – i. e., the *oikoumene*), which lay upon the sea from the Canary islands to the East had no sides turned downwards [i. e. extending] into the lower part of the sphere [of the earth] (*nulla latera habere per inferiorem partem sphaerae obtorta*".[146]

This meant that the Eurasian landmass extending eastwards from the Canary islands to China had no part of it extending into the southern hemisphere. GERALDINI records that he had intervened in the discussion and had remarked that this view did not correspond with the facts, that Nicolas de Lyra (i. e., Paul de Burgos) might know all about sacred theology, but that he knew nothing of cosmography.

[144] FERNANDO COLUMBUS, *Le Historie della vita e dei fatti di Cristoforo Colombo*, Venice, 1571. Modern edition: A cura di RINALDO CADDEO, con studio introduttivo, note, carte e incisioni, Milan, 1930, Vol. I, Chap. XII, pp. 106–7.

[145] BARTOLOMÉ DE LAS CASAS, *Historia de las Indias*, Lib. I, Cap. V, *Biblioteca de Autores Españoles*, Vol. XCV, Madrid, 1957, p. 111.

[146] ALEXANDER GERALDINI, *Itinerarium ad regiones sub aequinoctiali constitutas*, Rome, 1631, pp. 204–205.

> "For the Portuguese had sailed so far into the lower parts of the other [i. e. southern] hemisphere that, leaving ours behind, they had discovered another and that they had observed new stars under the sky of the Antipodes".[147]

The crucial question that was to decide the feasibility of COLUMBUS' project turned on the extent to which the *oikoumene*, generally recognized as circular in form, emerged from the Ocean. The Portuguese, in their discovery of the Cape of Good Hope in 1488, three years before the Santa Fé meeting, had shown that the *oikoumene* stretched far to the south into the Ocean of the southern hemisphere. Such evidence could only help to shake the scholastic arguments of the Spanish cosmographers and, in lending weight to the belief in a narrow Atlantic, encourage COLUMBUS and his supporters.

That COLUMBUS had long before convinced himself that the Portuguese discoveries had invalidated the theory of the relation between land and water held by the scholastics, is shown by the Portuguese historian JOÃO DE BARROS' description of the Genoese's first confrontation with the cosmographers of the Portuguese King D. João II in 1483–84. COLUMBUS, writes BARROS, had remarked that since the Portuguese had discovered the Azores and a considerable part of the African coast, other islands and lands might well lie to the west,

> "... for Nature could not have been so disorderly in the make-up of the globe, for a greater share to have been given to the element of the water, rather than to that of the land for living creatures to live on."[148]

Hardly had the Santa Fé meeting ended with another rebuff for COLUMBUS, when suddenly the position of the Spaniards changed completely. Cardinal Pedro Gonzales de Mendoza, whom GERALDINI had dared to contradict in bringing forward the evidence of the Portuguese discoveries south of the equator, now accepted to

> "give audience to Columbus and saw that he was knowledgeable and well spoken and that he supported his ideas with sound reasoning. The Cardinal judged that he was a man of intelligence and very clever and having reached this conclusion, held him in high esteem and chose to support him [...] As a consequence of this he was listened to by the King and the Queen, who began to accord credit to his memoranda and petitions".[149]

There is nothing in the contemporary documents or the published chronicles which tells us how and why this sudden change occurred in the Spaniards'

[147] IDEM, *loc. cit.*

[148] JOÃO DE BARROS, *Asia*, Lisbon, 1552, Déc. I, Livro III, Cap. XI (p. 112 of the Coimbra, 1932 edition).

[149] GONZALO FERNÁNDEZ DE OVIEDO Y VALDÉZ, *Historia General y Natural de las Indias*, Seville, 1535, *Biblioteca de Autores Españoles*, Vol. CXVII, Madrid, 1957, p. 22.

attitude. On this crucial point we are reduced to conjectures and, as a possible conjecture, we suggest that it was the experience of the Portuguese in their African voyages that finally led the Spaniards to abandon the doctrine of the scholastics on the relation of earth and water, and to embrace that of Ptolemy and Marinus of Tyre.

During his first (1492) and second (1494) voyages, COLUMBUS was convinced that he had reached Asia. When, on his second voyage he proceeded to explore the southern coast of Cuba, he made his crew swear that they were in fact sailing along a part of the China coast.[150] On his third voyage (1498), however, he seems to have had second thoughts about the principles of Ptolemaic geography, entertaining for a moment a return to the scholastic model.

> "I have always read," he wrote, "that the world, land and water, was spherical and authoritative accounts and the experience which Ptolemy and all the others have recorded concerning this matter, so describe it and hold it to be [...] Now I have seen so great irregularity that as a result [...] I find that it is not round as they describe it, but that it has the shape of a pear which is everywhere very round except where the stalk is, for there it is very prominent ..."[151]

In the confused expression of his obscure thinking, COLUMBUS stated that he did not question that the eastern hemisphere was spherical,

> "... but I declare that this other hemisphere [the western] is as the half of a very round pear".[152]

In this momentary revision of the Ptolemaic view, he persuaded himself that the sphere of the water was eccentric to the heavens, for he stated that the ships, after passing west of the Azores, "went upwards rising gently to the sky".[153]

Although COLUMBUS ended by believing to his death that the lands he had discovered in the Atlantic were in fact a part of Asia, not all Spaniards shared this belief. ANTONIO DE NEBRIJA wrote that "the land found by the Catholic Kings 45° west of the Canaries is incorrectly termed India by some".[154] A few

150 "Information y testimonio de como el Almirante fue á reconocer la isla de Cuba quedando persuadido de que era tierra firme" in: MARTIN FERNÁNDEZ DE NAVARRETE, *Colección de los viajes y descubrimientos*, Madrid, 1825 – 1837, Vol. II, pp. 143 – 149.

151 CHRISTOPHER COLUMBUS, "Letter on his third voyage", in: CECIL JANE (ed.), *Select Documents illustrating the Four Voyages of Columbus*, London, Hakluyt Society, 1933, Reprint Nendeln, Kraus, 1967, pp. 28 – 30. On the pear-shaped earth according to COLUMBUS cf. now G. MORETTI, "The Other World and the 'Antipodes'. The Myth of the Unknown Countries between Antiquity and the Renaissance," below in this volume (*CTA* I 1) p. 281 with note 83.

152 IDEM, *loc. cit.*

153 IDEM, *loc. cit.*

154 ANTONIO DE NEBRIJA, *op. cit.*, sign. aa. i (v.).

years later, the founder of the University of Seville, RODRIGO DE SANTAELLA in the introduction to his translation of Marco Polo's travels published in Seville in 1503, declared roundly that those who believed that the islands (*sic*) recently discovered by the Catholic Kings were in India were mistaken.[155]

By the middle of the 16th century the term used by the Spaniards to describe the new lands discovered was, instead of "India", the "Indies", but whether they were attached or not to Asia remained a subject of discussion throughout the century.

The news of COLUMBUS' first voyage was printed in a letter published in Barcelona in 1493 and it went through some 18 editions in Latin, Spanish, Italian, German and English between 1493 and 1497. Yet as we shall see presently, a similar text reporting a Portuguese voyage into the southern hemisphere in 1501, allegedly written by AMERIGO VESPUCCI was to have a much greater impact on the European imagination.

XIV. The Portuguese voyage of 1501, its description allegedly by AMERIGO VESPUCCI *and its impact on European culture*

If in their exploration of the African coast, the Portuguese penetration far into the southern hemisphere had encouraged COLUMBUS in his rejection of the pseudo-Aristotelian model of the relation between the earth and the water and had served to strengthen his confidence in the Ptolemaic model, it was above all the Portuguese voyage of 1501 that, in revealing the long southward stretching coastline of the South American continent, produced clear proof, enunciated in a pamphlet allegedly written by VESPUCCI and spread throughout Europe by the printing press, that the scholastic model was baseless and that the southern hemisphere, instead of being covered by water, had in it in fact a large emerging landmass. This new reality, if it seemed to vindicate Ptolemy's view of the relation between the earth and the water, clashed however with his world map of the *oikoumene*, which showed no continent where the Portuguese had just discovered one.

The Portuguese voyage of 1501 followed on that of Pedro Álvares Cabral, the discoverer of Brazil, who, in 1500 with a fleet of 13 ships on their way to India, had struck out into the Atlantic in order to benefit from the prevailing winds and thereby round the Cape of Good Hope more rapidly. Cabral's fleet went so far west that it touched the South American continent, but he did not stay long enough to establish definitely whether the land he had discovered was an island or a continent.[156]

155 RODRIGO DE SANTAELLA, *Cosmographia breve introductoria en el libro de Marco Polo*, Seville, 1503, sign. A. ii. (v.).

156 Among the narratives of Cabral's voyage, PERO VAZ DE CAMINHA spoke of the land discovered, as "an island" (cf. JAIME CORTESÃO, *A Carta de Pero Vaz de Caminha*, São

Very little is known about the Portuguese voyage of 1501 and it is doubtful if we shall ever know for certain who commanded it.[157] It is claimed that his name was Gonçalo Coelho,[158] although others have proposed that of Fernão de Loronha[159] or that of André Gonçalves, also called Alfonso Gonçalves.[160]

According to the Spanish historian Francisco Lopez de Gómara, the purpose of the voyage was to seek a strait in the South American coastline leading into the Indian Ocean.[161]

The exact role of Amerigo Vespucci on the voyage is not clear, whether, as some have declared, he was the leader of the voyage, or a pilot, or just a merchant.[162] It has been asserted that Vespucci had been invited by the Portuguese to accompany them on account of his supposed expertise, later proven illusory, in calculating longitudes by the lunar distance method, which they hoped would enable them to demarcate the Tordesillas line.[163] This is a reasonable hypothesis.

In about the year 1503 there appeared, probably in Paris, a description in Latin of the 1501 voyage, entitled *Mundus Novus* and allegedly written by Vespucci. This little work of a few pages immediately became one of Europe's best sellers and reached a total of 60 editions by 1529, translated into German, French, Italian, Dutch/Flemish and Czech.[164] Of the enormous outpouring of scholarship on Vespucci and the writings attributed to him, a consensus of

Paulo, 1943, p. 241); Joan Faras, the cosmographer of the fleet said the natives made it understood that the land was "an island and that there were four" (cf. "Letter of Joan Faras", 1/5/1500, in: Joaquim Barradas de Carvalho, *La traduction espagnole du « De Situ Orbis » de Pomponius Mela par Joan Faras et les notes marginales de Duarte Pacheco Pereira*, Lisbon, 1974, p. 107). The pilot of the fleet however wrote that "the land is of great extent and we do not know if it is an island or *terra firma*, but because of its size, we rather believe it is *terra firma*" (cf. Anonymous relation in: Fracanzano da Montalboddo, *Paesi novamente ritrovati*, Vicenza, 1507, Chap. LXVI; facsimile edition of the Milan edition of 1508, Princeton, 1916, p. 78).

157 Damião Peres, *História dos Descobrimentos Portugueses*, Oporto, 3rd ed. 1983, p. 379.

158 Max Justo Guedes, "As primeiras expedições portugueses e o reconhecimento da costa brasileira", in: *Revista Portuguesa de História*, Vol. XIV (Coimbra, 1970), pp. 247 – 267.

159 Duarte Leite, *História dos Descobrimentos*, ed. V. Magalhães Godinho, Vol. 1, Lisbon, 1959, p. 602.

160 Moacyr Soares Pereira, *A Navegação de 1501 ao Brasil e Amerigo Vespúcio*, Rio de Janeiro, 1984, pp. 158 – 163.

161 Francisco Lopez de Gómara, *Historia General de las Indias*, Zaragoza, 1552, f. xlix (*Biblioteca de Autores Españoles*, Vol. XXII, Madrid, 1946, p. 211).

162 Damião Peres, *op. cit.*, pp. 158 – 163.

163 Rolando Laguarda Trias, *El hallazgo del Rio de la Plata por Amerigo Vespucci*, Montevideo, 1982, p. 102. (We do not necessarily agree with the author's conclusions implied in the title of the book.)

164 Cf. Rudolf Hirsch, "Printed Reports on the Early Discoveries and their Reception", in: Fredi Chiappelli (ed.), *First Images of America*, Vol. II, Los Angeles, University of California Press, 1975, p. 540. (According to Hirsch the first edition of the *Mundus Novus* can be dated 1502 – 1504.)

opinion now tends in the wake of the work of ALBERTO MAGNAGHI[165] to follow the view that the *Mundus Novus* is a re-worked version, with additions, by an unknown hand, either the merchant Giuliano Bartolomeo del Giocondo, or the architect Giovanni del Giocondo, of an authentic letter, known as "the Bartolozzi letter" written from Lisbon by VESPUCCI and dated to about 1502.[166]

In the same year as the *Mundus Novus*, there appeared another best seller in Europe, the little encyclopaedia of GREGOR REISCH, *Margarita Philosophica*, which went through 12 editions before 1550.[167]

REISCH repeated Buridan's theory of the relation between land and water, but he added a perspicacious condition that the diameter of the sphere of the earth must necessarily be greater than the radius of the combined sphere of the water and the sphere of the earth in order that the centre of magnitude of the sphere of the water lie on the diameter of the sphere of the earth between the latter's centre of gravity and its centre of magnitude, otherwise nothing of the earth's weight would keep it at or near the centre of the universe. This condition was no doubt an Aristotelian argument to counter the pseudo-Aristotelian doctrine that the sphere of the earth was only a tenth of the volume of the sphere of the water. COPERNICUS, as we shall see later, when he used an argument from Euclid to lay to rest forever the doctrine that the volumes of the elements were in a proportion of ten to one, no doubt had in mind this point of REISCH's.

REISCH's prudent condition for the acceptance of Buridan's theory nevertheless did not prevent his leaving the whole of the southern hemisphere under water as the theory entailed.

It was here that the *Mundus Novus* brought clear empirical evidence to controvert the scholastic doctrine. Of the new regions which had been found, wrote the author,

> "... we may rightly call them a New World [...] For this",

he added, addressing the scholastic doctrine,

> "transcends the view held by our ancients, in as much as most of them held that there is no *continentem* [i. e. mainland] to the south beyond the equator, but only sea which they named the Atlantic; and if some,"

he continued, evoking the Cratesian theory qualified by Christian myth,

> "did aver that a *continentem* [i. e. mainland) there was, they denied with abundant argument that it was a habitable land. But that this their opinion

165 ALBERTO MAGNAGHI, *Americo Vespucci*, 2nd ed. Rome, 1926, pp. 17 – 40.

166 ROLANDO LAGUARDA TRIAS, *op. cit.* (n. 163), Chap. V.

167 D. E. SMITH, *Rara Arithmetica*, London, 1908, p. 82. Cf. also R. RITTER VON SRBIK, *Die Margarita philosophica des Gregor Reisch. Ein Beitrag zur Geschichte der Naturwissenschaft in Deutschland*, in: *Denkschriften der Akademie der Wissenschaften zu Wien, Mathemat.-Naturwiss. Kl.*, vol. CIV, Vienna, 1941.

> is false and utterly opposed to the truth, this my last voyage has made manifest; for in those southern parts, I have found a *continentem* [i. e. mainland] more densely peopled and abounding in animals than our Europe or Asia or Africa ..."[168]

The author of the *Mundus Novus* went on to state that the Portuguese fleet had reached a latitude of 50 degrees S.[169], a figure which RAFFAELLO MAFFEI DA VOLTERRA, writing in 1506 and relying on information received from Lisbon, declared was not confirmed by the Portuguese.[170]

Another text describing all of VESPUCCI's real and alleged four voyages is known as the *Lettera di Amerigo Vespucci* and was published in Florence about 1505–6.[171] A Latin version of the same was published in 1507 in St Dié in the Vosges by a group of humanists encouraged and supported by the Duke of Lorraine, René II. The work published by this group was entitled *Cosmographiae introductio*. Although often attributed to MARTIN WALDSEEMÜLLER, the editor and commentator of the four voyages is now claimed to be MATTHIAS RINGMANN.[172] The *Cosmographiae introductio* went into four editions: St Dié, 1507, Strasburg, 1509, one published in Lyons between 1515 and 1518 and one published in German in Strasburg in 1509.[173]

It was in this work that the famous suggestion appears that the long coastline stretching southward into the southern hemisphere and discovered during the 1501 voyage be named "America".

> "Now a fourth part of the earth has been discovered by Amerigo Vespucci [...] In as much as both Europe and Asia received their names from women, I see no reason why anyone should justly object to calling this part Amerige, i. e., the land of Amerigo, or America, after Amerigo, its discoverer, a man of great ability."[174]

168 *Mundus Novus*, "Letter to Lorenzo Pietro di Medici", translation based on that of GEORGE TYLER NORTHUP, Princeton University Press, 1916, p. 1. TYLER translates *continentem* by "continent". We prefer "mainland"; see infra.

169 IDEM, p. 11.

170 RAFFAELLO MAFFEI DA VOLTERRA, *Commentariorum urbanorum libri XXXVIII*, Rome, 1506 (f. 138 [v.] of the Bâle, 1530 edition). Cf. VSEVOLOD SLESSAREV, "Raphael Maffei's contribution to the history of the Portuguese Discoveries", in: *Congresso Internacional de História dos Descobrimentos* (Lisbon, 1960), Actas, Lisbon, 1961, Vol. III, pp. 551–575.

171 Reproduced with a French translation in HENRI VIGNAUD, *Améric Vespuce (1451–1512)*, Paris, 1917, pp. 313–363.

172 FRANZ LAUBENBERGER, "The naming of America", in: *Sixteenth Century Journal*, Vol. XIII, No. 4 (1982), pp. 91–111.

173 ALBERT RONSIN, *Découverte et Baptême de l'Amérique*, Montréal, 1979, pp. 146–47.

174 MARTIN WALDSEEMÜLLER [= MATTHIAS RINGMANN], *Cosmographiae introductio*, St. Dié, 1507, facsimile edition with English translation by Prof. J. FISCHER and FRANZ VON WIESER, New York, 1907, new facsimile edition, University Microfilms International, Ann Arbor, Mich., 1977, p. XXX (facsim.), p. 70 (Engl. transl.). On WALDSEEMÜLLER now cf. D. WUTTKE, German Humanist Perspectives on the History of Discovery, 1493–1534, below in this volume (*CTA* I 2) (forthcoming).

Neither the *Cosmographiae introductio*, nor the *Mundus Novus* was published in Spain or Portugal. If the term "New World" (*Nuevo Mundo*) was used in these countries, though not to the same extent as *Indias Occidentales* or *Nueva España*, that of *America* was never used until the 18th century. A Spaniard, MICHAEL SERVETUS, seeking to maintain the prestige of COLUMBUS as the first discoverer, was to start a campaign against the use of the word "America". In 1535 he wrote that

> "they are utterly mistaken, those who seek to call this continent 'America', for AMERICO [VESPUCCI] visited the same land as COLUMBUS long after he did, and he went there, not like the Spaniards, but with the Portuguese to trade".[175]

The Spanish exasperation with VESPUCCI and the name he had left behind him continued to such a point that in 1627 Fray PEDRO SIMÓN proposed that the Royal Council of the Indies prohibit the use of all Ptolemaic atlases, maps, globes and sailing charts as well as all books in which the Indies were called "America".[176] Fray PEDRO SIMÓN's efforts were in vain; by the 18th century the name America had become accepted.

XV. The transformation of the meaning of the term orbis terrarum *in the light of the new Discoveries*

Once Europe found itself confronted with the new curiously profiled landmass in the western Atlantic, all the old models of classical geography were brought out anew to see if they fitted the unfamiliar entity any better than the then current ones. We shall examine the fortunes of each in turn to see how they were either rejected or adjusted to help insert the new phenomenon into the European classical tradition.

One of the first concepts to be transformed was that of the *orbis terrarum*. The discovery of new lands of immense size led to the collapse of the notion of a single inhabited *oikoumene* or *orbis terrarum* as a block of lands united together in continuous fashion.[177] Each "part" (*pars*) of the *orbis terrarum* was progressively to become accepted as a "continent".

[175] MICHAEL SERVETUS, in his Latin edition of Ptolemy's *Geography*, Lyons, 1535, text on the reverse side of the map of the New World, entitled *Oceani Occidentalis Terrae Novae Tabulae.*

[176] Fray PEDRO SIMÓN, *Noticias Historiales de las Conquistas de tierra firme en las Indias Occidentales*, Cuenca, 1627, partially reprinted under the title *Noticias Historiales de Venezuela*, Vol. I, with preliminary study by DEMETRIO RAMOS PEREZ, Caracas, 1963, pp. 55–56.

[177] On this concept cf. F. GISINGER, art. "Oikumene", in: PAULY's *Realencyclopädie der classischen Altertumswissenschaft*, Vol. XVIII, Pt. 2, Stuttgart, 1937, col. 2135 („*der Erde als zusammenhängender gemeinsamer Wohnstätte der Menschheit*").

The modern concept of a "continent" is not found in antiquity. The word *continens* refers only to the mainland of the *oikoumene*. In COLUMBUS's letter published in 1493, it appears with the meaning of "mainland": *continentem Chatay provinciam* (the mainland province of Chatay [= Cathay]).[178] Pseudo-VESPUCCI in the *Mundus Novus* (1503) used the word with the same meaning of "mainland":

> *in partibus illis meridianis continentem invenerim frequentioribus populis & animalibus habitatam quam nostram Europam, seu Asiam, vel Africam.*[179]
>
> ("... in those southern parts I have found a mainland more densely peopled and abounding in animals than our Europe or Asia or Africa").

VESPUCCI's authentic letter in Italian of 1502 from Lisbon, the so-called "Bartolozzi letter", on which the *Mundus Novus* is said to be based, uses the expression *terra ferma* – mainland.[180]

MATTHIAS RINGMANN in the *Cosmographiae introductio* (1507) used the word *continens*, though he designated the discovery of 1501 as an island:

> *Hunc in modum terra iam quadripartita cognoscitur. sunt tres primae partes continentes, quarta est insula ...*
>
> ("Thus the earth is now known to be divided into four parts. The first three parts are continuous [i. e. joined together] landmasses, while the fourth is an island ...").[181]

In 1511 the Italian PIETRO MARTYR D'ANGHIERA who had settled in Spain, wrote a history of the exploration of America in which the title *De Orbe Novo* clearly expressed his opinion of the nature of the new discovery. His work went into 8 editions in the 16th century.[182]

The first reasoned statement of the autonomous nature of the American continent was given by GONZALO FERNÁNDEZ DE OVIEDO Y VALDÉZ in his *Historia General y Natural de las Indias*, published in Seville in 1535.

> "Now it is well known that these Indies are in no way part of Europe, nor of Africa, from what I have said of their limits. And if they turn out

[178] CHRISTOPHER COLUMBUS, *De Insulis in Mari indico nuper inventis*, Barcelona, 1493, facsimilé edition Johnson Publishing Co., Murfreesboro, North Carolina, 1970.

[179] AMERICO VESPUCCI, *Mundus Novus*, Latin text reproduced by H. VIGNAUD, *Améric Vespuce (1451 – 1512)*, Paris, 1917, p. 305.

[180] AMERICO VESPUCCI, "Lettre à Laurent di Pier Francesco de Medicis" Lisbon, 1502, in: H. VIGNAUD, *Améric Vespuce (1451 – 1512)*, Paris, 1917, p. 409.

[181] MARTIN WALDSEEMÜLLER [= MATTHIAS RINGMANN], *Cosmographiae introductio*, St. Dié, 1507, facsimilé edition with English translation by Prof. J. FISCHER and FRANZ VON WIESER, New York, 1907, new facsimile edition, University Microfilms International, Ann Arbor, Mich., 1977, p. XXX (facsim.) p. 70 (Engl. transl.).

[182] PIETRO MARTYR D'ANGHIERA, *De Orbe Novo*, Seville, 1511, Alcalá 1516, 1530, Bâle 1533, Paris 1536, Antwerp and Bâle 1537, Cologne 1574, Paris 1587.

to be connected with one of these three, it must be with Asia. And this will only be settled when it has been verified that the farthest most eastern land to be found in Asia beyond China, or any other that may exist, or be even farther to the eastward, is linked up to the westernmost part of the mainland [*Terra firma*] of these our Indies. This part of the mainland lies to the west of New Spain, such as the latter is known here [in Spain]. As it has not so far been entirely discovered, it is not yet known if it ends in sea or continues as land, or if it is entirely surrounded by the Ocean, which I tend to believe. And my opinion, like that of others up till now, is that it is not part of Asia, nor is it joined to what cosmographers call Asia. Rather it is more likely that the mainland [*Terra Firma*] of these Indies is another half of the world, as big, or perhaps greater than Asia, Africa and Europe and that the whole world is divided into two parts and one half is that which the ancients called Asia and Africa and Europe (which they divided up as I have mentioned) and that the other part or half of the world is this one of our Indies. Thus Pietro Martyr was right to call it a New World, in view of the concepts of the ancients, and of what it seems they were ignorant of, and of which we are now aware."[183]

At the date at which OVIEDO wrote he could only offer conjectures concerning the possible separation of the new continent from Asia, and such was still to be the case when the Jesuit JOSÉ DE ACOSTA wrote his *Historia Natural y Moral de las Indias* published in Seville in 1590, in which he took the opposite view.

"What I have said is, as far as I am concerned, largely a conjecture in thinking that the New World (*nuevo orbe*) which we call the Indies is not entirely divided and separated from the other world (*otro orbe*) [i. e. the Old World]. And to give my opinion, I have held for quite a time that both mainlands (*la una tierra y la otra*) are at some point joined together and that they are continuous, or at least close together [...] Thus there are no contrary reasons, nor experience to go against my conjecture or opinion that the whole landmass (*tierra*) is joined together and is continuous at some point, or at least is closely brought together."[184]

ACOSTA's opinion may have been influenced by theological dogma as much as by geographical speculation[185] since, as we have seen earlier, there could

183 GONZALO FERNÁNDEZ DE OVIEDO Y VALDÉZ, *Historia General y Natural de las Indias*, Seville, 1535, Livro XVI, Proemio, *Biblioteca de Autores Españoles*, Vol. XCVIII, Madrid, 1959, p. 86.

184 JOSÉ DE ACOSTA, *Historia Natural y Moral de las Indias*, Seville, 1590, Lib. I Cap. XX. Mod. ed. *Biblioteca de Autores Españoles*, Tome LXXIII, Madrid, 1954, p. 33. – On Acosta now see F. DEL PINO, "Renaissance Humanism and the Origins of Ethnology: Father José de Acosta (1539 – 1600), Paradigm of Jesuit Anthropological Humanism," below in this volume (*CTA* I 2) (forthcoming) and M. MUSTAPHA, "Aristotle and the Government of the Indians: Acosta versus Las Casas," ibid.

185 The separation of America from Asia was only finally proved by Bering's voyage in 1741.

be no separate Creations or Salvations. FRANCISCO LOPEZ DE GOMARA in his *Historia General de las Indias* (1552) discussed the matter at length. After quoting the Stoics, Aristotle, Alexander the Great, the Bible and the Fathers of the Church, GOMARA ended by noting that other worlds should be understood as being

> "*orbes* or parts of the earth, such as Pliny and other writers call Scandinavia and Taprobana [Ceylon]. [...] Although I believe there is only one world, I shall often mention two in this my work, to give variety to my terms (*variar de vocabulos*) in treating a single subject and in order to make myself better understood in calling the Indies, of which I am writing, a New World."[186]

XVI. The revival of the Cratesian theory as a model to help yield understanding of the nature of the American continent and to guide future explorers

The first use of the Cratesian theory and the term "*antichthones*" to situate the new lands across the Atlantic was by the Spanish humanist ANTONIO DE NEBRIJA in his little cosmography published about 1503.

> "Of the remaining part of our hemisphere and of the region opposite inhabited by the antichthones, nothing certain has been reported to our ancestors. But such is the boldness of the men of our time, that it will soon occur that they will bring us a description of this land, either that it is made up of islands, or that it is a mainland (*continentis*), of which a large part of its coastline has been revealed by our navigators, principally the part situated opposite the recently discovered islands (Hispaniola and Isabella) [= Santo Domingo/Haiti and Cuba] and the other neighbouring ones."[187]

NEBRIJA's source for the Cratesian theory was Martianus Capella.[188]

With the printing of Cleomedes' work *De Contemplatione excelsorum orbium disputatio*, Brescia, 1497, the Cratesian theory was given a new lease

[186] FRANCISCO LOPEZ DE GOMARA, *Historia General de las Indias*, Zaragoza, 1552, Mod. ed. *Biblioteca de Autores Españoles*, Vol. XXII, Madrid, 1946, pp. 157–58.

[187] ANTONIO DE NEBRIJA, *In Cosmographiae libros introductorium*, Salamanca (entre 1497 and 1504, probably about 1503), sign. a. ij (v.). Cf. FRANCISCO RICO, "El Nuevo Mundo de Nebrija y Colón", in: VICTOR GARCÍA DE LA CONCHA, *Nebrija y la introducción del Renascimiento en España*, Salamanca, 1983, p. 173.

[188] Cf. W. H. STAHL and RICHARD JOHNSON, *Martianus Capella and the Seven Liberal Arts*, Vol. II, *The Marriage of Philology and Mercury*, *Records of Civilization*, Vol. LXXXIV, New York, Columbia University Press, 1977, p. 226.

of life in the period of the Renaissance.[189] The first to re-use it with Cleomedes' terms was RAFFAELLO MAFFEI DA VOLTERRA in his rambling encyclopaedia published in Rome in 1506, which went through eleven editions in the 16th century.[190]

> "By God's providence and by the felicity of their name, the Christians have opened up our globe (*orbe*) in length and in breadth [...] To the west the limits of the world have, through the virtue of the Kings of Spain, been thrust back in many places beyond the Fortunate Isles [Canary Islands]. Ships have crossed as far as our Perioecians and have discovered islands. [An allusion to the Spanish voyages to America]. To the south, the sailors of the King of Portugal have sailed in their voyages beyond the Tropic of Capricorn and have reached the Antoecians [An allusion to the discovery of the Cape of Good Hope]."[191]

The pertinence of the Cratesian model in helping to situate the newly discovered lands may seem strange to us, but it had far more immediate impact on the minds of Renaissance readers than the new maps that were then appearing and that seemed incomprehensibly unfamiliar.

MAFFEI's text was to be later copied word for word by the Portuguese PEDRO MARGALHO, in his *Physices Compendium* (1520).[192]

After MAFFEI in Rome, the Swiss humanist JOACHIM VADIANUS, who had settled in Vienna, noted that Macrobius in his treatment of the Cratesian model had been mistaken in declaring that there could be no communication between the four Cratesian islands.

> "Who can doubt" he wrote, "that at the tip of Africa are the Antoecians of our climate? If we can go to them, what prevents them from coming to us, as we see occurring everyday."[193]

As the situation of the new lands became better known on the maps, so the Cratesian model evolved. In PETER APIAN's *Cosmography* of 1524, the four Cratesian "islands" became first of all abstract geometrical "points"; then, developing his reasoning, he situates them without reference to the Medieval *oikoumene*:

[189] Cf. mod. ed. Cléomède, *Théorie élémentaire* (*De motu circulari corporum caelestium*), ed. and translated by RICHARD GOULET, Paris, 1980, p. 95.

[190] RAFFAELLO MAFFEI DA VOLTERRA, *Commentariorum urbanorum libri XXXVIII*, Rome, 1506 (f. 4 [v.] of the Bâle, 1530 edition); cf. on this text G. CARACI, "Americo Vespucci al recente congresso di storia delle scoperte di Lisbona (4 – 12 Settembro, 1960)", in: *Memorie Geografiche*, Vol. VII, Rome, 1961, pp. 142 – 144.

[191] RAFFAELLO MAFFEI DA VOLTERRA, *op. cit.*, *loc. cit.*

[192] PEDRO MARGALHO, *Physices Compendium*, Salamanca, 1520, f. vii (r.).

[193] JOACHIM VADIANUS, *Habes lector: hoc libello, Rodolphi Agricolae ... ad J. Vadianum ...*, Vienna, 1515, sign. C. i (v.) – C. ii (r.). On VADIANUS cf. GEORG GEILFUS, *Joachim v. Watt genannt Vadianus als geographischer Schriftsteller*, Winterthur, 1865.

> "The whole earth is divided into four parts. Authors such as Cleomedes and others, place on each meridian and on each point on any meridian four habitats having each in relation to the others a given proportion, of which the first is the one that we inhabit. Likewise, for the first habitat one may take any place or point on the whole terrestrial globe (*terrarum orbe*). A second habitat will be those called the Perioecians, that is, those who live round behind us (*circumcolae*). A third habitat is that of the Antoecians, that is those who live opposite and across from us (*anticolae*). The fourth and last region or habitat is that of the men called Antichthones or Antipodes."[194]

APIAN's little book went into 34 editions in the 16th century in Latin, French, Dutch and Spanish.

Towards the middle of the 16th century, the Cratesian model appears in reverse form: Instead of serving to situate the new lands, the latter, now well enough known on maps, were used in an attempt to explain the model, which had by this time played out its usefulness. The Spanish historian FRANCISCO LOPEZ DE GOMARA wrote thus in 1552:

> "Besides the Antipodes, there are others, called Perioecians (*Parecos*) and Antoecians (*Antecos*). For included in these three names are all the inhabitants of the world. The Antipodes are those whose feet tread the earth directly opposite one another, such as those of Guinea and Peru. The Antoecians [should be Antipodes] of the Spaniards and the Germans are those who live about the Rio de la Plata and the Patagonians who live around Magellan's strait [...] For us Spaniards, our Perioecians are those who live in New Spain and in Cibola and thereabouts and those who live in Chile [sic!]."[195]

GOMARA shows a poor understanding of the Cratesian model completely mixing up the order of his Cratesian populations. From this point on, the Cratesian model disappears from geographical texts.

If the Cratesian model was used for a time to help situate the newly discovered lands, it also served at the beginning of the 16th century to guide explorers of the new continent in their search for straits which would lead them into the Indian Ocean. If, according to the Cratesian model, there were two landmasses in the western Atlantic, one in the northern hemisphere and one in the southern, there should theoretically be a strait separating them near the equator and also "a north-west passage" and a "south-west passage".

COLUMBUS, during his fourth voyage in 1502 – 1503, sought a strait in the region of Panama, which he called "the province of Ciguare". Ciguare, he

[194] PETER APIAN, *Cosmographus liber*, Landshut, 1524, Chapter XVI, f. 54.

[195] FRANCISCO LOPEZ DE GÓMARA, *Historia General de las Indias*, Zaragoza, 1552, Mod. ed. *Biblioteca de Autores Españoles*, Vol. XXII, Madrid, 1946, p. 160.

said, has the sea on the other side of it and would be only ten days distance from the Ganges in India.[196]

In 1508, the King of Spain sent Juan Díaz de Solis and Vicente Yañez Pinzon to "... discover [in the Carribean] the channel or open sea (*mar abierto*) which you are going to look for and which I want to be sought".[197] On 20 December 1512, the King wrote once more insisting that "... the Gulf of Hispaniola [i. e. the Caribbean] should finally be discovered and that it should be learnt if there was a strait there as some have believed up till now."[198]

The strait was never found; it had to be dug. As early as about 1542, Alonso de Santa Cruz proposed to the King that a canal be cut in the region of Panama to Nombre de Dios.[199]

Several early attempts were made to discover a passage round the southern extremity of the South American coastline. The first was that of a Portuguese voyage of 1511 – 12, under the command of Diogo Ribeiro, with as pilot João de Lisboa.[200] The principal source for the voyage is found in a pamphlet in German with the title of *Neuen Zeytung ausz Pressilg Landt*, printed, with no indication of place, about 1514.[201] It was during this voyage that the Rio de la Plata was discovered and thought at first to be the much searched for strait until it was found to contain fresh water.

The second attempt to find a southwest passage into the Indian Ocean was that of Juan Díaz de Solis in 1515. He was sent to discover the "back of South America" (*descubrir las espaldas de Tierra Firme*).[202] He reached the Rio de la Plata, where he was killed by Indians; his crew was able to return home.[203]

The third and finally successful attempt was Magellan's. During his voyage of 1519 – 22, he discovered the strait which bears his name and which led him into the Pacific. Although he was killed by the inhabitants of a Pacific island, his ship and crew were able to return to Spain circumnavigating the globe.[204]

196 Christopher Columbus, "Letter on his fourth voyage", in: Cecil Jane (ed.), *Select Documents Illustrating the Four Voyages of Columbus*, London, Hakluyt Society, 2nd series, Vol. LXX (1933), p. 82.

197 Document in José Toribio de Medina, *Juan Díaz de Solis*, Vol. II, Santiago de Chile, 1897, pp. 29 – 30.

198 Document in Idem, Vol. I, Santiago de Chile, 1897, pp. cxxi – cxxii.

199 Alonso de Santa Cruz, *Islario*, in: Mariano Cuesta, *Alonso de Santa Cruz y su obra cosmografica*, Vol. II, Madrid, 1984, p. 286.

200 F. M. Esteves Pereira, "O Descobrimento do Rio de la Plata", in: Carlos Malheiro Dias (ed.), *História da Colonização Portuguesa do Brasil*, Vol. II, Oporto, 1923, pp. 351 – 90, and Rolando Laguarda Trias, *El Predescubrimiento del Rio de la Plata por la expedición portuguesa de 1511 – 12*, Lisbon, 1973, Chap. X.

201 Anon., *Tidings out of Brazil*, English translation by Mark Graubard with facsimile reproduction of the original text and commentary and notes by John Parker, Minneapolis, University of Minnesota Press, 1957.

202 José Toribio de Medina, *op. cit.*, Vol. I, p. ccxxxvi.

203 Idem, p. cclx – cclxi.

204 Jean Denucé, *Magellan. La question des Moluques et la première circumnavigation du globe*, *Académie Royale de Belgique, Classe des Lettres, Mémoires*, Vol. IV, fascicule III,

The north-west passage was also sought for from an early date.

The Portuguese Corte Real brothers are said to have sought it in their voyage of 1500.[205]

In 1524 – 25 Esteban Gomez, a Portuguese in the service of Spain, was sent by Charles V to seek the strait which it was believed would lead into the Pacific and enable him to reach Cathay. What he thought was the strait he had discovered turned out to be only the Penobscot River on the Canadian coast leading out into the Bay of Fundy.[206]

Another attempt in 1524 was that of GIOVANNI DA VERRAZANO, an Italian in the service of François I, King of France. VERRAZANO wrote that

> "My intention [...] was to reach Cathay on the extreme eastern coast of Asia, but I did not expect to find such an obstacle of new land as I have found and if for some reason I did expect to find it, I estimated that there would be some strait to get through to the Eastern ocean. This was the opinion of all the ancients, who certainly believed that our Western Ocean [i. e. the Atlantic] was joined to the Eastern Ocean of India [= Indian Ocean] without any land between. Aristotle supports this theory with arguments of various analogies, but this opinion is quite contrary to that of the moderns, and has been proved false by experience."[207]

VERRAZANO's reference to Aristotle concerns the latter's remark in the *Meteorologica* that "the Ocean severs the habitable land and prevents it forming a continuous belt round the globe."[208]

VERRAZANO sailed up the North American coastline from the shores of North Carolina to New Foundland. His brother GEROLAMO made a map of the voyage in which the Hatteras sand spit is shown as a narrow neck of land separating the Atlantic from the Pacific Ocean. The latter, in reality only

Brussels, 1911 and JOSÉ TORIBIO DE MEDINA, *El Descubrimiento del Océano Pacífico*, Santiago de Chile, 1914.

205 FRANCISCO LOPEZ DE GOMARA, *Historia General de las Indias*, Zaragoza, 1552, *Biblioteca de Autores Españoles*, Vol. XXII, Madrid, 1946, p. 177. On the Corte Reals' voyage cf. ANTÓNIO GALVÃO, *The Discoveries of the World*, Lisbon, 1563, English translation, London, 1601. Reprint with Portuguese and English texts, Hakluyt Society, 1st series, Vol. XXX, London, 1862, Facsimile reprint New York, Burt Franklin, n. d., pp. 96 – 97, and DAMIÃO PERES, *História dos Descobrimentos Portugueses*, 3rd ed., Porto, 1983, pp. 349 – 361.

206 ALONSO DE SANTA CRUZ, *Islario (1542 – 60)*, in: MARIANO CUESTA, *op. cit.*, II, p. 285. English Translation in H. P. BIGGAR, *Les précurseurs de Jacques Cartier, 1497 – 1534*, Ottawa, 1913, p. 193. Cf. also D. B. QUINN, *North America from Earliest Discovery to First Settlement. The Norse Voyages to 1612*, New York, 1977, p. 160.

207 GIOVANNI DA VERRAZANO (1524), in: LAWRENCE C. WROTH, *The Voyages of Giovanni da Verrazano, 1524 – 28*, New Haven, Connecticut, Yale University Press, 1970, p. 131 (Italian text) and p. 142 (English translation).

208 Aristotle, *Meteorologica*, Bk. II, Chap. 5, 362 b, translated by H. D. P. LEE, Loeb edition, Cambridge, Mass. and London, 1978, p. 183.

Pamlico sound, is shown on his map as of enormous extent filling almost the whole of present day Canada.[209]

Many other attempts to find the northwest passage were to follow throughout the 16th century and after.

XVII. The revival of the Ptolemaic theory of the oceans enclosed in basins as a model for the continuous nature of the coastline of the American continent

By the year 1500 three strips of coastline had been discovered in the western Atlantic: COLUMBUS' discovery of the mainland in the Caribbean, Cabral's discovery of a part of the coast of Brazil and the Corte Reals' discovery of Newfoundland.[210]

The theory that these strips of coastline were linked in a continuous manner is first expressed in a letter addressed to the Signoria of Venice, dated 18th Oct 1501 by PIETRO PASQUALIGO:

> "They [the Portuguese] also believe," he wrote from Lisbon, "that they [Newfoundland and Labrador] are joined to the Andilie [i. e., Antilles] which were discovered by the Spanish monarchs and to the land of the parrots (*terra dei papagà*), [i. e. Brazil] recently discovered by the ships of King of Portugal on their way to Calicut [in India]."[211]

Thirteen years later another confirmation of this notion appears, in which the American coastline is seen as not only continuous but as passing south of the tip of Africa to join up with the southeast Asian peninsula. GIOVANNI DA EMPOLI wrote thus from Lisbon on the 22 August 1514:

> "The land of Santa Cruz, called Brazil, has not yet been [entirely] discovered, for it is supposed that the Antilles of the King of Castile as well as the land of the Corte Reals [Newfoundland] link up with the land of Malacca and are one and the same with it."[212]

The existence of land extending in an east-west direction south of the Cape of Good Hope is shown on a sketch map of the early 16th century by

209 GEROLAMO DA VERAZANO, World Map, 1529, Vatican Library.

210 On the Corte Real brothers, cf. DAMIÃO PERES, *História dos Descobrimentos Portugueses*, 3rd edition, Porto, 1983, pp. 349 – 361.

211 Reproduced in HENRI HARRISSE, *Les Corte-Real et leurs voyages au Nouveau-Monde d'après des documents nouveaux ou peu connus tirés des archives de Lisbonne et de Modène, suivi du texte inédit d'un récit de la troisième expédition de Gaspar Corte-Real etc.*, Paris, 1883, pp. 209 – 10.

212 GIOVANNI DA EMPOLI, "Lettera mandata da Giovanni da Empoli a Leonardo suo padre, Lisbon, Aug. 22, 1514", in: *Archivio Storico Italiano*, Serie I, Tome III, Appendix, Florence, 1846, pp. 78 – 79.

Alesandro Zorzi. It concerns a Portuguese voyage commanded by António de Saldanha that left Lisbon for India on March 14, 1503.[213] On the sketch of this southern land, identified as part of Brazil, appears the following legend:

> "Land seen by the Portuguese 600 [leagues] to the south [not necessarily to the exact south of the Cape of Good Hope, as will be seen in the following quotation from Enciso]. It is called Brazil."[214]

This enigmatic note was later repeated by MARTÍN FERNÁNDEZ DE ENCISO in his *Summa de Geographia* (1519). ENCISO admitted that no one had ever seen this southern land, but he nevertheless placed it in 42° Lat. S.

> "This Cape of Good Hope lies to the west of this land, called southern (*austral*) and from this Cape of Good Hope to this southern (*austral*) land there are 450 leagues. This southern (*austral*) land lies in 42° and at 600 leagues from Cape St Augustin [in Lat 8° on the Brazilian coast] [...] Nothing more is known of this land than that which has been seen from ships, for no one has landed there."[215]

Already some ten years previously the Portuguese soldier-sailor-cosmographer DUARTE PACHECO PEREIRA had gathered together these reports concerning the lands discovered in the Western and Southern Atlantic and on the basis of them he worked out a general theory of Ptolemaic inspiration, but borrowed, as he himself admitted, from JACOB PEREZ DE VALENCIA.

> "Therefore we must consider," wrote Pereira, "how the philosophers who treated of this matter said that the earth is completely surrounded by the sea, bringing themselves to believe that the whole of our world, [...] is made into an island by the surrounding waters [...] but some of the modern doctors have held different, even contrary opinions. For they sought to prove by the authority of Holy Scripture and able arguments opposed to those of the ancients that the earth is larger than the sum of all the waters, which are placed in the depths and concavity of the earth and are surrounded by the earth; as to which we must note the words of Jacob, Bishop of Valencia, an excellent scholar and master of sacred

[213] Cf. on this voyage A. TEIXEIRA DA MOTA, *A Viagem de António de Saldanha em 1503*, *Agrupamento de História da Cartografia Antiga*, *Secção de Lisboa*, Vol. LXIV, Lisbon, 1971.

[214] ALESSANDRO ZORZI, Collezione Alberico, Biblioteca Magliabecchiana, XIII, Vol. 84 (= Collez. Alberico, Vol. 4: Voyage of Antonio de Saldanha), f. 169 (v.). Reproduction of the relevant part of the manuscript with a transcription in S. CRINÒ, "Schizzi cartografici dei primi anni della scoperta dell'America", in: *Rivista Marittima*, Rome, 1930, Supplementary fascicule, Nov. 1930, pp. 51 – 52, fig. 21. That the sketch was made by ZORZI and not by BARTOLOMÉ COLUMBUS, as has been claimed, is shown by ROBERTO ALMAGIÀ, "Intorno a quattro codici fiorentini e ad uno ferrarese dell'erudito Alessandro Zorzi", in: *La Bibliofilia*, Vol. XXX (1936), pp. 313 – 347.

[215] MARTÍN FERNÁNDEZ DE ENCISO, *Suma de Geographia*, Seville, 1519, sign. f. v. (v.).

> theology, who in his commentary on the Psalter and with reference to Psalm 104 [...] says that the waters are all set within the concavity of the earth and that the earth is far larger than all the waters."[216]

PEREIRA does not seem to have absorbed Ptolemy's doctrine directly on this point and he makes no reference to the fact that Ptolemy is at least in part the source of PEREZ's theory.

For PEREIRA the whole coastline of America, or rather the sections explored up until then, tended to confirm PEREZ's theory.

> "There has been found and explored," he wrote, "a very large landmass with many large islands adjacent, extending to 70° North of the Equator, toward the Arctic pole [...] and to 28½° on the other side of the Equator towards the Antarctic pole; and such is the spread of its greatness and its length that on either side its end has not been seen or known, so that it is certain that it circles all around."[217]

Although PEREIRA admitted ignoring the northern and southern limits of this land, he was nevertheless certain that it "circles all around". This certainty, inspired most probably from his reading of PEREZ, he inferred from the fact that the land in question was always found at a constant distance on a radius from the *oikoumene*, which appeared to him very small in contrast to the immensity of the Ocean and to the land beyond it.

> "Thus we have learnt", he continues, "that, if from the shores and coast of Portugal or from any other point of Europe, Africa or Asia, we sail across the Ocean due west, that is in terms of seamanship through 36° of longitude which are equivalent to 648 leagues, counting 18 leagues to a degree,[218] (and there are places where this distance is somewhat greater), we find this land [...] Now since this land that lies over there is so great and on this side we have Europe, Africa and Asia, it is obvious that the Ocean lies in the middle between these two landmasses and is 'mediterranean'. Hence we can say that the Ocean does not surround the earth as the philosophers have declared, but rather the earth surrounds the sea, and that it lies in its hollow and centre. And from this I conclude that the

216 DUARTE PACHECO PEREIRA, *Esmeraldo de Situ Orbis* (c. 1508), edited by EPIFÂNIO DA SILVA DIAS, Lisbon, 1905. Reprint Lisbon, 1975, pp. 21–22. English translation by G. H. T. KIMBLE, Hakluyt Society, 2nd series, Vol. LXXIX, London, 1936, Reprint Nendeln, Kraus, 1967, pp. 10–11 (not followed at every point by us).

217 IDEM, Portuguese edition, p. 23, English translation p. 12.

218 PEREIRA's value for the length of a degree of the meridian is unusual but closer to the real value than the two others commonly found at this period: 16⅔ and 17½. Cf. A. TEIXEIRA DA MOTA, "Bartolomeu Dias e o valor do grau terrestre", in: *Congresso Internacional de História dos Descobrimentos* (Lisbon, 1960), Actas, Lisbon, 1961, Vol. II, pp. 299–309.

> Ocean is nothing other than an enormous lake in the hollow of the earth and that the earth and the sea together make up a single roundness."[219]

PEREIRA's description of the Atlantic appears to be a transfer from Ptolemy's representation of the Indian Ocean as a closed sea. The Portuguese voyages of BARTOLOMEU DIAS (1488) and of VASCO DA GAMA (1497) had shown Ptolemy to be wrong and that the Indian Ocean was connected to the Atlantic. PEREIRA retained the Ptolemaic principle applying it to the Atlantic itself and to the lands beyond. His new image of the relations of land and water can be seen graphically illustrated on the map of LOPO HOMEM of 1519 (pl. IV).[220]

In stating that "the earth and the sea make up a single roundness" PEREIRA showed how the experience of navigators had helped to swing opinion away from the scholastic theory of the spheres of the earth and the water with different centres, in favour of Ptolemy's definition of the land and the water having the same continuous surface.

Later, after Magellan had found and passed through the strait which bears his name, PEREIRA's image of a continuous coastline collapsed as a geographical model.

XVIII. The final collapse of the pseudo-Aristotelian doctrine of the two spheres of the water and of the earth with non-coinciding centres and the emergence of the concept of the terraqueous globe

The first clear recognition of the clash between the experience of the Discoveries of the Portuguese in the southern hemisphere and the Aristotelian doctrine of the elements appears in 1506 in the work of RAFFAELLO MAFFEI DA VOLTERRA, which went into ten editions in the 16th century.

After having, as we have seen above, resuscitated the Cratesian model to situate the extension of the African continent in the southern hemisphere, as well as the presence of new lands in the Western Atlantic, MAFFEI adds that the navigators of the King of Portugal have shown that the torrid zone is inhabited and that they have "clearly revealed the false opinion of the poets and of Aristotle himself on the subject". He then notes that the discovery of the new lands had disproved the pseudo-Aristotelian doctrine that the proportions of the elements are in a ratio of 10:1.

> *... neque sane aliud credere par est quam apparens, quod aquae locus non esset si decuplo terram ex ipsius philosophi sententia superet.*[221]

[219] DUARTE PACHECO PEREIRA, *op. cit.*, Portuguese text, pp. 23 – 24, English translation, pp. 12 – 13 (not closely followed by us).

[220] LOPO HOMEM, Mappaemundi, 1519, Département des Cartes et Plans, Bibliothèque Nationale, Paris. Reproduced in A. CORTESÃO & A. TEIXEIRA DA MOTA, *Portugaliae Monumenta Cartographica*, Lisbon, Vol. I, 1960, Plate 16.

[221] RAFFAELLO MAFFEI DA VOLTERRA, *Commentariorum urbanorum libri XXXVIII*, Rome, 1506 (f. 4 [v.] of the Bâle edition of 1530).

("And it is certainly not reasonable to believe other than what is apparent [i. e. observed by the navigators] because [in view of their experience] there would be no place for the water if, as the philosopher [= Aristotle] declares, it is ten times that of the earth.")

We have already seen that the doctrine of the 10:1 ratio was not Aristotle's, but that of his followers. Yet the impact of MAFFEI's remark should be gauged against the statement of a typical teacher of Aristotelian philosophy who taught at Padua in the 15th century: GAETAN DE TIENE (died 1465), who in his commentary of Aristotle's *Meteorologica* wrote that

"... the air is ten times that of the water, just as the fire is ten times that of the air and the water is ten times that of the earth [...] though others have a different opinion for which I do not care (*de quo non est cura*)."[222]

The whole of MAFFEI's lines were plagiarized by the Portuguese scholar PEDRO MARGALHO in his work *Physices Compendium*, published in Salamanca in 1520. MARGALHO however added the following significant words of his own in which we see a cogent argument, clearer than those of PEREZ DE VALENCIA and JEROME MÜNZER and drawn from the experience of sailors, in defence of Ptolemy's doctrine in the *Geography* that the earth and water had the same continuous surface.

... quapropter immixta esse duo aquae et terrae elementa constat, nec dari in aqua gradum, ut isti dicunt, cui ex adverso non resideat terrae gradus et haec duo elementa equari quando cuicumque parti unius residet pars alterius e contrario probabile videtur.[223]

("This is why it is agreed that the two elements of water and of earth are intermixed and that it does not seem probable, as they [i. e., the navigators of the King of Portugal] confirm, that there exists a degree on the water to which there does not correspond in [symmetrical] opposition a degree on the land, and it seems likely that these two elements are equal, given that to any one of the parts of the one there corresponds in [symmetrical] opposition a part of the other.")

There is no known document to tell us when the Portuguese navigators had perceived this fact and what had been the procedures by which they had verified such a conclusion. MARGALHO is the first scholar to confront scholastic Aristotelian doctrine with the experience of sailors.

PEDRO NUNES, the famous Portuguese mathematician, was later to repeat, in a marginal note to his vernacular translation of Sacrobosco's *Sphere*, MARGALHO's remark, though without mentioning his name.

[222] GAETAN DE TIENE, *Libri Metheororum ... cum commentariis*, Padua, 1476, Lib. I, Cap. III (no pagination).

[223] PEDRO MARGALHO, *Physices Compendium*, Salamanca, 1520, f. vii (r.).

"These two elements of water and earth together make up one sphere as is proved from the eclipses of the moon [...] It can also be proved that this is so because the same number of leagues or of miles corresponds to a celestial degree equally on land (*terra*) as at sea. This could not be so if both did not together make up one sphere [...] The ancients made much difficulty over this for they held that the sea was much greater than the land (*terra*), but the recent voyages of discovery have shown, as appears in the descriptions of the Portuguese, how the surface of the earth is greater than that of the sea."[224]

The single sphere of the earth and the water, much later to be given the name of the "terraqueous globe", was defined by the Spaniard MARTÍN FERNÁNDEZ DE ENCISO in the following terms in his *Suma de Geographia* published in Seville in 1519. His definition is an important modification of that in Sacrobosco's *Sphere*, which provided him with his model of the Aristotelian arrangement of the elements.

"The sphere which is commonly called the world (*mundo*), is divided into two regions, one etherial and celestial; the other region is that of the elements subject to change. And the region of the elements is divided among four elements which are earth, air, water and fire. The earth is in the centre and after it comes the water which surrounds it all about. But one must remember that the water and the earth together make up one body in which is the earth on one hand and the water on the other and both together make up the centre [of the world] and one cannot [suffice] without the other, because the earth does not surround the water nor does the water surround the earth and after the earth and the water comes the region of the air which lies all round the earth and the water ..."[225]

The influence of the Discoveries, in forcing this fundamental modification of Aristotelian doctrine, was to spread slowly through Europe in the face of entrenched resistance from Aristotelian scholars, above all in the universities.

In the German cultural area the first developed challenge to the scholastic Aristotelian doctrine concerning the earth and the water appears in Vienna in 1515, after MAFFEI in Italy, but before the Spanish and Portuguese scholars just quoted. The occasion was an exchange of letters between two humanists: RUDOLF AGRICOLA (1490–1521) and JOACHIM VADIANUS (1484–1551). VADIANUS, a Swiss settled in Vienna, sought to reassure his friend, troubled by the clash between the Church's doctrine on the Antipodes (in particular the views of Lactantius and St Augustine) and the observed experience of the navigators of the Renaissance.

"If you understand," wrote VADIANUS to AGRICOLA, "that the whole mass of the inhabited earth is round, as both the mathematicians and the

[224] PEDRO NUNES, *Tratado da Sphera*, Lisbon, 1537, facsimilé edition by J. BENSAUDE, Munich, 1915, sign a. iiii (r.).

[225] MARTÍN FERNÁNDEZ DE ENCISO, *Suma de Geographia*, Seville, 1519, sign. A. iij (r.–v.).

physicists agree, conceive of it as if it were a sort of clod, that emerges from the immense ocean on one side and that together with the water under which it is partially hidden and from which it partially emerges (*partim latet, partim eminet*), comes out to be one single roundness."[226]

VADIANUS here wrestles with the problem of how to define the arrangement of the two elements of earth and water, if it is agreed that they are not eccentric to one another. Compared with that of ENCISO, his definition is more accurate. He next proceeds to attack the holders of the scholastic position.

"It might seem thus, Rudolf, to those who believe that on one side the earth emerges from the water like a hilltop rising above the flat surface of a lake or even like an apple floating in it, as some of the ancients have thought. But, by Hercules, to imagine this is the work of a dull mind, given that, in reality, things are very different. As the many facts of experience have shown and as George Peurbach [1423–61], a German very clever at mathematics, has recorded in his collected writings on the *Sphere* of Sacrobosco, the earth together with the waters which surround it, forms a globe in such a way [...] that it emerges like a small clod out of a lake or out of a swamp and that, together with the waters, it follows a curved globe-like shape as far as [its surface] extends."[227]

VADIANUS' reference to the "many facts of experience" may probably be to knowledge of the Discoveries which had reached him through Pseudo-VESPUCCI's letters, to which he alludes. PEURBACH's work, quoted by VADIANUS, does not appear to have been printed and no trace of a manuscript appears in ZINNER's list.[228]

With the exception of VADIANUS, who became Rector of Vienna University,[229] most of those who challenged the Aristotelian doctrine of the scholastics were outside the universities. They tended to be concerned with astronomy and mathematics and it often occurred that both these subjects were handled by doctors of medicine.

Such was the case of JEAN FERNEL (1497–1558), a doctor of medicine, whose *Cosmotheoria*, published in Paris in 1528 and dedicated to King D. João III of Portugal, remains one of the most original contributions to cosmography in the French Renaissance. After having given a condensed

226 JOACHIM VADIANUS, *Habes lector: hoc libello Rudolphi Agricola ... ad I. Vadianum ...*, Vienna, 1515, sign. B. ii (r.). There were three other editions, Vienna, 1518, and 1522 and Bâle, 1557.

227 IDEM, sign. B iij (r.).

228 ERNST ZINNER, *Verzeichnis der astronomischen Handschriften des deutschen Kulturgebietes*, Munich, 1925, pp. 241–243.

229 GEORG GEILFUS, *Joachim von Watt, genannt Vadianus, als geographischer Schriftsteller*, Winterthur, 1865, p. 5.

account of Ptolemy's proofs of the sphericity of the earth from observations of the pole star and of eclipses, adding that the proof from eclipses does not wholly satify him, FERNEL declares that

> "... we need to examine more carefully the reasoning of certain philosophers, most of them modern ones, who are convinced that the centre of magnitude of the earth is different from the centre of the universe [...] If, they say, the surface of the earth were at an equal distance from the centre of the universe, the flooding of the waters would immediately cover the whole earth. They are accustomed to add another argument. Given that, because of the wetness it receives, the side of the earth submerged under the waters is heavier than the other side which is out of the water and is continuously dried by the sun, it is therefore the former side which must lie closer to the centre of the universe, while the latter one must lie farther away from it; and that the centre of gravity of the earth must be the same as the centre of the universe, while its centre of magnitude must lie toward its less heavy side. But there is no doubt whatever that this derives from a false hypothesis such as I shall briefly demonstrate. First of all, all the philosophers and Aristotle in particular in Book II of the *De Caelo*, prove with arguments that are above all natural ones, that the inhabited side of the earth together with the waters which go together with it (*connexo*), forms one single convex surface whose centre coincides with the centre of the universe. Therefore, since the writings of learned men and the reliable evidence of navigators have shown us, scattered everywhere in the sea, innumerable islands and sandbanks that follow almost the same convexity with the sea, it is obvious that these places are not less distant from the centre of the universe than is the inhabited side of the earth. And the great number of places that they occupy in the sea amply shows that the part of the earth that is covered by the waters can be said to be at the same distance from the centre of the universe as the other part. These elevated places which, as we have said, lie in the sea, are like the mountains by the valleys, surrounded by hollows which receive the waters [...].
>
> One must thus agree that the earth looks like a wooden globe in which there are many hollows in which water can gather."[230]

FERNEL's interpretation of Aristotle, wielded to refute Buridan's doctrine, leaves one perplexed, for what is clearly a Ptolemaic demonstration of the relation of earth and water can hardly be substantiated by the references to the passages in Aristotle to which he refers. Was FERNEL trying to cover his tracks, criticizing the scholastics and at the same time invoking fidelity to the "Philosopher" out of fear of the Aristotelians? Though he placed the experience of navigators on the same level as the "writings of learned men", he clearly used the "reliable evidence" of the former to give conclusive cogency to his argument.

[230] JEAN FERNEL, *Cosmotheoria*, Paris, 1528, sign. B. i (v.).

There is no evidence that Fernel's book was known to Copernicus and it is commonly agreed that the general outlines of Copernicus's heliocentric theory were formulated in the first decade of the sixteenth century before Fernel's book was published.

In Copernicus we find the clearest and most convincing refutation of the scholastic doctrine of the spheres of the earth and the water and in particular of the 10:1 ratio in the proportion of the elements.

Chapter 3 of Book I of Copernicus' *De Revolutionibus Orbium Coelestium*, published in Nuremberg in 1543, is subtitled as follows: "How earth forms a single sphere with water."

> "... we should not heed certain Peripatetics who declared that the entire body of water is ten times greater than all the land. For, according to the conjecture which they accepted, in the transmutation of the elements as one unit of earth dissolves, it becomes ten units of water. They also assert that the earth bulges out to some extent as it does because it is not of equal weight everywhere on account of its cavities, its centre of gravity being different from its centre of magnitude. But they err through ignorance of the art of geometry. For they do not realize that the water cannot be even seven times greater and still leave any part of the land dry, unless the earth as a whole vacated the centre of gravity and yielded that position to water, as if the latter were heavier than itself. For, spheres are to each other as the cubes of their diameters. Therefore, if the earth were the eighth part to seven parts of water, earth's diameter could not be greater than the distance from [their joint] centre to the circumference of the waters. So far are they from being as much as ten times greater [than the land].
>
> Moreover, there is no difference between the earth's centres of gravity and magnitude. This can be established by the fact that from the ocean inward, the curvature of the land does not mount steadily in a continuous rise. If it did, it would keep the sea water out completely and in no way permit the inland seas and such vast gulfs to intrude. Furthermore, the depth of the abyss would never stop increasing from the shore of the ocean outward, so that no island or reef or any form of land would be encountered by sailors on the longer voyages. But it is well known that almost in the middle of the inhabited lands barely fifteen furlongs remain between the eastern Mediterranean and the Red Sea. On the other hand, in his *Geography* Ptolemy extended the habitable area halfway around the world. Beyond that meridian, where he left unknown land, the moderns have added Cathay and territory as vast as sixty degrees of longitude, so that now the earth is inhabited over a greater stretch of longitude than is left for the ocean. To these regions, moreover, should be added the islands discovered in our time under the rulers of Spain and Portugal, and especially America, named after the ship's captain who found it. On account of its still undisclosed size it is thought to be a second group of inhabited

> countries. There are also many other islands, heretofore unknown. So little reason have we to marvel at the existence of antipodes or antichthones. Indeed, geometrical reasoning about the location of America compels us to believe that it is diametrically opposite the Ganges district of India.
>
> From all these facts, finally, I think it is clear that land and water together press upon a single centre of gravity; that the earth has no other centre of magnitude; that since earth is heavier, its gaps are filled with water; and that consequently there is little water in comparison with land, even though more water perhaps appears on the surface."[231]

COPERNICUS uses Euclid's theorem in Book XII, Proposition 18[232] that the volumes of two spheres are in the proportion of the cube of their diameters, to invalidate Buridan's synthesis of Aristotle's physics and Archimedes' principle of the centre of gravity of a body. His procedure is a development of that in GREGOR REISCH's *Margarita Philosophica*, as we have seen.

The example of the small isthmus separating the Mediterranean from the Red Sea is a further cogent and original geographical argument against Buridan's natural-philosophical theory. Finally COPERNICUS uses the evidence of the Discoveries and especially of the discovery of America to maintain that "land and water press together upon a single centre of gravity" and that there "is little water in comparison with land". If the first statement is in conformity with the doctrine of Aristotle himself, the second is clearly in contradiction with it.

It is possible that the formulation of the heliocentric hypothesis depended for COPERNICUS on his reaching a point where he was sufficiently freed from the weight of Aristotelian tradition and this point was provided by the experience of navigators showing the untenability of the scholastic doctrine of the spheres of the water and of the earth. The resulting proof of the water and the earth making up a single globe with a common centre of magnitude and of gravity enabled COPERNICUS to present the earth as a planet in his system, something which adherence to the scholastic doctrine had previously made impossible.

All those who, in the 16th century, were to adopt the Ptolemaic definition of the relation between land and water, necessarily felt themselves obliged to justify their choice to the Aristotelians in terms which could force the latter's attention.

231 NICHOLAS COPERNICUS, *De Revolutionibus Orbium Coelestium*, Nuremberg, 1543, edited by JERZY DOBRZYCKI, translation and commentary by EDWARD ROSEN, Warsaw – Cracow, 1978, Book I, Chapter 3, pp. 9 – 10. On COPERNICUS's allusion to America cf. EDWARD ROSEN, "Copernicus and the Discovery of America," in: *Hispanic American Historical Review*, Vol. XXIII (1943), pp. 367 – 71.

232 Cf. Euclid, *The Thirteen Books of Euclid's Elements*, translated with an introduction and commentary by Sir THOMAS L. HEATH, Vol. III, New York, N. Y., Dover Publications, 1956 (repr. of 2nd rev. ed., Cambridge, 1926), pp. 434 – 437.

Such was what CASPAR PEUCER (1525 – 1602) attempted in his *Elementa doctrinae de circulis coelestibus et primo motu*, published in Wittenberg in 1551 and reprinted five times in the 16th century.

> "That the earth and the waters that surround it and penetrate it, together make up a single globe, bound by one and the same convex surface – the earth here opening to allow the waters into its hollows and elsewhere protruding and hugging in the waters between its shores as if the latter were barriers, so that it becomes the place where living creatures can live – is first of all proven by the voyages during which a great number of vast lands have been discovered in all the directions of the compass, lands whose existence has been clearly established and it has been thus proven that the low lying parts of the earth are not, as some have imagined, encircled by the waters like a girth while the higher parts stick out, and that the earth does not float like an apple on the waters, nor is the lower hemisphere plunged in the waters while the upper one emerges out of them. But it is clear that the earth makes up, together with the waters, a single body in the form of a globe, the two elements mutually embracing one another, while maintaining that which distinguishes each from the other, certain parts of the earth being raised up and the others hollowed out and filled with the waters."[233]

PEUCER's definition is obviously a more careful attempt than that of Ptolemy's, i. e., "that the continuous surface of the earth and the seas forms one sphere",[234] for it takes into account the necessity of keeping the elements distinct from one another, something with which Ptolemy as an astronomer was not concerned.

At the end of the century, the Professor of mathematics at the Jesuit College in Rome, CHRISTOPHER CLAVIUS (1537 – 1612) formulated at length a definition of the terraqueous globe in his commentary on Sacrobosco's *Sphere*, which was to go into 18 editions between 1570 and 1618.

CLAVIUS, after summarizing the scholastic doctrines concerning the spheres of the earth and the water, noting both the providentialist and the naturalistic solutions, then reiterates the same objection raised by both MARGALHO and NUNES that one would have to

> "... recognize that to each degree on the surface of the sea, there would have to correspond a greater number of stades or of miles than to a degree on the surface of the earth [...] But all the navigators affirm the opposite, for they have often observed by experience that there are the same number of stades or of miles in a degree on the surface of the earth as in a degree on the surface of the sea."[235]

233 CASPAR PEUCER, *Elementa doctrinae de circulis coelestibus et primo motu*, Wittenberg, 1551, sign. vii (r.).

234 Ptolemy, *Traité de Geographie*, Greek text and French translation by the Abbé HALMA, Paris, 1828, Bk. I, Chap. 2, p. 11.

235 CHRISTOPHER CLAVIUS, *In Sphaeram Ioannis de Sacrobosco commentarius*, Rome, 1581, pp. 117 – 118 (first augmented edition).

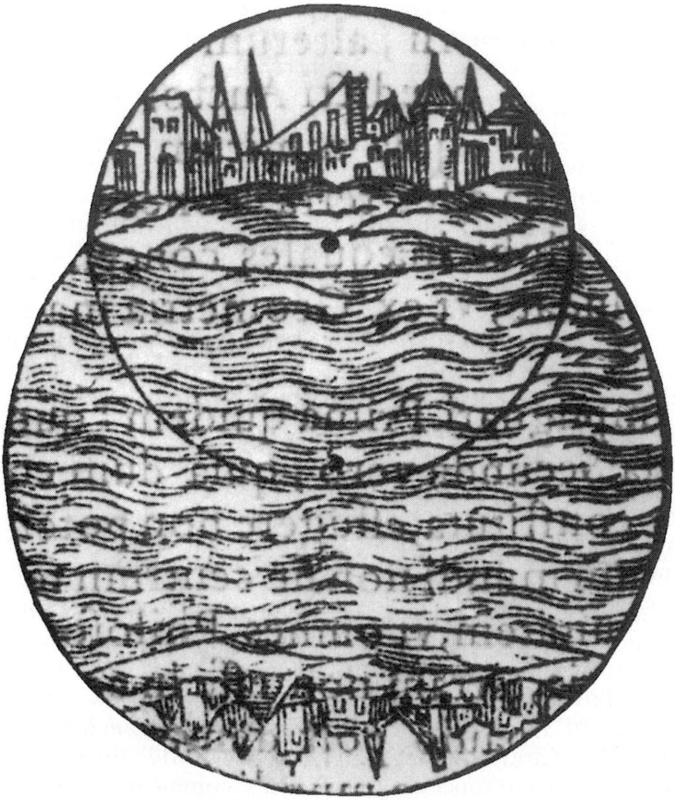

Fig. 8. CLAVIUS, *In Sphaeram Ioannis de Sacrobosco commentarius*, Lyon, 1593, p. 134

CLAVIUS, like his predecessors, gives no details of how this conclusion was reached. The Jesuit mathematician then proceeds to re-define the proofs of the sphericity of the earth and the water as given by Ptolemy in the *Almagest*.[236]

> "Secondly, basing ourselves on a large number of observations of astronomers, we shall show that the earth and the water have one and the same convex surface and that as a consequence they have the same centre. Thus for a city situated 15 degrees farther to the east than another, the sun and the stars rise, reach the middle of the heaven and set, one hour earlier, and for a city situated 30 degrees farther to the east, they do so two hours earlier and this occurs along any stretch of the earth lying in an east – west direction in which these cities may be found, given that they

[236] Ptolemy, *Almagest*, translated by G. J. TOOMER, London, 1984, pp. 40 – 41.

lie on the same parallel. Now experienced sailors have verified that the same thing happens at sea and in the Ocean. Navigating toward shores situated to the westward, as for example from Portugal to America or New Spain, and in particular to that province called Florida, after having advanced through 15 degrees, they noticed by clear signs, especially at the time of a lunar eclipse, that the sun and the other stars rose and set one whole hour earlier in Portugal, and they noticed that the same thing happened in the same proportion across the Ocean when they sailed from east to west. Now this could not occur unless the convex surface of the sea were uniformly continuous with the convex surface of the earth, as is well known to all geometricians. If indeed the sea were to rise slowly to form a sort of bulge or mount such as those imagine who hold the opposite opinion, for these navigators, after they had run through several degrees, the sun would rise earlier for them than when they were on shore. By similar reasoning, if someone said that the earth's [surface] spread downwards bit by bit, it would be impossible to maintain the proportional difference between the rising and the setting of the sun and of the other stars. Since this is not so, it is clear that from east to west, the earth and the water form at any point, one and the same convex surface. Furthermore, for example, if someone were to proceed by land from north to south or from south to north, he would notice that after having travelled through a whole degree, the Arctic pole would have dropped one degree, and if he travelled through two degrees he would find that it had dropped two degrees, and so on proportionately. Now this has also been observed at sea. When then one starts sailing from north to south, as for example from Portugal where the height of the pole is 40°[237] towards the Canary islands, after having sailed through a whole degree, one finds the height of the pole to be 39° and so on proportionately. One observes the opposite sailing from south to north, for example from the above-mentioned islands to Portugal or from Portugal to England. This is an evident sign that the water and the earth have the same convex surface from north to south, so that neither the earth nor the sea rises more [than the other] but both elements lie at an equal distance from the centre of the earth. If this were not so, one could not in any way be agreed on the above proportion in the variation of the height of the pole.

Since therefore there is absolutely no difference between the convex surface of the water and the convex surface of the earth, both from east to west and from north to south, there remains no reason to doubt that one single globe is made up of both elements. This reasoning had such an effect on the mind of a certain person, who held the opposite opinion (as I learnt from reliable people who knew him well) that, at his own expense and provided with various mathematical instruments, he sailed in several

[237] The latitude of Lisbon is in reality 38° 44′ N.

different directions to see from experience if the proportion, which we have described as having been observed in the rising and the setting of the stars and in the height of the pole, was true or invented by the astronomers, and having realized that it was absolutely exact, he gave up his previous wrong ideas and embraced the right way of thinking."[238]

Although CLAVIUS refers to recognizable regular voyages in the Atlantic, he gives no precise description of the procedures by which the conclusions he enunciates were verified in practice. His proofs, however carefully formulated, are very largely circular and the observation of angular distance is not explicitly related to terrestrial distance, although one may suspect that when CLAVIUS speaks of someone having "travelled through a whole degree" he is probably referring to a sailor's estimation of miles or leagues run, converted into degrees.

Yet for all its defects, this massive definition of what was later to be called the "Terraqueous Globe" was to put an end to any further rearguard resistance from the Aristotelians. The term, however, as far as we can determine, was only first used by a Frenchman: PHILIPPE LABBÉ in his *Géographie Royale*, Paris, 1646. He spoke of the "*Globe terraqué*" as "the earth and the water united together in the centre of the universe."[239]

The Jesuit GIOVANNI BATTISTA RICCIOLI, in his monumental work, *Almagestum Novum*, defined the terraqueous globe in singularly non-Aristotelian terms:

"*Terraqueum hoc, in quo degimus Elementum; hoc est Terra et Aqua compositum.*"[240] ("This terraqueous element in which we live, that is to say, composed of earth and water").

There is here a confusion of the elements which PEUCER, as we have seen, had tried to avoid.

It was in these terms that the Medieval doctrine of the spheres of the water and of the earth ultimately saw its demise.

XIX. Conclusion

Until the recovery of Ptolemy's *Geography* by the West, during the Renaissance, the models of world geography inherited by the Middle Ages from antiquity, were basically of two varieties. One, derived from empirical data collected from visual experience, aimed at a direct mapping of the inhabited world as known to the Greeks (the *oikoumene*). We have called it the Homeric representation.

[238] CHRISTOPHER CLAVIUS, *op. cit.*, p. 123.
[239] PHILIPPE LABBÉ, *Géographie Royale*, Paris, 1646, p. 3.
[240] GIOVANNI BATTISTA RICCIOLI, *Almagestum Novum*, Bologna, 1651, Vol. I, p. 47.

2. T & O Map from Isidore of Seville, *Etymologies*, Ms XIIIth century, Bibliothèque Municipale, Douai, Ms 319, f. 101 v°

I

PLATE II

3. Miniature representing the Antipodes in St Augustine, *The City of God*, Ms dated 1478, Bibliothèque Municipale, Nantes, Ms Fr. 8, f. 163 v°

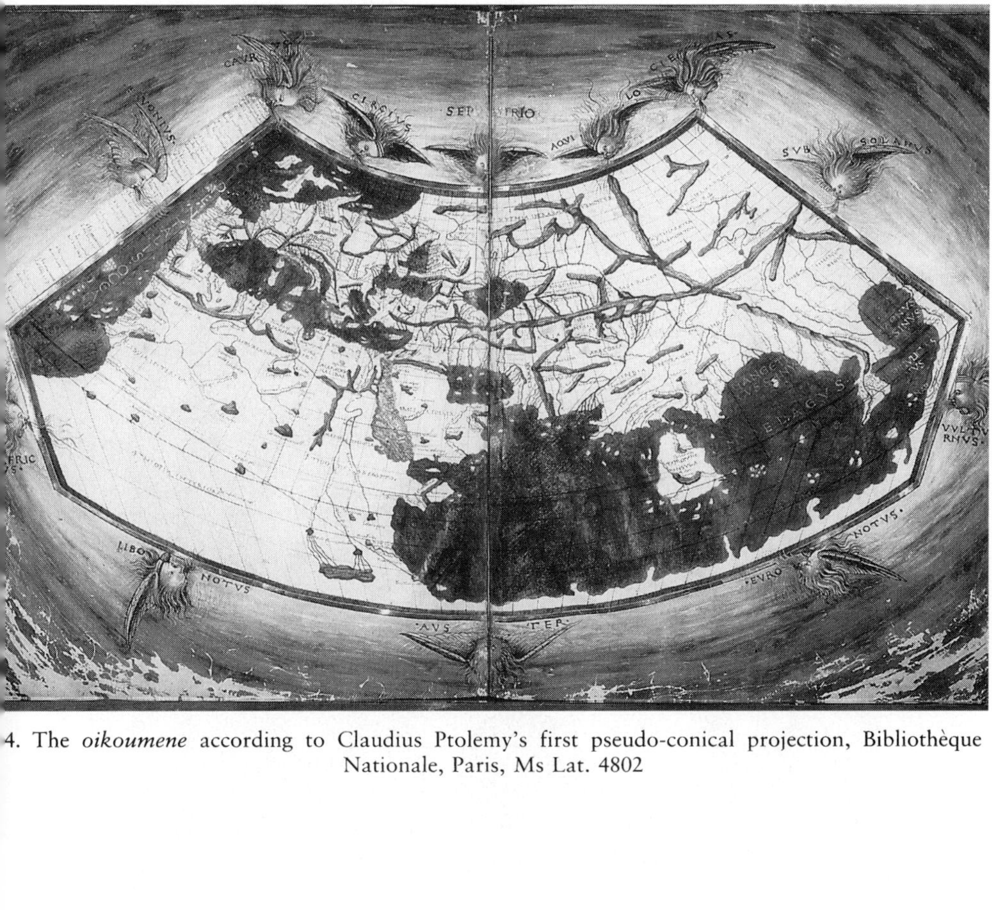

4. The *oikoumene* according to Claudius Ptolemy's first pseudo-conical projection, Bibliothèque Nationale, Paris, Ms Lat. 4802

PLATE IV

7. Planisphere by the Portuguese map-maker Lopo Homem (1519)

The other was theoretically based and founded on hypothetico-deductive conclusions drawn from the observation of celestial phenomena. The central feature of the second model was the postulate of a spherical earth first enunciated by Pythagoras and later elaborated by Aristotle with the aid of his doctrine that the weight of the parts of the earth determined its spherical shape.

The attempt of the Greek Crates of Mallos to achieve a synthesis of the two models by placing four small landmasses on the surface of a spherical globe largely covered by water, was later taken up by the Neoplatonists Martianus Capella and Macrobius (Vth century) and through them transmitted to the early European Middle Ages where it was widely adopted, after being modified by theologians to conform to Christian dogma.

Except for a brief resurgence during the Renaissance, the Cratesian model was abandoned after the recovery in the West of Aristotle's works in the late 12th century and the spread of his doctrines through Europe in the 13th. European scholars, conscious that Aristotle's concentric arrangement of the four elements of earth, water, air and fire, according to their relative densities, had apparently led to his placing the earth entirely under water, sought to make his doctrine conform to observed reality.

They first introduced the action of a Divine Providence in the form of a perpetual miracle pushing aside the waters to enable the dry land to appear; then later in the 14th century, they turned to Archimedes' principle of the centre of gravity of a body (a principle not in contradiction with Aristotle's thought), to postulate two centres in the sphere of the earth, one of gravity and the other of magnitude, the former eccentric to the second. Making the centre of gravity coincide with the centre of magnitude of the water and of the universe, they thus made the earth emerge out of the water in the form of a circular slightly convex disc which was seen to be equivalent to the Homeric representation of the *oikoumene*.

Scholars of the Italian Renaissance were able, at the beginning of the 15th century, to recover, from Constantinople, manuscripts of Ptolemy's *Geography* unknown to the Middle Ages. Although this work represented in detail only the *oikoumene* known to antiquity, it possessed the novel advantage of situating the latter on a grid designed to cover the whole of the globe of the earth.

The spread and application of Ptolemy's principles of mathematical geography coincided with the first voyages of Discovery into the Atlantic by the Iberian peoples in the second half of the 15th century. The experience of the maritime discoveries in providing confirmation of Ptolemy's doctrine of the relation of the water and the earth as a single continuous convex surface, was to play a significant role in the collapse of Aristotle's doctrine of the elements. The new concept of the "terraqueous globe" emerged to take its place and it was on the framework of this concept that the outline of the continent of America was constructed as an entirely new entity that no model from Classical Antiquity had been able to foresee.

List of Illustrations

II

The Atlantic in European Cartography and Culture from the Middle Ages to the Renaissance*

The Atlantic in history can be represented in three successive phases leading from a period in which myth transmitted from Antiquity dominated, to the modern period in which systematic reconnaissance of the Ocean and its shores was carried out by Italian, Portuguese and Spanish navigators.

During the earliest phase, which runs from the early Middle Ages to the 13th century, the Atlantic appears as a domain held to lie 'outside' the known world, an area that man was forbidden by powerful myths to enter. In the 7th century A.D. Isidore of Seville declared the 'Ocean [i.e.the Atlantic] to be unlimited in extent and untraversable'.[1]

In a later phase which runs from the 13th to the middle of the 15th century, there appear the first nautical charts of the Atlantic towards the end of the 13th century.[2] The Ocean is shown with scattered islands whose names vary and who positions change from chart to chart and whose coastlines beyond the Mediterranean and Western Europe are indistinct or unrepresented. In this Atlantic, myth and reality are inextricably intertwined. Real islands are shown side by side imaginary ones and historically atttested voyages are described in contexts in which myths concerning the impossibility of navigating such seas are simultaneously evoked.

In the more modern phase starting about mid 15th century, the islands begin to appear regularly from chart to chart in positions which they occupy on modern charts. Little by little the representation of the Atlantic becomes organized into a

---

* First published in Italian in Guglielmo Cavallo, ed., *Cristoforo Colombo e l'Apertura degli Spazi*, Istituto Poligrafico e Zecca dello Stato, Libreria dello Stato, (Rome, 1992), Vol. I, pp. 427–448. I am grateful to Professor Cavallo for authorizing publication of the original English text of the article, to which notes have been added, not present in the Italian version. The latter was accompanied by reproductions of maps and charts not reproduced here, but for which others have been substituted.

1 Isidore of Seville, (c. 570–636), *De Natura Rerum*, ed. with Fr. trans. by Jacques Fontaine, (Bordeaux, 1960), p. 307.

2 The first extant nautical chart known is the so-called Pisan Chart, (c. 1290), in the Bibliothèque Nationale, Paris; reproduced in Youssouf Kamal, *Monumenta Cartographica Africae et Aegypti*, Tome IV, fasc. 1, (Cairo, 1936), f. 1137; reprint, (Frankfurt, 1987), Vol. 5, pp. 114–115.

systematic and scientific cartography. The ancient interdictory myths disappear or are transformed into new ones proclaiming the enlargement of the known world.

Let us examine the earliest phase in greater detail in order to define it better.

The Myth of the Pillars of Hercules

It was above all the myth of the Pillars of Hercules, situated at the straits of Gibraltar, which symbolized not only for Medieval Europe, but also for Islam, the interdiction for man to penetrate into the Atlantic. The bounds set by Hercules take, as we shall see, the form sometimes of statues and sometimes of pillars.

In his commentary on the work of astrology by Claudius Ptolemy, the *Tetrabiblos*, the Arab astronomer Ali ibn Ridwan (998–1061/69), wrote that in the straits of Gibraltar (*passagium Herculis*), were islands called *Albalidet*, and that on one of these islands is,

> 'a statue called Catez, which holds in its hands keys with which it shows that beyond it there lies no inhabited place, and these islands are three in number, all three of them of the same size and on each one there is a statue [...] and that it is there that passes the limit of inhabited places to the West'.[3]

Catez here refers to Cádiz, once an island separated from the mainland. The text was translated from the Arabic into Latin by John of Seville in 1142 and from Latin into Spanish by Juan Gil (c. 1350–52).[4] It was thus well-known in the Iberian penisula.

The same myth of Hercules appears in the *Crónica Geral de Espanha* (1344) a Portuguese version of the Spanish *General Historia* of Alphonso X (begun in 1272).

> 'The first place which Hercules came to, after passing from Africa into Spain, was an island [Cádiz] situated where the great Ocean enters the Mediterranean. [...] He there built a very high tower and placed upon it an image [=statue] of copper very finely made, which looked towards the east and it had in its right hand a large key, as if it wished to open a door, while it held its left hand up in the air and extended in the direction of the East, with on its palm this inscription: "Here are the boundaries of Hercules"'.[5]

---

3 Ali ibn Ridwan, *Liber Ptholomei quattuor tractatuum cum Centiloquio ejusdem Ptholomei: et commento Haly...*, (Venice, 1484), Tract. 2, Cap. 3, (f. 32^r^, of the Venice 1493 edn of the Latin trans. of 1260 by Egidius Thebaldus).

4 See Guy Beaujouan, *Manuscrits scientifiques médiévaux à l'Université de Salamanque*, (Bordeaux, 1962), p. 65 and Idem. *L'astronomie dans la péninsule ibérique à la fin du Moyen Age*, Agrupamento de Estudos de Cartografia Antiga, série separatas XXIV, (Coimbra, 1969), p. 11.

5 *Crónica Geral de Espanha de 1344*, ed. Luís Lindley Cintra, (Lisbon, 1954), Vol. II, Cap. VI, pp. 19–20. For the slightly different Spanish text, see extract in F. Rubio Alvarez, 'Andanzas de Hercules por España segun la *General Historia*', in *Archivo Hispalense*, Vol. XXV, (1956), p. 51.

The statues, or Pillars, of Hercules appear on many *mappaemundi* of the Middle Ages and they are usually set in the straits of Gibraltar.[6] But as European navigators began to penetrate the Atlantic, the Pillars of Hercules accompanied them, receding along the Atlantic coast or withdrawing to the neighbourhood of the islands in the midst of the Ocean, thus marking the limits reached on successive voyages and emphasizing the new enlarged extent of the known world. By the early 15th century, they had receded as far as Cape Nun on the African coast opposite the Canary Islands. Referring to the voyage of D. Fernando de Castro to the Canary Islands in 1425, a Portuguese navigator, Diogo Gomes, reported around 1495 that at Cape Nun,

> 'Hercules set up Pillars carrying this inscription: "Whoever sails beyond Cape Nun has uncertain chances of returning" [in Portuguese in the Latin text: *Quem passa o Cabo de Não tornará si ou não*]. Hercules here ended his voyage fearing that on account of the strong currents he would not be able to return'.[7]

The current compressed between the Canary Islands and Cape Juby (Lat. 27° 57' N) just south of Cape Nun can be quite strong.[8] It was a force to be reckoned with, but was not unsurmountable. Yet allied to the interdictory myth of Hercules, both were enough to require fifteen attempts over twelve years before sailors sent by Prince Henry of Portugal were bold enough to round Cape Bojador beyond Cape Juby. This feat was accomplished by the Portuguese navigator Gil Eanes in 1434. Prince Henry's biographer Gomes Eanes de Zurara details in his *Chronicle of Guinea* (1460), the arguments that the Portuguese sailors had previously opposed to the Prince, their master.

> 'How shall we, they said, pass beyond the boundaries laid down by our ancestors and what good can bring the Prince the loss of our souls together with our bodies, since we shall be consciously our own murderers'.[9]

In a letter dated 22 Oct. 1443, the Prince's brother the Regent D. Pedro wrote noting that before Gil Eannes voyage the lands beyond Cape Bojador had never been charted by direct observation on nautical charts or on *mappaemundi*, except according to the fantasy of map-makers. The Prince had been the first to have a marine chart made of the coast beyond Cape Bojador.[10]

---

6 See for example Ranulf Higden, *Mappemond* (1363), British Library: Royal Ms 14. C. ix, ff. 1^v – 2^r; reproduced in J.B. Harley and David Woodward, *The History of Cartography*, Vol. I, (Chicago, 1987), Plate 15 and Fig. 18.67.

7 Diogo Gomes, 'De Prima Inventione Gujnee', in Valentim Fernandes, *Códice Valentim Fernandes*, ed. José Pereira da Costa, Academia Portuguesa de História, (Lisbon, 1997), p. 278.

8 See Raymond Mauny, *Les navigations médiévales sur les côtes sahariennes antérieures à la découverte portugaise (1434)*, Centro de Estudos Históricos Ultramarinos, (Lisbon, 1960), p. 3.

9 Gomes Eanes de Zurara, *Crónica dos feitos de Guiné*, Agência Geral das Colónias, (Lisbon, 1949), Vol. II, Cap. VIII, pp. 47–52 (states that it was believed that the force of the current was such that a ship could never return).

10 Letter from the Regent Prince Pedro dated 22 Oct. 1443, in *Monumenta Henricina*, Vol. VIII, (Coimbra, 1967), p. 107.

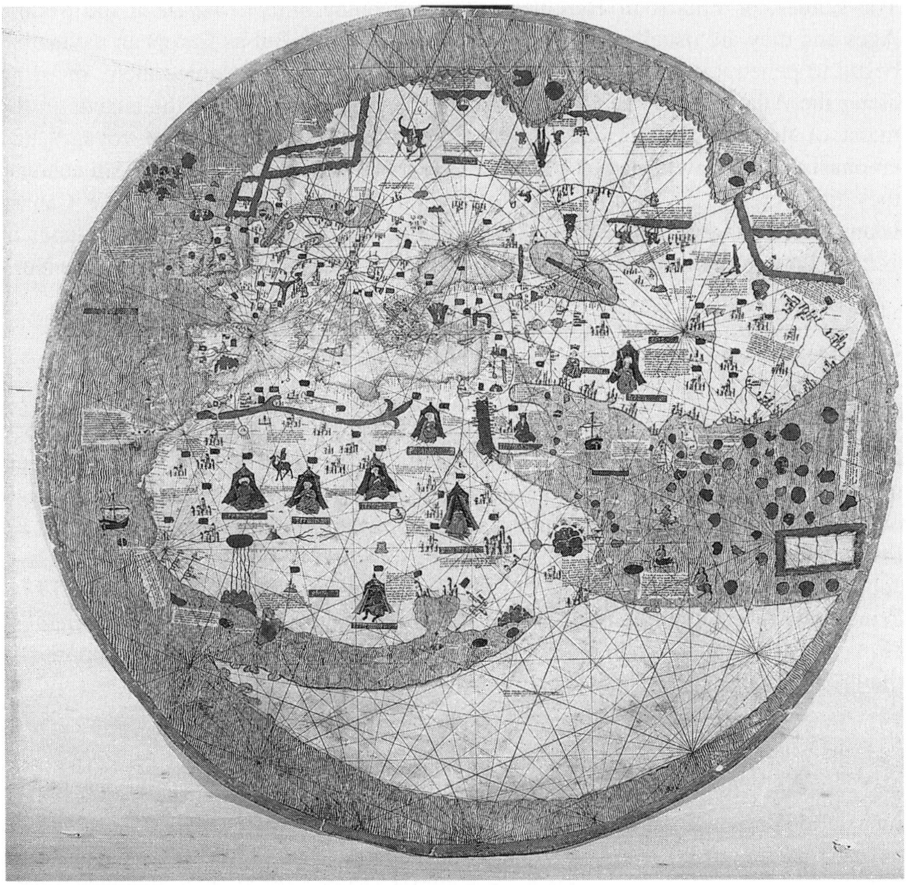

Fig. 1. Catalan World Map (c. 1450) Biblioteca Estense, Modena.

By the second half of the 15th century, in the wake of continued Portuguese exploration, the Pillars of Hercules had receded as far as Cape Verd. The Catalan World Map (c. 1450), has a legend which places the islands of Cadiz in front of Cape Verd: *Illa de Cades asi posa ercules duas colunas* (Island of Cadiz, where Hercules set up two pillars).[11]

On Fra Mauro's World Map (1459) conserved in the Marciana Library in Venice, the Pillars of Hercules are placed at the same Cape, but Fra Mauro's accompanying legend introduces a doubt already hinting at the decline of the myth of Hercules and his Pillars:

> 'I have often heard it said that here [at Cape Verd] there is a Pillar with a hand which has an inscription forbidding one to go further. But I would like the Portuguese, who sail this sea, to say if what I have heard is true for I hesitate to affirm it.'[12]

Hercules' Pillars not only receded down the African coast, but also withdrew into the Atlantic to the neighbourhood of the present day island of Madeira, as can be seen on an anonymous Venetian chart (c. 1430). Beside an island representing Cadiz, on which are placed the Pillars of Hercules, are written the words: '*Ysola gades* [=*Cadiz*] [...] *coloni misi Ercolis quando compasa lo mondo per arti di strolezia*' (the island of Cadiz [...] Hercules here set up pillars when he measured the extent of the [known] world by astronomical methods). Two other legends appear in close proximity to the island with the pillars: *Ysola di legname* (= the present day island of Madeira – *legname* = Italian for 'wood' – *Madeira* = Portuguese for 'wood'); *Ysola di Porto Santo* (the present day island of Porto Santo close to Madeira).[13]

The Pillars of Hercules also appear on another much older chart of 1367, where they were long associated with the 'mythical' island of Antilia, situated near the Azores. This is the chart of Francesco and Domenico Pizzigani at present conserved in the Biblioteca Palatina at Parma. Near an island placed far into the Atlantic at the westenmost extremity of the chart, lies the following legend as transcribed by K. Kretschmer in 1892:

> 'Here stand the statues which are before the shores of Antilia, of which some are on the shore itself. They are there for the safety of sailors, for it is just so far that

[11] For reproduction of the map see G.H.T. Kimble, *Memoir on the Catalan World Map of the R. Biblioteca Estense at Modena,* Royal Geographical Society, (London, 1934). For the transcription of the legend, see F.L. Pullé & M. Longhena, 'Illustrazione del Mappamondo catalano della Biblioteca Estense di Modena', in *Atti del Sesto Congresso Geografico Italiano*, (Venice, 1908), Vol. II, p. 360.

[12] For a facsimile reproduction in colour of the map, with transcriptions of the legends, see R. Almagià & T. G. Leporace, *Il Mappamondo di Fra Mauro*, (Rome, 1956), Tavola XIII, p. 32.

[13] Anonymous Venetian Chart (c.1430), Correr Museum, Venice, Port. 38, (Morosini Gatterburg Collection 469). Hercules' use of astronomical methods to measure the extent of the world would seem to be an exaggeration attributable to any hero founder.

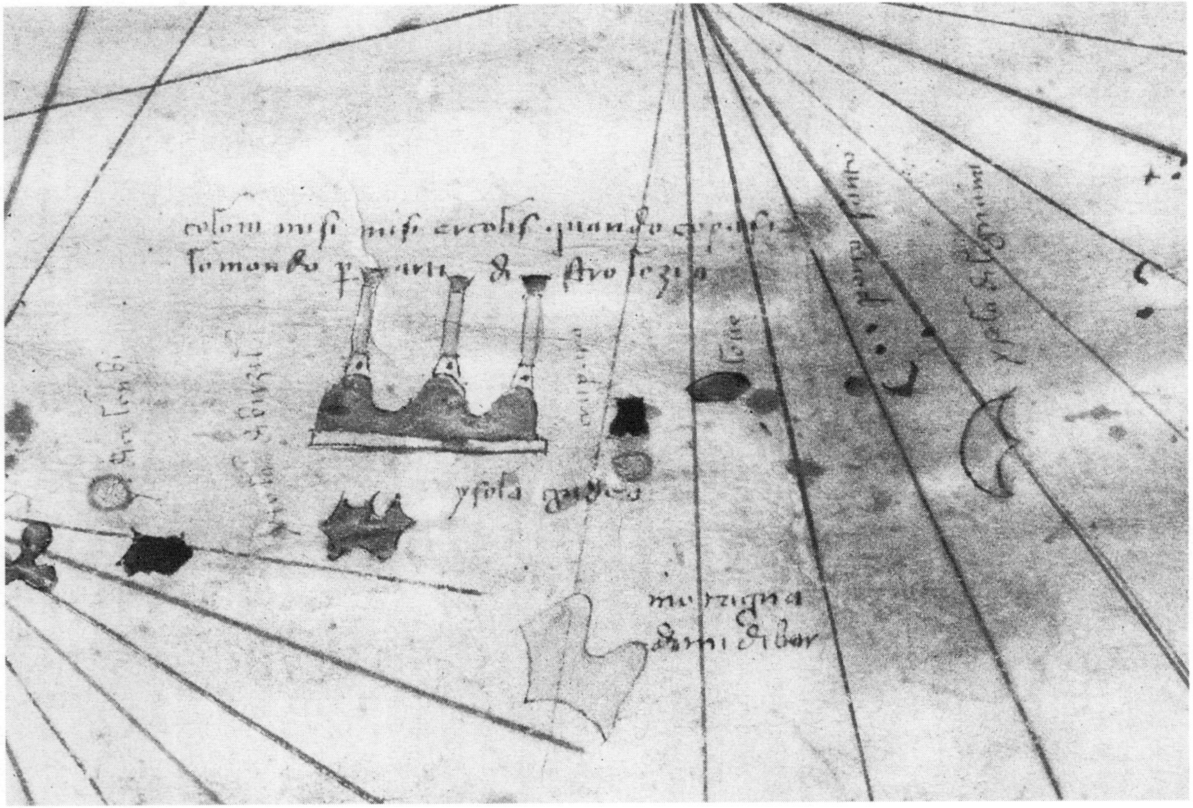

Fig. 2. Anonymous Venetian Chart, (c. 1430), Correr Museum, Venice, Port. 38, (Morosini-Gatterburg Collection 469). The Pillars of Hercules are shown in the vicinity of the island of Madeira (*Ysola di legname*).

these seas can be navigated and beyond these statues the sea is *sorde* (curdled) where navigators cannot go'.[14]

The word *sorde* may be an early reference to the Sargasso Sea or *Mar de Baga* as it is written on André Bianco's chart of 1436,[15] *baga* being a Portuguese word referring to the little grape-like bulbs of the floating seaweed (*fucus natans* and *fucus bacciferus*).[16]

In 1954 a new reading of the first lines of the legend was proposed by Richard Hennig:

> '*Hoc socit (?) statuas, quae fuirant ante tenplum Arc ... ulis quarum quae in fundo ad securantes homines navigantes, quare est fusum ad ista maria quousque possint navigare; et foras porrecta statua est mare sorde quo non possunt intrare nautae*'.[17]
>
> 'Here stand (?) the statues which are in front of the temple of Hercules some of which are on land. They are there for the safety of navigators, for this is the limit to which they can navigate these seas and beyond these statues the sea is soiled (*sorde*) where sailors cannot penetrate'.

The name 'Antilia' would thus in fact be a mistaken reading for 'Hercules'. Nevertheless, the island of Antilia, enlarged and rectangular in shape was to continue as a feature of Atlantic cartography throughout the 15th century.

The Canary Islands in European Culture and Cartography

In the 14th century the Atlantic was still for Europeans a region that was not rationally part of the known world. The Italian humanist Francesco Petrarch (1304–1374), when evoking, in his work *De Vita Solitaria* (1356), the voyage of Lanzarote Malocello to the Canary Isands in 1336, uses, in order to situate these islands, expressions such as *extra orbem* (outside the world) and *disiunctos mundi angulos* (corners separated from the world).[18]

---

14 K. Kretschmer, *Die Entdeckung Americas*, (Berlin, 1892), p. 197.

15 André Bianco's Chart (1436) in Biblioteca Marciana Venice. For reproduction of the chart and transcription of the legend: *questo xe mar de baga*, see Vincenzio Formaleoni, *Nautica dei Veneziani*, (Venice, 1783), p. 47.

16 See the note on the *Mar de Baga* in *Archivo dos Açores*, Vol. I, (1878), p. 193. The Portuguese term *Sargaço* appears for the first time in the work of the Portuguese António Galvão (1555), *Tratado [...] dos diversos & desvayrados caminhos*, (Lisbon, 1555), Hakluyt Society edn, (London, 1862), 1st ser. Vol. XXX, (1862); reprint, (New York, Burt Franklin, s.d.), p. 82. (Port. text and Eng. trans.). On the limits of the Sargasso Sea, see Charles C. Dixon, 'The Sargasso Sea', *Geographical Journal*, Vol. LXVI, (1925), pp. 435–442.

17 See Richard Hennig, *Terrae Incognitae*, (Leiden, 1956), Vol. IV, p. 292 and Armando Cortesão, 'A Carta náutica de 1424', in *Esparsos*, (Coimbra, 1975), Vol. III, pp. 106–108.

18 Francesco Petrarch, 'De Vita Solitaria', in *Opera quae extant omnia*, (Basle, 1581), Lib. II, Sect. VI, Cap. 3, p. 277.

Such was the impression of the Italians of these islands that even in the first third of the 15th century, Piero Quirino, who had spent forty days in their neighbourhood in 1431, could write of them that they were ' ... *luoghi incogniti & spaventosi a tutti i marinari massimamente della parte nostre* ... ' (unknown places terrifying to all sailors and particularly to our own).[19]

In the cartography of the early 14th century, the Canary Islands do not appear on the chart of Angelino Dalorto (1325/30),[20] but they do appear on that of Angelo Dulcert (1339).[21] The presence of the Canary Islands on the latter chart is held to be a reflection of the spread of knowledge of Lancerote Malocello's voyage. The voyage of Lanzarote Malocello (1336), in which both Italian and Portuguese sailors participated, was the first of a regular series of voyages which led to the Canary Islands becoming progressively integrated into European consciousness and cartography.[22]

The Islands of the Azores on 15th century nautical charts

The Canary Islands were to become more rapidly integrated into the cartography of the 14th century than the other Atlantic islands such as the Azores, where myth and reality remained intertwined into the 16th century.

Much ink has flowed in discussions on how and when the Azores were discovered. The general consensus of scholars supports the view that the Azores were probably discovered by navigators returning from the Canary Islands and who were carried out into the Ocean, either by accident or deliberately, by the prevailing circular wind system of the North Atlantic, which at the latitude of the Azores then took them back on a straight easterly course to Portugal.[23]

---

19 Piero Quirino, 'Viaggio del Magnifico Messer Piero Quirino gentilhuomo Venitiano', in G.B. Ramusio, *Delle Navigationi et Viaggi*, Vol. II, (Venice, 1559), ff. 144r–150v.

20 Corsini Collection, Florence; reproduced in Youssouf Kamal, *op. cit.*, Tome IV, fasc. II, f. 1197; reprint, (Frankfurt, 1987), Vol. V, pp. 218–219.

21 Bibliothèque Nationale, Paris; reproduced in Youssouf Kamal, *op. cit.*, Tome IV, fasc. II, f. 1222; reprint, (Frankfurt, 1987), Vol. V, pp. 266–267. Gonçalo de Reparaz entertains, with H. Winter, the conjecture that both cartographers were one and the same person. See Gonçalo de Reparaz, 'Les sciences géographiques et astronomiques au XIVe siècle dans le Nord-Est de la Péninsule Ibérique et leur origine', in *Archives internationales d'histoire des sciences*, No 1, Oct. (1947), pp. 453–457.

22 On this voyage, see Charles Verlinden, 'Lanzarotto Malocello et la découverte portugaise des Canaries' in *Revue belge de philologie et d'histoire*, Vol. XXXVI, 3, (1958), pp. 1173–1209.

23 The first description the 'wide arc' sailing technique or *volta ao largo* for return from the African coast appears relatively late in the work of Gaspar Frutuoso (1522–1591). In his *Saudades da Terra*, written between 1586 and 1590, he stated that, 'some of those whom Prince Henry sent to explore the coast of Africa, did what those do today when returning from India or from São Tome [Island], La Mina [the Gold Coast], Arguim and Cape Verd. Carried by the winds which blow [along the West

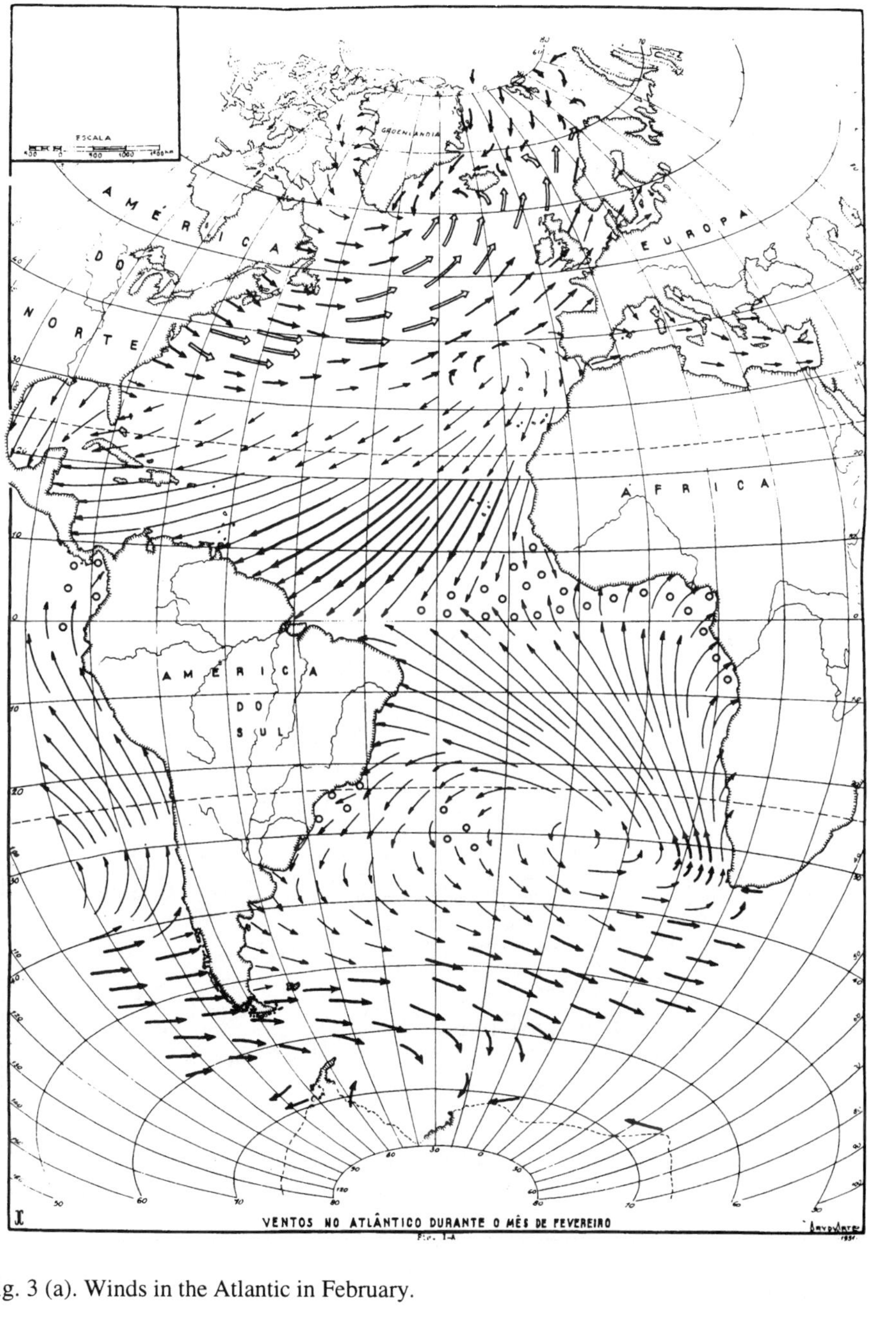

Fig. 3 (a). Winds in the Atlantic in February.

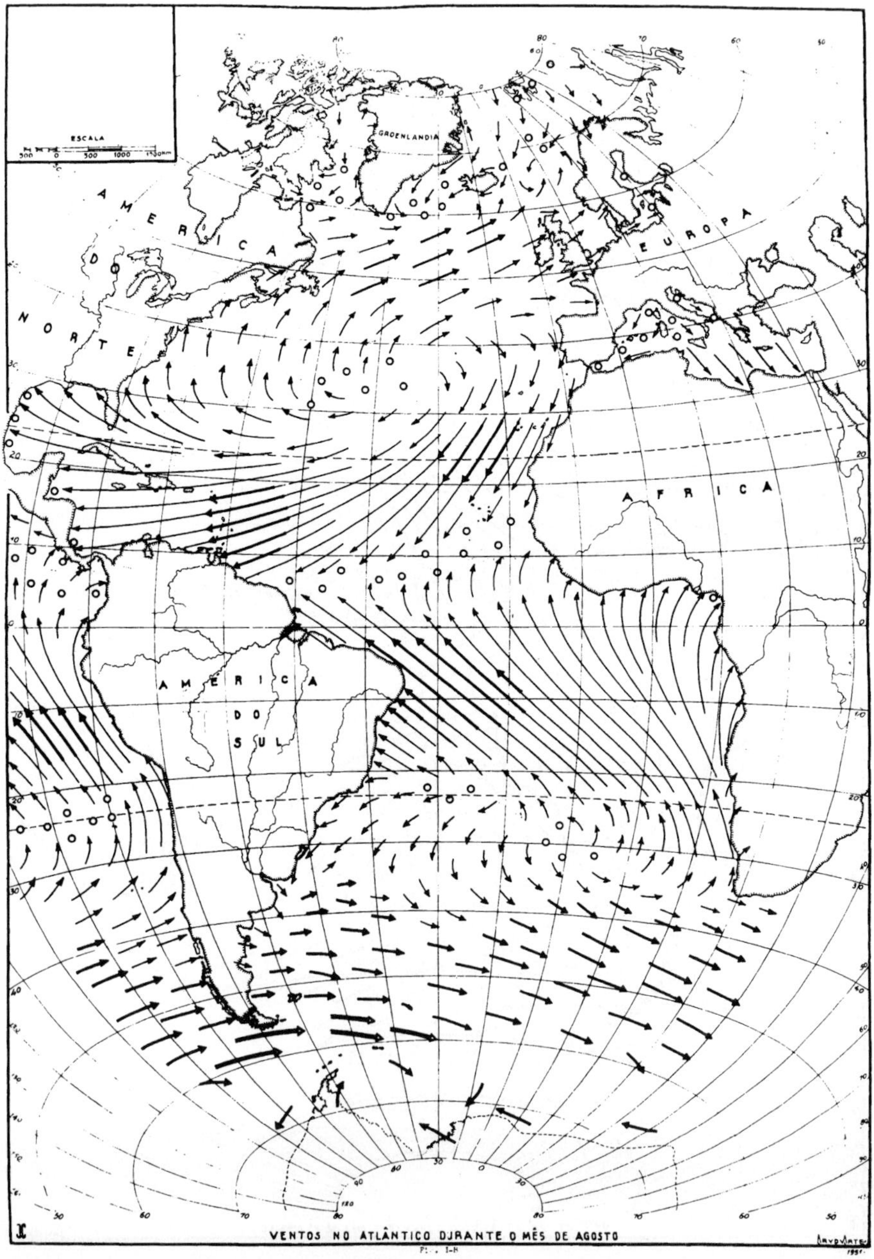

Fig. 3 (b). Winds in the Atlantic in August.

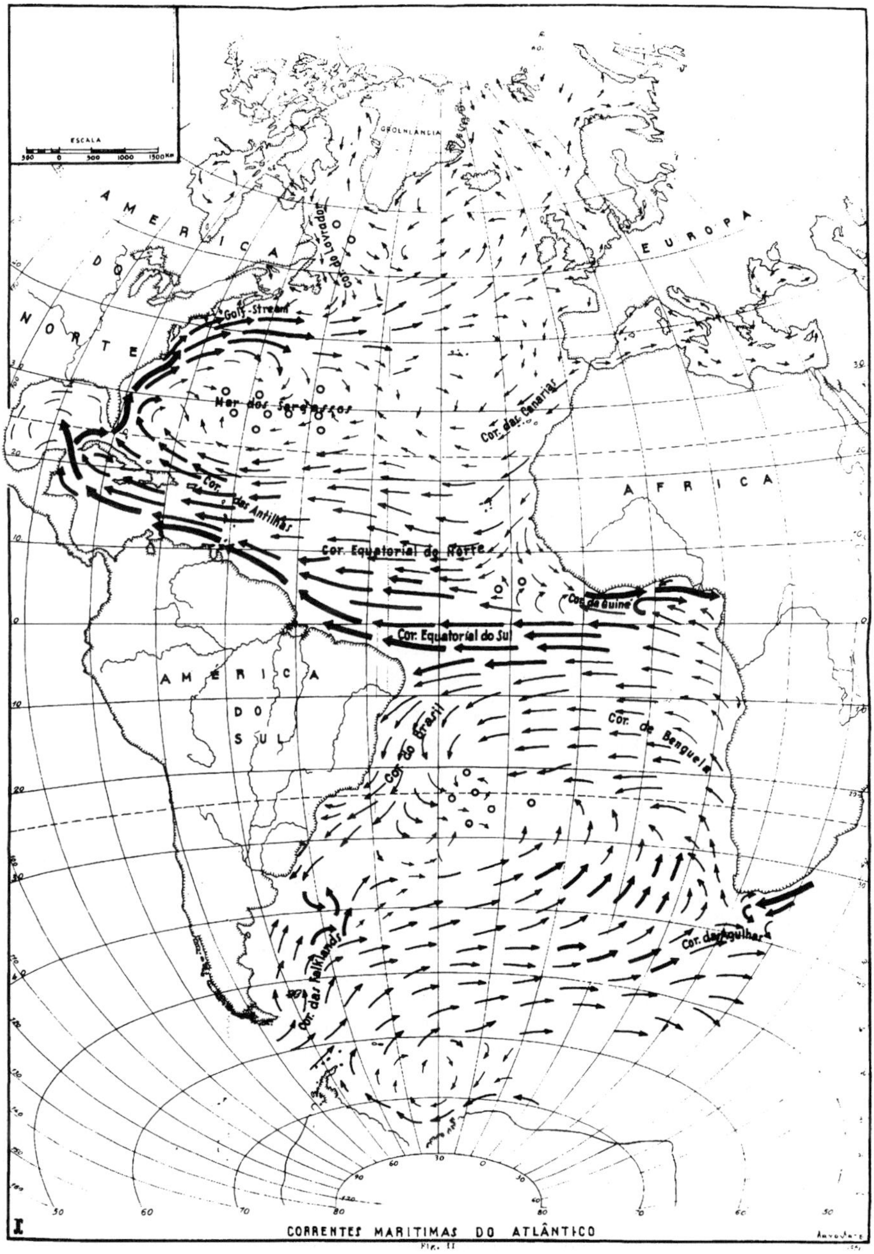

Fig. 3 (c). Currents in the Atlantic.

On Italian and Majorcan charts of the 14th century, islands scattered over the Western Atlantic with Italianized names associated with birds, sea lions or rabbits, are claimed to represent the discovery of the Azores by Italians, or Catalans, but no documentary evidence exists of Italian or Majorcan navigators having reached them. Only from the first third of the 15th century, after their documented discovery by the Portuguese (on a Majorcan chart), do the Azores begin to appear in their correct places with their modern Portuguese names of Christian saints.

In 1435 Alonso de Cartagena, the Bishop of Burgos, protesting in favour of Spanish rights to the Canary Islands, could still quite seriously write of the mythical island of Brazil, 'which is said to be in the West, opposite Lisbon and which is uninhabited, but which navigators can find only very rarely'.[24]

In an old rutter of sailing directions, probably dating from the 15th century, contained in a manuscript of the mid-16th century, known as the *Livro de Marinharia de João de Lisboa*, is a list of 'undiscovered' islands (*Ylhas não descobertas*), with the distances in leagues between them and other known islands. For example: *Dorsey* [=*Zorzi*, Italian for 'George'] *e a ilha dos braçiles leste-oest 50 leguas* (The island of Saint George [one of the islands of the Azores] distant 50 leagues from the island of Brazil).[25] The mythical island of Brazil, often shown on maps of the 15th century, is stated by Armando Cortesão to correspond to the present day island of Terceira.[26]

In 1427 the Azores were formally 'discovered' and, except for the two more western islands of Flores and Corvo discovered later about 1452,[27] they were integrated into Atlantic cartography and appear henceforth in their true positions. The discoverer was a Portuguese navigator, whose name appears, though only decypherable with great difficulty, on the chart of a Majorcan, Gabriel Valseca, (1439),[28] The legend concerning the discoverer reads: *Aquestes illes foram*

African coast], they steered out toward these islands [Azores] to seek the fresh regular winds in this [part of the Atlantic]. Whereas along the coast they are not as regular, nor do they blow as strongly, serving rather more for outward voyages than for return ones, away from the coast they change direction and out in the Ocean they blow towards these islands [Azores]. On the basis of information which his sailors brought him and from past tradition known to him, [...] it was not by guesswork, but from sound knowledge that he surmised from information from his sailors that these islands were in the Ocean and he had them explored judging them to really exist.' See Gaspar Frutuoso, *Saudades da Terra*, edn, (Ponta Delgada, 1963), Livro Sexto, Cap. I, pp. 7–8.

24 Alonso de Cartagena, 'Allegationes factas [...] super conquesta Jnsularum Canarie contra Portugalensses ... ' (1435), in João Martins da Silva Marques, ed., *Descobrimentos Portugueses*, Vol. I, (Lisbon, 1944), p. 315.

25 João de Lisboa, *Livro de Marinharia*, ed. Jacinto Ignácio de Brito Rebelo, (Lisbon, 1903), p. 121.

26 See Armando Cortesão, *História da Cartografia Portuguesa*, (Coimbra, 1970), Vol. II, p. 44. An English edition of this work was published simultaneously at Coimbra.

27 See Damião Peres, *História dos Descobrimentos Portugueses*, 3rd edn, (Oporto, 1983), pp. 72–77.

28 The chart is in the Museu Maritim in Barcelona; reproduced in A. Cortesão *op. cit.*, Vol. II, p. 151, Fig. 77.

trobades per diego de silues pelot delrey de Portugal an lay Mccccxxvij (These islands were discovered by Diogo de Silves, a pilot of the King of Portugal in the year 1427).[29]

From 1439 on, the Azores became systematically colonized by Portuguese and Flemish settlers, and this entailed regular voyages to and from them. However the 'mythical' Azores of the earlier charts continue to be shown simultaneously on charts throughout the 15th century, often arbitrarily strung out on a more or less vertical line running from north to south, such as on part of the chart showing the western Atlantic of the Majorcan Jacobo Bertran (1482).[30]

The invention of nautical astronomy by the Portuguese and its rôle in the exploration of the Atlantic

From about the middle of the 15th century, Portuguese navigators began to make observations of the height of the Pole Star with the aid of a quadrant in order to measure relative distances sailed north and south along the Atantic coast of Africa. This did not in any way yet imply the calculation of latitude as such, and in any case the nautical charts of the 15th century carried no coordinates of latitude and longitude.[31]

Without astronomical navigation, at least of this rudimentary sort, experienced Portuguese mariners have contended that the Azores could not have been discovered and still less revisited on successive voyages.[32]

As the Portuguese navigators, in the course of their explorations down the African coast, approached the equator, observation of the Pole Star became more and more difficult as it declined closer and closer to the horizon before finally disappearing. Penetration into the southern hemisphere thus required a new method of celestial navigation and it was here that the Portuguese achieved a major breakthrough in about 1485 by turning to the observation of the meridian height of the sun with the aid of a nautical astrolabe developed for the purpose. From a medieval tradition in the Iberian peninsula inherited from Arab astronomers, tables were developed showing the sun's declination for the day and the year of the observa-

---

29 On the successive attempts at decyphering his name, see Damião Peres, *op. cit.* pp. 65–71.

30 In the Archivio di Stato, Florence; reproduced in Armando Cortesão, *op. cit.*, Vol. II, p. 175.

31 See Luís de Albuquerque, 'A navegação astronómica', Chapter VII, in Armando Cortesão, *op. cit.* pp. 225–371. See also reprint of chapter VII, by the Comissão Nacional para as Comemorações dos Descobrimentos Portugueses, (Lisbon, 1988).

32 The mariners are quoted by Damião Peres, 'A acção do Infante D. Henrique no desenvolvimento da ciência', in *Congresso Internacional de História dos Descobrimentos*, Lisbon, 1960, *Actas*, Vol. I, (Lisbon, 1961), p. 111.

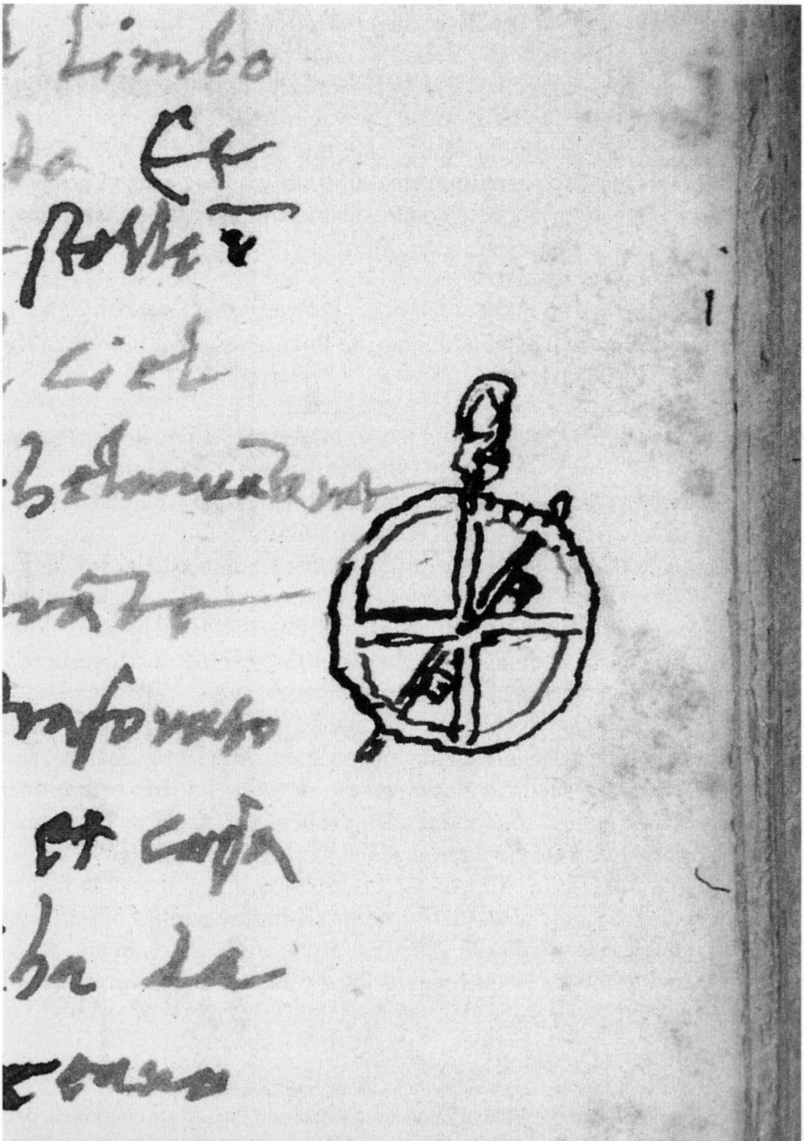

Fig. 4. Earliest drawing of a nautical astrolabe in [Alessandro Zorzi]. 'Informatiõ hauuto Io Alexandro da Portogalesi. 1517. in Veneza', Ms. Magl. XIII, 80, fl. 136r, Biblioteca Nacional, Florence.

tion. With the use of these tables, it became possible to calculate latitudes and these were later integrated into the nautical charts of the 16th century.[33] Longitude however remained until the 18th century an intractable problem from an astronomical point of view and the only solution in the 16th century consisted of the use of hour glasses, whose accuracy was very limited.

The cartography of the Portuguese explorations of the coast of Africa in the South Atlantic

The progressive reconnaissance by the Portuguese of the west African coast is remarkably illustrated in the two oldest Portuguese charts known, that of an anonymous cartographer dated to between 1471 and 1482,[34] and that of Pedro Reinel, dated to between 1484 and 1492.[35]

On Pedro Reinel's chart, the cartographer, having failed to foresee that the African coast extended so far southward, placed the rest of the coast showing the Gulf of Guinea down to the mouth of the Congo River, within the area showing the Sahara desert.

The cross at the mouth of the Congo River, named the *Padrão de São Jorge*, set up in Lat. 6° 04' S., by the Portuguese navigator Diogo Cão during his voyage of 1483, is the first of the marble pillars surmounted by a cross, which the Portuguese began, from this point onwards, to set up along the African coast to mark the limits reached in their voyages. That these crosses were intended to continue the function of the Pillars of Hercules in a new Christianized form is explained by the Portuguese historian and humanist João de Barros. In a triumphant exalted tone, Barros declaims in his *Panegyric to the Infante D. Maria*, written in 1555:

> 'I shall not speak of the Pillars of Hercules, on the island of Cádiz, which he set up at our very doorstep, as it were, to mark the extreme limits of the whole earth and which in our time have been effaced from human memory and thrust into silence and oblivion, by others much nobler, which have been set up by the King of Portugal in the most distant eastern parts of the world, and which have brought more benefit to his Majesty for being those on which Christ placed his shoulders, than did those of Hercules, by which so many souls have been lost'.[36]

---

[33] On the development of navigation using observation of the meridian height of the sun, see chapter VII by Luís de Albuquerque, in Armando Cortesão, *op. cit.*

[34] In the Biblioteca Estense, Modena; reproduced in Armando Cortesão & A. Teixeira da Mota, *Portugaliae Monumenta Cartographica*, (Coimbra, 1960), Vol. I, Pl. 2. and by A. Fontoura da Costa, *Uma Carta Náutica Portuguesa, Anónima de 1471*, (Lisbon, 1940).

[35] In the Archives Départementales de la Gironde, Bordeaux; reproduced in *Ibid.* Vol. V, Plates 521/522. On the dating of this chart and the previous one, see Alfredo Pinheiro Marques, 'The dating of the oldest Portuguese Charts', in *Imago Mundi*, Vol. 41, (1989), pp. 87–97.

[36] João de Barros, 'Panegírico da Infanta D. Maria', in Idem. *Panegíricos*, ed. M. Rodrigues Lapa, (Lisbon, 1937), p. 173.

O BRASÃO DE DIOGO CÃO SEGUNDO UM LIVRO DE ARMARIA DE 1575, O "TESOURO DE NOBREZA" DE FRANCISCO COELHO, REI-DE-ARMAS DA ÍNDIA

Fig. 5. Coat of Arms awarded to Diogo Cão for the discovery of the African coast on his first voyage. Two *padrões* surmounted by Christian crosses are shown with the obsolete and collapsed Pillars of Hercules above them. From Francisco Coelho, *Tesouro de Nobreza, Livro de Armaria*, (1575).

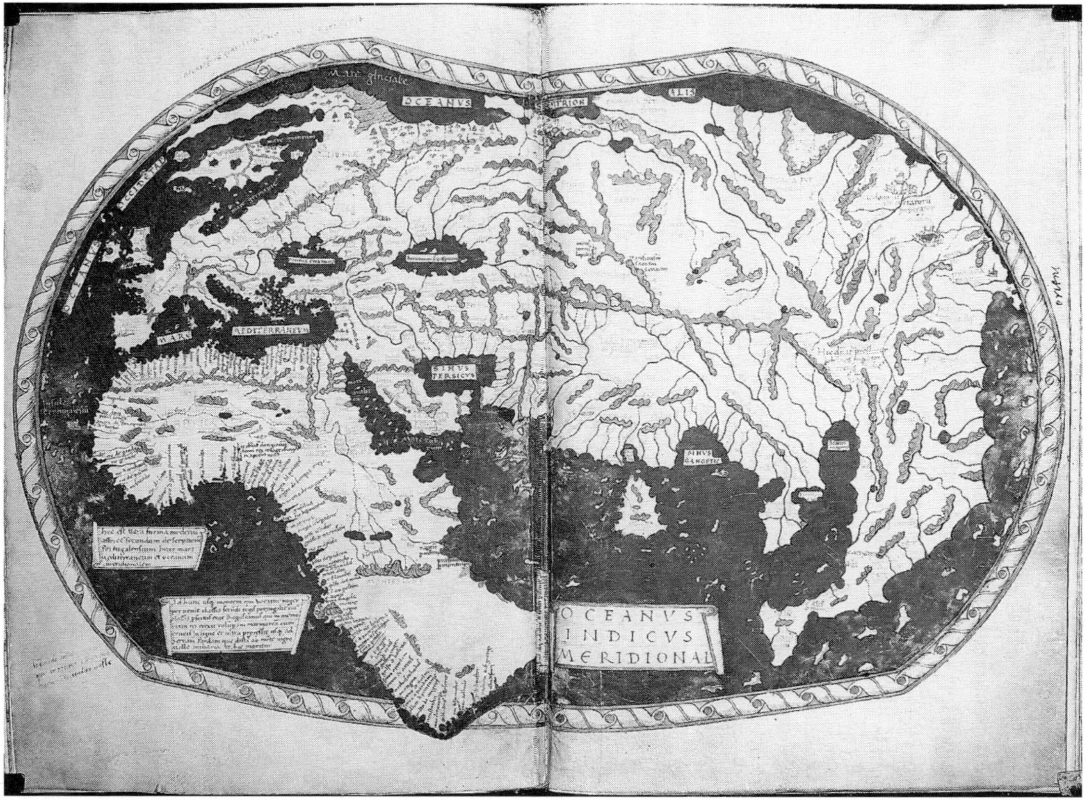

Fig. 6. Henricus Martellus, Map, (c. 1490), British Library, Ms Add. 15760, ff. 68^v–68^r. Reproduced by permission of the British Library.

Cão's second voyage (1485–86), destined to achieve the circumnavigation of the African continent, was a desastrous failure. The coast continued ever southward and Cão, finally dispairing of finding its end, returned after having reached a place he named the *Serra Parda*. The success of his first voyage had led to his being ennobled and awarded a coat of arms showing on it two *padrões*; for his inability to find the hoped for access to the Indian Ocean he was disgraced and his second voyage is not mentioned in any chronicle.[37]

The World Map of Henricus Martellus (c. 1490),[38] a synthesis of Ptolemy's world map and of contemporary nautical charts, shows the limit of Cão's furthest voyage. A legend placed opposite Cape Negro, north of the *Serra Parda*, (Lat. 15° 42' S.) reads:

> 'The mountain called *Monte Negro*, was reached by the fleet of the King of Portugal John II; the fleet was commanded by Diogo Cão, who to commemorate the fact, set up a marble column with a cross on it and the said fleet went on to the *Serra Parda*, (today *Ponta dos Farilhões*, Lat. 22° 10' S.) which is distant 1000 miles from the *Monte Negro* and there it went no further'.

The succeeding voyage of Bartolomeu Dias, the discoverer of the Cape of Good Hope in 1488, is also recorded on the Martellus Map. Dias, after rounding the Cape of Good Hope and having seen the coast turn northwards and northeastwards, went no further and turned back at a place on the south-east African coast he named the *Penedo das Fontes* (today Kwaai Hoek on the Western Province coast in Lat. 33° 38' S.). The legend on the Martellus Map reads: 'In the course of their last voyage in 1489 [=1488], the Portuguese reached the *Penedo das Fontes* ["Fountain Rock"]'. There Dias set up his last pillar, the *Padrão de São Gregório,* since recovered by Eric Axelson in 1938.[39]

Although the profile of the African continent was not fully established until the voyage nine years later of Vasco da Gama, it was Bartolomeu Dias who was the

---

[37] Damião Peres, *História dos Descobrimentos Portugueses, ed. cit.*, pp. 206–218.

[38] In the British Library, Add. 15760, ff. 68^{v}–68^{r}.

[39] See Eric Axelson, 'The Dias Voyage, 1487–1488: toponymy and *Padrões*', in *Revista da Universidade de Coimbra*, Vol. XXXIV, (1988), pp. 29–55; see also A. Teixeira da Mota, *Bartolomeu Dias, Discoverer of the Cape of Good Hope*, (Lisbon, 1988), 32p. 1st edn in Portuguese, (Lisbon, 1955); see also illustrations of the *padrões* in Damião Peres, *op. cit.*, Plates XXIX, XXX, and XXXV and pp. 238–245. Later in their discovery of the East the Portuguese continued to set up *padrões* as they proceeded. In the neighbourhood of Calicut, see João de Barros, *Ásia*, Déc. I, Liv. VI, Cap. IX, (Lisbon, 1552), facsimile of the (Coimbra 1932) edn, (Lisbon, 1988), p. 159. On the island of Ceylon (Sri Lanka), see Fernão Lopes de Castanheda, *História do Descobrimento & Conquista da India pelos Portugueses*, Liv. II, Cap. XXIII, (Lisbon, 1552), edn (Coimbra, 1924), p. 262. Their most distant one was on the coast of China in the vicinity of Macau, see João de Barros, *Terceira Década da Ásia,* (Lisbon, 1563), Liv. VI, Cap. II, facsimile edn, (Lisbon, 1992), f. 160^{r}. On the Portuguese *padrões* in general see A. Teixeira da Mota, 'Padrões dos Descobrimentos', in Idem, *Mar, Além Mar*, (Lisbon, 1972), pp. 47–51.

first to demonstrate the real possibility for European navigators to sail from the Atlantic into the Indian Ocean.

The cartography of the Atlantic on the eve of Columbus' first voyage and the earliest maps showing his discoveries

While the Portuguese were systematically pushing their way round Africa in the belief that this was the only possible route to reach India, Columbus, then resident in Portugal, had become convinced that the shortest route to the Orient was across the Atlantic. This conviction seems to have come to him from reading the enthusiastic claim of the feasibility of reaching Asia by sailing westwards, as expounded in a letter accompanied by a map addressed by the Florentine cosmographer Paolo dal Pozzo Toscanelli to the King of Portugal in 1474. The text of the letter has come down to us,[40] but the map has disappeared. Among the various attempts to reconstruct the map from elements in the text of the letter, that of Hermann Wagner in 1894, has been held to be the most plausible.[41] The map proposes a representation of the Atlantic between the Iberian peninsula and China on a rectangular cylindrical projection based on that of Marinus of Tyre. Both parallels and meridians are straight lines intersecting at right angles, such as appear on the chart of the Scandinavian regions by Claudius Clavus, drawn as early as 1427.[42] On the Marinus projection Toscanelli introduced features from the medieval nautical charts of the Atlantic, (e.g. the island of Antilia) and elements concerning the Far East taken from the travels of Marco Polo and Nicolò di Conti. Toscanelli was also able, by adopting Marinus of Tyre's value of 225° for the extension of the *oikumene* (which he preferred to Ptolemy's 180°) and a module of 66 2/3 miles to a degree, to postulate an exact distance between Spain and China. Toscanelli's map, unlike, on the one hand, the Mediterranean nautical charts, or, on the other hand, the world map in Ptolemy's *Geography* known to Europe only in the 15th century, at once both assumed the sphericity of the earth and remained practically usable by mariners.

Although Columbus possessed and used many maps and charts, none used to prepare his first voyage have come down to us. In 1940 Roberto Almagià offered a clever argument in which he maintained that Columbus was influenced by a type of

---

40 Reproduced by N. Sumien, *La correspondance du savant florentin Paolo dal Pozzo Toscanelli*, (Paris, 1927).

41 See Hermann Wagner, 'Die Rekonstruktion der Toscanelli Karte vom Jahre 1474', in *Nachrichten der K. Gesellschaft der Wissenschaft zu Göttingen*. Phil.-Hist. Kl. (1894), pp. 208–312.

42 In Atlas of Guillaume Fillastre, Ms 441, Bibliothèque de Nancy; reproduced in A.E. Nordenskiold, *Facsimile Atlas*, (Stockholm, 1889); reprint, (New York, 1973), p. 49.

world map such as the one of Henricus Martellus in the British Library.[43] Almagià did not then know of the existence of another of Martellus' maps now in the Yale Library. The British Library map has no scale of latitudes and longitudes, but the Yale map has both. On the Yale map, the distance separating Europe from Asia coincides, according to R.A. Skelton who was able to examine the map carefully, almost exactly with that given by Toscanelli in the description of his map.[44] Thus it can be said that at the end of the 15th century, a type of world map with coordinates of latitude and longitude was circulating in Europe and provided a mathematical framework on which enterprizing navigators like Columbus could conceive their voyages of exploration.

The only map made by Columbus himself of his discoveries, to reach us, is a little sketch map showing the northwest coast of Hispaniola (Santo Domingo), drawn in 1492.[45] The earliest maps showing his discoveries in the context of the world map, are to be found in a Florentine manuscript of the early 16th century. They are three sketches attributed to Columbus' brother Bartolomeo Columbus, and they relate to Columbus' fourth voyage.[46] Roberto Almagià has shown however that the sketches are not by Bartolomeo Columbus, but by the Italian scholar Alessandro Zorzi.[47] The sketch of the Atlantic usefully illustrates the partly verified and partly conjectured configuration of the lands of the western Atlantic as perceived in Europe at the time.

The earliest known nautical chart that shows Columbus' discoveries in relation to the Atlantic is that of one of his pilots, Juan de la Cosa (1508).[48] In the isthmus of Central America, where Columbus had hoped to find a strait leading him into the Indian Ocean, the cartographer placed a figure representing the navigator, in the guise of Saint Christopher carrying Christ on his shoulders as a child to convey Christianity to the New World. The representation of a greatly extended American coastline, both northwards and southwards is, with one exception, largely conjectural and probably corresponds to a principle of Ptolemy's *Geography* according to

---

43 Roberto Almagià, 'I mappamondi di Enrico Martello e alcuni concetti geografici di Cristoforo Colombo', in *La Bibliofilia*, Vol. XLII, (1940), pp. 288–311;

44 R.A. Skelton, in Marcel Destombes, *Mappemondes A.D. 1200–1500*, (Amsterdam, 1964), pp. 229–234; see also Ilaria Luzzana Caraci, 'Il planifero di Enrico Martello della Yale University e i fratelli Colombo', in *Rivista Geografica Italiana*, Annata LXXXV, fasc. 2, (Giugno, 1978), pp. 132–143.

45 Reproduced in Kenneth Nebenzahl, *Atlas of Columbus and the Great Discoveries*, (Chicago, New York, San Franscisco, 1990), p. 26.

46 In Biblioteca Nazionale Centrale, Florence; reproduced by Sebastiano Crinò, 'Schizzi cartografici inediti dei primi anni della scoperta dell'America', in *Rivista maritima*, (Rome, 1930), Vol. IX, Suppl. fasc. (Nov. 1930), pp. 18–20.

47 Robert Almagià, 'Intorno a quattro codici fiorentini e ad uno ferrarese dell'erudito Alessandro Zorzi', in *La Bibliofilia*, Vol. XXXVIII, (1936), pp. 313–347.

48 In the Museo Naval, Madrid; reproduced in Kenneth Nebenzahl, *op. cit.*, Plate 10.

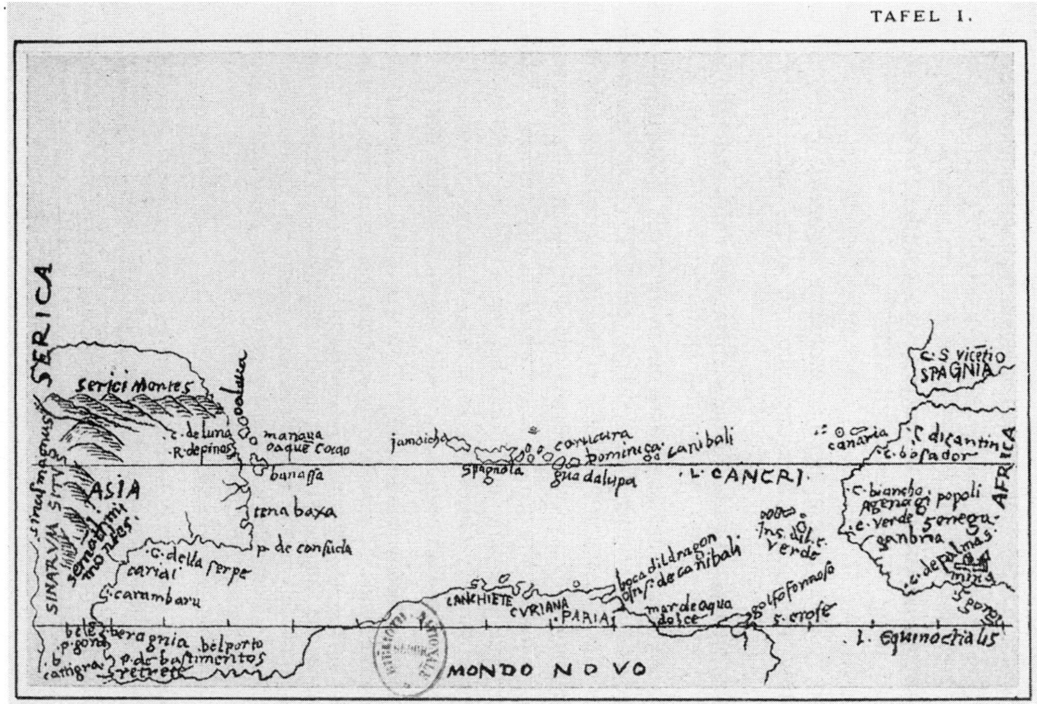

Fig. 7. Third of Alessandro Zorzi's three sketch maps (wrongly attributed to Bartolomeo Colombus) showing *AFRICA,* the north coast of Brazil (*MONDO NOVO*) with *ASIA* attached to the latter. Ms. Magl. XIII, 'Alberico', Vol. II, 81, in Biblioteca Nazionale, Florence: 'Informatione avuta da Bartolomeo Colombo'. Reproduced from F. von Wieser, 'Die Karte des Bartolomeo Colombo über die vierte Reise des Admirals', in *Mittheilungen des Instituts für Oesterreichische Geschichtsforchung,* Erganzungsheft IV, (1893).

which the oceans were lakes surrounded by landmasses. The exception is the length of coastline running east-west in the northern part of the chart on which are written a number of legends, one of which reads: *Mar descubierto por ingleses* (Sea discovered by the English). Five English flags are placed along this stretch of coast. The legend and the flags reflect John Cabot's voyage of 1497 and refer to an area of coast explored by him in the neighbourhood of Newfoundland and Nova Scotia.[49]

The impact in Europe of Amerigo Vespucci's voyage under Portuguese command in 1501 along the east coast of Brazil

In 1501 Amerigo Vespucci accompanied a Portuguese voyage down the east coast of Brazil, probably in search of a passage through to the Spice Islands of the Moluccas.[50]

According to the analysis of authoritative scholars, the ships, under the command of a Portuguese Gonçalo Coelho, reached to about 25° Lat. S., then drew away from the coast to advance further southward out in the ocean.[51] Amerigo Vespucci described the voyage in a letter from Lisbon, dated 1502 and known as the 'Bartolozzi letter'. In it he stated that the fleet finally reached a latitude of 50° S. This authentic letter was then re-worked and added to by an unknown hand, and published under the title of *Mundus Novus,* of which the first edition is dated Paris, c. 1503 or 1504.[52] The little pamphlet at once became a best seller across the whole of Europe and went into sixty editions by 1529 in six languages.[53]

What chiefly impressed European readers was Vespucci's claim that,

> '... the new region which we have found [...] we may rightly call a New World. [...] For this transcends the view held by our ancestors, in as much as most of them hold that there is no continent to the south beyond the equator, but only sea, which they named the Atlantic. [...] But that this their opinion is false and utterly opposed to the truth, this my last voyage has made manifest; for in those southern parts I have found a continent more densely peopled and abounding in animals than our Europe or Asia or Africa'.[54]

---

49 See J.A. Williamson, *The Cabot Voyages and Bristol Discovery under Henry VII*, Hakluyt Society, Second Series, Vol. CXX, (London, 1961), Fig. 2.

50 As alleged by Francisco López de Gómara, *Historia General de las Indias*, (Zaragoza, 1552), f. xlix^r; Bibliotheca de Autores Españoles, Vol. XXII, (Madrid, 1946), p. 211.

51 Max Justo Guedes, 'As primeiras expedições portugueses e o reconhecimento da costa brasileira', in *Revista Portuguesa da História*, Vol. XIV, (Coimbra, 1970), pp. 247–267.

52 Alberto Magnaghi, *Americo Vespucci*, 2nd edn (Rome, 1926), pp. 17–40.

53 On the *Mundus Novus* and the number of its editions, see Rudolf Hirsch, 'Printed Reports on the Early Discoveries and their Reception', in Fredi Chiapelli ed., *First Images of America*, Vol. II, (Los Angeles, 1976), pp. 537–560.

54 Pseudo-Vespucci, *Mundus Novus*, edn (Augsburg, 1504), reproduced by Henri Vignaud, *Améric Vespuce, (1451–1512)*, (Paris, 1917), p. 305.

Other sources concerning the voyage, place doubt upon Vespucci's claim that the fleet reached as far as Lat. 50° S. Raffaello Maffei da Volterra in his encyclopedia *Commentariorum urbanorum libri XXXVI*, published in Rome in 1506, noted, on the basis of his contacts with the Portuguese Cardinal Jorge da Costa in Rome that the Portuguese had not confirmed the figure of 50° Lat S.[55] Although many imaginative explanations have been offered, its origin still remains enigmatic.

The reservations of the Portuguese regarding Vespucci's claim passed unnoticed by European readers; Vespucci's text was widely read and commented on both by humanists and university scholars as an example of recent experience to be opposed to the authority of scholastic culture. Across the whole of Europe, the evidence of Vespucci's voyage was used as a weapon to attack the reigning Aristotelian science of the universities.

The expression 'a New World' was not used by Vespucci in the 'Bartolozzi' letter and the anonymous editor of the *Mundus Novus* was not the first to use the term. It had already in 1493 been used with reference to Columbus in a letter of the Italian humanist Pietro Martyr d'Anghiera: 'Columbus, the illustrious discoverer of a New World'.[56] But as Pietro Martyr's correspondence was only published in 1530, his remark did not have the impact of the expression attributed to Vespucci.

In 1507, the text of four of Vespucci's real and alleged voyages (one of them in 1497 is considered apocryphal) came into the hands of a group of humanist scholars in the little town of St Dié in the Vosges. They printed it under the title of *Cosmographiae Introductio*.[57] Often attributed to Martin Waldseemüller, the real editor of the text has been shown to be Matthias Ringmann.[58] In the introduction Ringmann suggested calling the new continent found by Vespucci 'America' in honour of the Florentine voyager.

Martin Waldseemüller, the cartographer of the group, produced in 1507 and 1516, two celebrated world maps to illustrate the vast transformation of knowledge brought about by the Portuguese voyages round Africa, Columbus' voyage across the Atlantic and the Portuguese voyage along the east coast of Brazil, in which Vespucci had participated. By means unknown to us, Waldseemüller obtained

---

55 See Vsevolod Slessarev, 'Raphael Maffei's contribution to the history of the Portuguese Discoveries', in *Congresso Internacional de História dos Descobrimentos, Lisboa, 1960, Actas*, (Lisbon, 1961), Vol. III, pp. 551–575.

56 Pietro Martyr d'Anghiera, *Opus Epistolarum*, (Alcalá, 1530), f. xxxiii^r.

57 Martin [Hylacomilus] Waldseemüller, The *Cosmographiae Introductio*, facsimile edn by Joseph Fischer and Franz von Wieser, (New York, 1907), University Microfilms International, (Ann Arbor, Michigan, U.S.A., 1977).

58 See Franz Laubenberger 'The naming of America', in *Sixteenth Century Journal*, Vol. XIII, No 4, (1982), pp. 91–111.

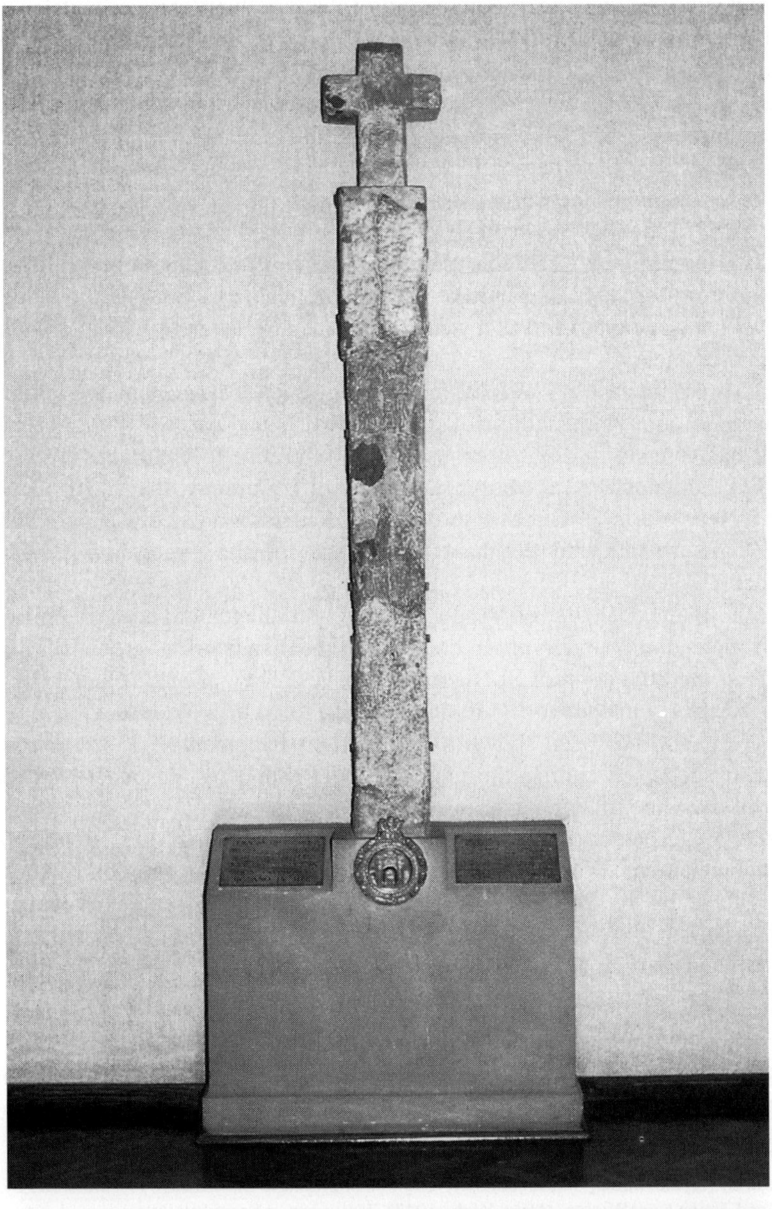

Fig. 8. Reconstructed fragments of the *Padrão* of São Gregório, recovered by Eric Axelson at Kwaai Hoek, and today in the Cullen Library, University of Witwatersrand, Johannesburg.

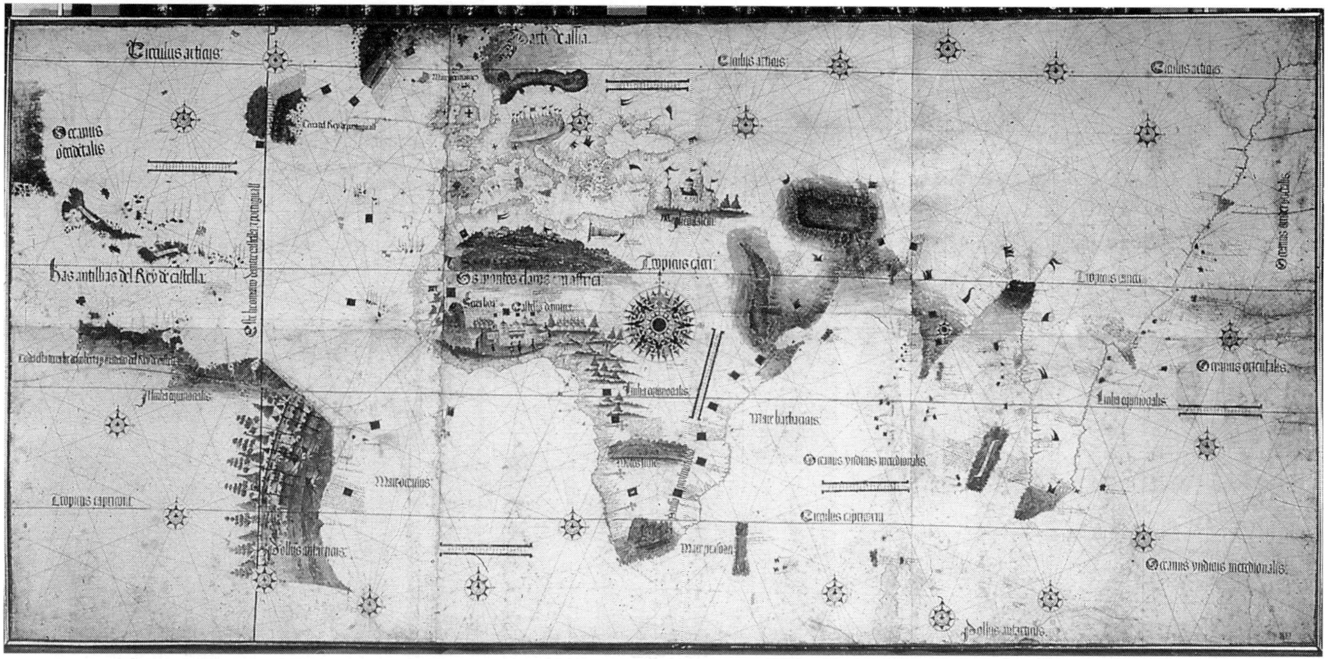

Fig. 9. The so-called 'Cantino Chart', (1502), in Biblioteca Estense, Modena. Copy of a lost Portuguese master chart obtained by Alberto Cantino, secret agent in Lisbon of the Duke Hercules d'Este.

copies of Portuguese nautical charts,[59] with the aid of which he sought to produce a new synthesis with the cartography of Ptolemy, the greatest cartographer of Antiquity.

In Waldseemüller's world map of 1507, the frame still remains essentially Ptolemaic. In that of 1516, the frame he chose was that of the traditional Mediterranean nautical chart. He was thus not able to resolve the conflict between the classical tradition of cartography and that of the nautical chart. Only later in the 16th century was this accomplished by the Dutch cartographer Mercator with his projection, which consisted of progressively increasing the spacing between the parallels of latitude, thus producing a map which gave a distorted image of reality, but which could be efficiently used by sailors.

The geographical revolution in European culture

If one compares the Catalan World Map of c.1450, with the nautical chart of Alberto Cantino of 1502, an Italian copy of an unknown Portuguese master chart,[60] it will be seen how completely the face of the known world had been transformed in the very short period of half a century. The frame of the medieval *mappaemundi*, traditionally round in shape, gave little or no idea of the extent of the surrounding ocean. The frame of the new marine world chart, rectangular in form, gave a clear impression of the immense extent of the area of the sea in relation to land.

Already in 1498, the Italian humanist Angelo Poliziano praised the Portuguese King D. João II, for ' ... having discovered other lands, another sea, other worlds, indeed other stars and for having made them emerge from the eternal shades of ancient chaos to the light of day'.[61]

Of Columbus, the Genoese Agostini Giustiniani wrote in his Psalter (1516), that, ' ... Columbus had in our day, by his own energy, explored in a few months, more land and sea than almost all other mortels in the ages that are gone'.[62]

Among the features of medieval geography which the Discoveries had trans-

---

[59] He appears to have had access to the chart (or a copy of it) of Nicolo Caveri (c.1504–05), an Italian copy of a Portuguese chart. It is today in the Bibliothèque Nationale, Paris; reproduced in Kenneth Nebenzahl, *op. cit.*, Plate 13.

[60] In the Bibliotheca Estense; reproduced in Kenneth Nebenzahl, *op. cit.*, Plate 11. On Cantino's chart see Armando Cortesão, *Cartografia e Cartógrafos Portugueses dos séculos XV e XVI*, (Lisbon, 1935), Vol. I, pp. 142–151.

[61] Angelo Poliziano, letter to King D. João II, in *Omnia Opera Angeli Politiani*, (Venice, 1498), sign. n.iii^{r}.

[62] Agostini Giustiniani, *Psalterium*, (Genoa, 1516), reproduced in *Raccolta di Documenti e Studi pubblicati dalla R. Commissione Colombiana*, Parte III, Vol. II, (Rome, 1893), p. 245.

formed, was the torrid zone astride the equator, alleged since Antiquity to be uninhabitable by man on account of the extreme heat. Experience had shown on the contrary that it was densely populated by blacks, 'who form' wrote the Portuguese Diogo Gomes, about 1495, 'such a multitude as can hardly be believed'.[63]

The island of Taprobana (the present day Sri Lanka), reported by Ptolemy in Antiquity to be of great size and extending below the equator, had been shown to be in reality of very modest proportions.[64] Instead another 'island' – some said 'another world' – called Brazil, had been shown to exist, unmentioned by Ptolemy and unsuspected by Europeans. The German printer resident in Lisbon, Valentim Fernandes, did not hesitate in 1503 to identify Brazil and its inhabitants with the Antipodes,[65] despite what Saint Augustine had written of them that they were uninhabited, since there was no record that any of the sons of Adam had ever reached them.[66]

Contrary to the medieval belief that the southern hemisphere contained fewer stars than that of the northern, Vespucci's voyage with the Portuguese fleet of 1501, wrote Jean Dullaert, a professor at the University of Paris, in a commentary on Aristotle's *Meteorologia*, (1514), had proved just the opposite. The stars in the southern hemisphere were, he said, in reality greater in number and brighter.[67]

The Portuguese humanist and historian João de Barros wrote in 1532 that if Ptolemy, Strabo and Pomponius Mela were to return with their

> 'mediocre writings' modern cosmography ' ... would make them embarrassed and ashamed, and they would perceive that the parts of the world which they had not dealt with were greater in extent than the three parts into which they had divided it. And the most embarrassed would be Ptolemy, on account of the graduation of his maps, since as soon as he deals with regions beyond Alexandria [his home] he represents them with the same licence which Horace allows to painters and poets'.[68]

Other humanists saw the geographical discoveries as a Promethean restitution to mankind of the heritage lost by Adam and Eve after the Fall. The Swabian mathematician Simon Gryneus, wrote in 1532 of the navigators of the Discoveries as

---

63 Diogo Gomes, 'De Prima Inventione Gujnee', in Valentim Fernandes, *op. cit.* p. 284.

64 See João de Barros, *Terceira Década da Ásia,* (Lisbon, 1563), Liv. II, Cap. I, facsimile edn (Lisbon, 1992), f. 26ʳ.

65 Valentim Fernandes, 'Acto Notarial' (1503), ed. António Alberto de Andrade, 'O Auto Notarial de Valentim Fernandes (1503) e o seu significado como fonte histórica', in *Arquivos do Centro Cultural Português*, Paris, Vol. V, (1972), p. 540.

66 Saint Augustine, *City of God*, Loeb edn (London, 1965), Bk. XVII, Chap. 9, pp. 49–51.

67 Jean Dullaert, *Habes humanissime lector librorum metheororum Aristotelis facilem expositionem & questiones super eodem...*, (Paris, 1514), sign. C.iʳ.

68 João de Barros, *Ropica Pnefma*, (Lisbon, 1532), sign. D.iijʳ, facsimile edn with modern transcription, notes and study by I.S. Révah, (Lisbon, 1952), Vol. I, p. 51.

'... invincible and truly divine souls, who are seen to claim dominion over land and sea as the heritage of their first parents [i.e. Adam and Eve]'.[69]

In conclusion the opinions of the humanists may be summed up in the remark of the Spanish humanist Luis Vives in 1531: 'Truly, humankind has seen its orb revealed to it in full'.[70]

---

69 Simon Gryneus, *Novus Orbis*, (Basle, 1532), edn (Basle, 1537), Introduction, (no pagination).

70 Luis Vives, *De Disciplinis libri XX*, (Antwerp, 1531), Dedicatory Epistle to King D. João III, f. A.ij^r, (*plane generi humano suus orbis patefactus*).

III

The alleged nautical school founded in the fifteenth-century at Sagres by Prince Henry of Portugal, called the 'Navigator'[1]

The statement that in order to implement his program of maritime discovery, Prince Henry of Portugal (1394–1460) had founded at Sagres in the early fifteenth century a nautical school to which he attracted experts in astronomy, navigation and cartography, has appeared in recent works published in connection with the quincentenary of Columbus' first voyage. The assertion has already for long been the subject of heated controversy among Portuguese historians of which the best and most authoritative have amassed respectable evidence to show that no such institution ever existed.[2] In his *Atlas of Columbus and the Great Discoveries* (1990), destined for a wide audience, Kenneth Nebenzahl writes:

> 'By 1420, Henry had founded his renowned school for mathematicians, astronomers, navigators, cartographers, and instrument makers at Sagres.'[3]

In a sumptuously produced work, *Voies océanes de l'ancien aux nouveaux mondes*, (1990), destined for a much narrower audience and prefaced by Emmanuel Le Roy Ladurie, Mireille Pastoureau states:

> 'From 1413 [sic] the Infante [= Prince Henry] had settled at Sagres, on a wild flat-topped headland overlooking a bay near Cape Saint Vincent, not far from Lagos at the extreme tip of Portugal. He was to transform this austere place into a dockyard and attract to it all those who could help him succeed in his plans: astronomers, captains and map-makers.'[4]

---

1 We put 'Navigator' in inverted commas since the term 'Henry the Navigator' was first used successively by two German historians Heinrich Schaefer and Gustav de Veer and then later taken up as the title of their books by the English historians Henry Major in 1868 and Raymond Beazley in 1895. Only secondary Portuguese historians, following the above writers, have called him the 'Navigator'.

2 The most recent critic of the 'Sagres School' was the late Luís de Albuquerque, *Dúvidas e Certezas na História dos Descobrimentos Portugueses*, Lisbon, 1990, Part I, Chap. II.

3 Kenneth Nebenzahl, *Atlas of Columbus and the Great Discoveries*, Rand McNally, New York, Chicago and San Francisco, 1990, p. vi.

4 Mireille Pastoureau, *Voies océanes de l'ancien aux nouveaux mondes*, Preface by Emmanuel Le Roy Ladurie, Editions Hervas, Paris, 1990, p. 33.

We shall discuss the alleged dockyard later. In other less recent writers, similar allusions to a centre for specialists at Sagres appear. The learned Samuel Morison showed a certain caution, but he nevertheless declared that:

> 'Prince Henry the Navigator established around 1420 a center for exploration and hydrography at Sagres on Cape Saint Vincent [sic].'[5]

Joseph Needham, in a recent volume of his *Science and Civilization in China*, threw caution to the winds, asserting the existence of,

> ' . . . a scientific citadel [sic] on Sagres Promontory [where Prince Henry] welcomed and entertained for long periods astronomers, cartographers, cosmologists and sea-captains.'[6]

The 'welcoming and entertaining' both of nationals and foreigners is indeed referred to by Zurara the Infante's fifteenth century biographer, but he makes no mention of astronomers, cartographers and cosmologists [sic].[7]

Scepticism about the existence of an organized centre for the study and teaching of nautical science had already started among Portuguese scholars at the beginning of this century. By 1945 Francisco Fernandes Lopes was to denounce four features of the myth of the 'Sagres School': the alleged 'palace' of Prince Henry at Sagres, the 'school of cosmography', the 'observatory' and the 'dockyard'.[8] In the 1940's one of Portugal's most distinguished historians, Duarte Leite, a university professor of mathematics and a diplomat, examined the whole question of the Sagres settlement in a series of articles mainly published in daily newspapers. Professor Magalhães Godinho gathered these together and published them in 1959–60 in book form with notes, additions and corrections in the light of more recent scholarship. These two little known volumes remain one of the finest contributions to the history of Portuguese cartography and nautical science in this century.[9] The sources analysed by Duarte Leite in his discussion of the Sagres settlement can be divided into those of three distinct periods: writers who wrote during the Prince's life-time, those of the sixteenth century and those of later centuries up to the nineteenth. Leite saw in writers of the second group the beginnings of a deliberate distortion of the past guided by the desire to build up Prince Henry into a great figure. In writers of

---

[5] S. E. Morison, *The European Discovery of America. The Northern Voyages 500–1600*, New York, Oxford University Press, 1971, p. 94.

[6] Joseph Needham, *Science and Civilization in China*, Vol. IV, Part 3, Cambridge, 1971, p. 504.

[7] Gomes Eanes de Zurara, *Crónica dos feitos de Guiné*, ed. A.J. Dias Dinis, Lisbon, 1949, Vol. II, Chap. 4, p. 19.

[8] Francisco Fernandes Lopes, *Terçanabal e a 'Escola de Sagres'*, Lisbon, 1945, 38 pp.

[9] Duarte Leite, *História dos Descobrimentos, Colectânea de Esparsos*, organização e notas e estudo final de V. Magalhães Godinho, 2 vols, Lisbon, 1959 & 1962. Duarte Leite's discussion of Sagres appears in his chapter on Prince Henry in Vol. 1, pp. 67–265. Magalhães Godinho's notes on the same subject appear in Vol. 2, pp. 438–511.

the eighteenth and nineteenth centuries he saw the myth appear in all its full blown exaggeration. Having provided evidence of the progressive growth of the myth, Leite, carrying scepticism to the limit, applied the rigorous criterion that only writers of the first group, who had lived in Henry's life-time could be considered reliable. Six in number, they were all shown to make no mention whatever of any school or scientific activity at Sagres.[10]

While admiring the minutely detailed quality of Leite's scholarship, one hesitates to follow him all the way in his systematic brushing aside of all writers on the mid-sixteenth century, one of which, as it happened, escaped his vigilant eye.[11] We propose then to approach the Sagres problem in the light of the oldest sources presently available, giving a critical hearing to those of the mid-sixteenth century and, but not beyond, the beginning of the seventeenth, and dividing our analysis into four parts: (a) the dating and construction of Prince Henry's settlement at Sagres and the origin of the claim that a nautical school existed there; (b) evidence for the presence of experts in Prince Henry's household; (c) evidence of Prince Henry's own scientific capacities.

The dating and construction of the Sagres settlement and the origin of the claim that a nautical school existed there

Contrary to what is often repeated, Prince Henry did not settle at Sagres after his return from the taking of Ceuta in Morocco in 1415.[12] Only in 1443 did the Portuguese Regent D. Pedro grant Prince Henry the right to a piece of land a league in circumference around the Cape and the right to build on it a settlement or settlements.[13] As this grant was renewed in 1448,[14] it would appear that no construction had so far started by then. According to José Manuel Garcia, it was only after 1452 that the Prince really settled in the Algarve and only from 1457, in the last three years of his life, that he concerned himself directly with Sagres.[15] When the chronicler Gomes Eanes de Zurara wrote his *Crónica dos feitos de Guiné*, finished in

---

10 Duarte Leite, *op. cit.* Vol. I, pp. 131–155.

11 Damião de Góis, *Fides, religio moresque aethiopum*, Louvain, 1540.

12 The source for this date is late: the 16th century Portuguese historian Damião de Góis, *Crónica do Príncipe D. João [II]*, Lisbon, 1567, ed. Graça Almeida Rodrigues, Lisbon, 1977, Chap. VII, p. 20.

13 *Monumenta Henricina*, Vol. VIII, Coimbra, 1967, Doc. 62, p. 107.

14 Ibid. Coimbra, 1969, Vol. IX, Doc. 194, p. 309.

15 José Manuel Garcia, *Sagres*, Câmara Municipal de Vila do Bispo, 1990, pp. 17–18. (Also in English and German). This richly illustrated little book lacks footnotes, but the documents referred to can be readily found in the *Monumenta Henricina*. Garcia notes (p. 51) that so difficult was it to get voluntary settlers established at Sagres that the Prince resorted to granting pardons to banned criminals in return for their willingness to settle there.

1453, he noted of the settlement being built by Prince Henry, that there were, when he wrote, only the walls, which he acknowledged were 'well fortified' (*de boa fortelleza*).[16] In his letter in the form of a will, Prince Henry wrote, on the 29 September 1460, two months before he died, that he '. . . had ordered a settlement (*vila*) to be built on the Cape [...] called *Terçanabal* [= Sagres] and which I have now called *Vila do Infante*.'[17] There is thus no evidence that just before his death, 13 November 1460, there was very much more at Sagres than what is shown on Drake's map of 1587 and certainly nothing that could be called a *palácio*. In an earlier letter written by the King D. Afonso V, 10 April 1451, Prince Henry's '*Vila*' is called *Terçanavall*,[18] (proof that construction had already been started at that date). Much speculation has taken place on the origin and meaning of the word *Terçanabal*. D. Francisco Manuel de Melo (1660), hazarding a specious etymology, found the origin of the word in the Venetian term for a dockyard: '*Darcena* or *Arsenal*'.[19] Hence arose the legend that there had been a dockyard at Sagres. *Terçanabal*, it is otherwise agreed, is a word of Arabic origin and Magalhães Godinho's suggested explanation seems plausible enough: *Tarf-an-nabl* – 'Cape of the Arrow' or Tarf-es-annabl – 'Shelter of the Arrow'.[20]

The Prince had sought, so he declared in his will of 15 September 1460, to make his settlement a refreshment station, where food and other things could be bought, since many ships often sheltered there.[21] The Genoese, his chronicler Zurara had heard it said, held the Prince's project in high esteem.[22] The Sagres peninsula has in fact on either side of it, two bays, the one to the eastward offering shelter from strong and frequent north winds which prevent rounding the Cape and one to the westward side which offers shelter from the less frequent easterly winds.[23] Prince Henry's decision to settle at Sagres seems more likely to have been dictated by the above consideration than by reasons which we shall examine below, such as the

---

16 Gomes Eanes de Zurara, *Crónica dos feitos de Guiné*, ed. A.J. Dias Dinis, O.F.M., Lisbon, 1949, Vol. II, Cap V. p. 31.

17 *Monumenta Henricina*, Vol. XIV, Lisbon, 1973, Doc. 1, pp. 1–2. Polemic has centred on the correct identification of the Cape referred to, but consensus holds that it is in fact the modern Sagres. Cf A. Fontoura da Costa, 'Vila do Infante antes Terça Nabal e Sagres depois', in *Arquivo Histórico da Marinha*, Vol. I, 1933–36, pp. 25–56.

18 *Ibid.* Vol. XI, Coimbra, 1970, Doc. 36, pp. 51–52.

19 D. Francisco Manuel de Melo, *Epanáphoras de Vária História Portuguesa*, Lisbon, 1660, facsimile reprint n.p.n.d. [Portugal, 1975], p. 314. See also Francisco Fernandes Lopes, *op. cit.* p. 14.

20 V. Magalhães Godinho, in Duarte Leite, *op. cit.* Vol. II, pp. 443–444. That the word could be read as *Tarf-Anabal*, 'Hannibal's Cape' is an explanation to be set aside.

21 *Monumenta Henricina* Vol. XIV, Coimbra, 1973, Doc. 1 p. 1.

22 Gomes Eanes de Zurara, *op. cit. ed. cit.* Vol. II, Cap. V, p. 31.

23 A. Fontoura da Costa, *art cit.* pp. 25–32.

need for clear skies to observe the stars,[24] or to seek a life of contemplation away from the bustle of crowds.[25]

The earliest mention of a 'School of Navigation' at Sagres appears in the work of Samuel Purchas (1577–1628), *Hakluyt Posthumus*, London, 1625.

> '[Prince Henry]' wrote Purchas, ' ... also from Majorca caused one Master James, [= Mestre Jacome], a man skilful in Navigation, and in Cards [= Charts] and Sea Instruments, to be brought into Portugal, there at his charge, as it were, to erect a School of Marinership, and to instruct his countrymen in that Mysterie'.[26]

Purchas' source for the first part of his paragraph is undoubtedly the Portuguese historian João de Barros, but the expression 'School of Marinership' is nowhere to be found in Barros' text (given below).[27] There remains the possibility of another source upon which Purchas might have drawn, though this is unlikely. From Purchas the notion of a 'School' spread into continental scholarship through the Abbé Prevost's *Histoire Générale des Voyages*, Paris, 1746,[28] and thence back into Portugal.[29]

The Presence of experts on nautical matters in Prince Henry's household

The only source for the presence of experts in Prince Henry's household concerns the Majorcan cartographer *Mestre* Jacome. The Portuguese soldier-cosmographer Duarte Pacheco Pereira, born about the middle of the fifteenth century, describes, in his *Esmeraldo de Situ Orbis* (c. 1508), how Prince Henry,

> ' . . . sent to Majorca for *Mestre* Jacome, a skilled maker of charts and by many gifts and favours brought him to these realms, where he taught his skill to men who in turn taught men who are alive at the present time.'[30]

The sixteenth century Portuguese historian João de Barros repeats this information in rather different terms, which would seem to show, as Duarte Leite has remarked,[31] that his source was not the same as Pereira's.

---

24 Damião de Góis, *Fides, religio moresque aethiopum* ... Louvain, 1540, sign. A.iv. (r.).

25 Alvise da Cadamosto, 'Navigatione' in G. B. Ramusio, *Delle Navigationi et Viaggi*, Venise, 1550, Vol. I, F. 105 (v.). This remark only appears in Ramusio's text and not in the earlier editions of Cadamosto or in the oldest manuscript known. See below.

26 Samuel Purchas, *Hakluyt Posthumus*, London, 1625, Book II, Chap. 1, § 2 p. 5.

27 João de Barros, *Ásia*, Lisbon, 1552, Déc. 1, Liv. I, Cap. XVI, ed. Coimbra, 1932, p. 61.

28 Abbé Prevost, *Histoire Générale des Voyages*, Tome I, Paris, 1746, p. 5 n. (a).

29 Duarte Leite, *op. cit.* Vol. I, p. 162.

30 Duarte Pacheco Pereira, *Esmeraldo de Situ Orbis*, trans by G.H.T. Kimble, Hakluyt Society, 2nd Series, No. LXXIX, London, 1937, facsimile reprint Kraus Reprint Ltd, Nendeln, 1967, pp. 100–101.

31 Duarte Leite, *op. cit.* Vol. I, p. 390.

> 'Prince Henry' wrote Barros, ' ... sent to the Island of Majorca for *Mestre* Jacome, a man expert in the art of navigation and in the making of charts and instruments and in so doing it cost him much effort to bring him to Portugal to teach his expertise (*sciencia*) to Portuguese practitioners of that art.'[32]

The word 'instrument' in the context of the time, could mean anything from an astronomer's instrument to a magnetic compass or a graphic representation such as the *Toleta di Martoloio*, so that there is here no unequivocal proof that the 'instruments' concerned were necessarily astronomical.

There is no evidence that the Majorcan cartographer was ever brought to Sagres, if anything only to Lisbon. Leite, in a later study, went to the point of denying that *Mestre* Jacome was ever in Prince Henry's service.[33] This seems unwarrantedly extreme and Magalhães Godinho writes that the Majorcan's presence in Portugal cannot be rejected *in limine*.[34] In 1930, Gonçalo de Reparaz, in a long and thorough study, brought together evidence to suggest that *Mestre* Jacome could be identified with the well-known Majorcan map-maker Jaime Ribes, whose original Jewish name before conversion was Jafuda Cresques, son of Abraham Cresques, the author of the famous Catalan Atlas of 1375, at present in the Bibliothèque Nationale, Paris. Jafuda Cresques, wrote Reparaz, probably came to Portugal between 1420 and 1427.[35] Magalhães Godinho pushes this dating forward to 1434–35, adding that the Majorcan would have reached Portugal when he was about 75 years of age.[36] It has recently been claimed that Jafuda Cresques died in 1410 and therefore could not have served Prince Henry as an adviser, since the latter's enterprise of discovery only began much later.[37]

If the identification of *Mestre* Jacome with Jafuda Cresques remains open to doubt,[38] there seems no reason to question the evidence of Duarte Pacheco Pereira and of João de Barros that a Majorcan cartographer, whatever his name, was brought to Portugal so that he could transmit his skill to Portuguese pilots and this in no way implies that the latter did not already have nautical charts of some kind.

---

32 João de Barros, *op. cit. ed. cit.* Déc. I, Liv. I, Cap. XVI, p. 61. *Sciencia* in the period had a much wider meaning than the modern word 'science' – it being more equivalent to 'knowledge' or 'expertise'.

33 Duarte Leite, *op. cit.* Vol. I, p. 175.

34 V. Magalhães Godinho, in Idem. Vol. I, 408.

35 Gonçalo de Reparaz, 'Mestre Jacome de Malhorca, cartógrafo do Infante' in *Biblos*, Vol. VI, 1930, fasc. 3–4, pp. 163–186.

36 V. Magalhães Godinho, note in Duarte Leite, *op. cit.* Vol. I, p. 409.

37 Cf A. Pinheiro Marques, 'Realidades e mitos da ciência dos descobrimentos quatrocentistas (a propósito da "Escola de Sagres") e do célebre Mestre Jaime de Maiorca)' in *Congresso Internacional Bartolomeu Dias e a sua época, Actas*, Oporto, 1989, Vol. II, pp. 347–361. Marques relies on jaume Riera Sans in his facsimile edition of the Catalan Atlas, *El Atlas Catalan de Cresques Abraham*, Barcelona, Diáfora, 1975, pp. 14–21.

38 According to a document in the Portuguese National Archives published by Armando de Sousa Gomes, in *Memórias do Congresso do Mundo Português*, Lisbon, 1940, Vol. I, pp. 647–5, a *Mestre*

Prince Henry's own scientific capacities

Finally we come to the question of Prince Henry's own competence in astronomy, mathematics and cartography, a competence which was vigorously denied by Duarte Leite.

Our earliest reference (published after Leite wrote), comes from André do Prado, a Franciscan monk, in a manuscript work dating from the middle of the fifteenth century. André do Prado praised Prince Henry in the following terms:´

> ' ... in astris videlicet cognoscendis ardua, in acquerendis terras grandia, in gerandis bellis strenua, in maria perlustrando forcia ... '[39]

That Prince Henry spent efforts in studying the stars (his other proficiencies do not concern us here) is no proof that he was competent in making astronomical measurements, or that he could put his knowledge to use in navigation. Astrology might have been all that he was concerned with. However, as we shall see, the theme of his interest in the stars recurs in the next writer to be examined, whose existence apparently escaped the notice of Duarte Leite. The Portuguese humanist and historian Damião de Góis, in a little work on the religion of the Ethiopians published at Louvain in 1540, wrote as follows of Prince Henry:

> 'Of all the sons of King D. João I, [Prince] Henry was the cleverest, particularly in mathematics. To wholly give himself up to the study of the stars, and to measure accurately their movements, he lived on the Sacred Promontory, called Saint Vincent [sic]. He chose this place because the sky there is rarely overcast and observation of the movement of the stars, with the instruments he used for the purpose, is not hindered by intervening clouds.'[40]

This is the first source that declares that Prince Henry was proficient in mathematics, a subject about which his biographer Gomes Eanes de Zurara says not a word, though the latter's silence on this point is no absolute proof. That the Prince chose to settle on the Sagres promontory because the sky there was rarely overcast is scarcely born out by fact, since the sky over the promontory is more often cloudier than at nearby Lagos, though, according to one local authority, when not overcast, visibility at Sagres is better than at Lagos.[41] Góis remains the only six-

---

Jacome is known to have rented lands near Lisbon in 1426. This still does not prove any connection with Jafuda Cresques as A. Cortesão appears to suggest in his *História da Cartografia Portuguesa*, Lisbon, 1970, Vol. II, p. 95.

39 André do Prado, *Horologium fidei*, Biblioteca Vaticana, Ms 1062 (sec. XV), reproduced by Fernando Felix Lopes, 'A volta de Fr André do Prado' in *Colectânea de Estudos*, 2e série, Ano II, No 1, Braga, 1951, pp. 121–131. Cf also Mário Martins, 'O Diálogo entre o Infante D. Henrique e Fr André do Prado', in *Revista Portuguesa de Filosofia*, Tome XVI, fasc. 3, 1960, pp. 281–295. This study however adds nothing concerning the context in question to the article of Fernando Felix Lopes.

40 Damião de Góis, *Fides, religio moresque aethiopum*, Louvain, 1540, sign. A.iv. (r).

41 Information from Snr José de Dcus Vieira Rodrigues, Mayor of the village of Castelejo on the Atlantic coast near Sagres.

teenth century source which states that Prince Henry undertook measurements of the movements of the stars (*stellarum*) with instruments. The fact that Góis wrote in the middle of the sixteenth century should not, in our opinion, preclude, as Duarte Leite would have it, such testimony from all consideration, but it still does not prove that the Prince's activity was directly connected with nautical astronomy.

Another sixteenth century reference to Prince Henry's interest in astronomy appears in Ramusio's edition of the account of the voyage of a contemporary of the Prince's, the Italian navigator Alvise da Cadamosto, who had reached the region of Guinea in 1455–56. Cadamosto wrote:

> 'Although in the study of the science of the movements of the heavens and of astronomy, I can greatly praise him, nevertheless I pass over all references to it.'[42]

In the first printed edition of Cadamosto's text in Fracanzano da Montalboddo's little work *Paesi nouamente ritrouati* Vicenza, 1507, the passage in question is reduced to the following:

> 'Although of the [Prince's] qualities, many noteworthy things may be said, I pass over all reference [to them]'[43]

The same lines appear exactly as above in the earliest known manuscript copy of Cadamosto's voyages preserved in the Marciana Library in Venice.[44] Duarte Leite attributed the discrepancy between the two editions to Ramusio's own gratuitous additions, such as he has been found guilty of in other cases. Leite saw here the beginnings of the sixteenth century 'build-up' of Prince Henry as a national hero.[45] There is however the possibility that Ramusio used a different and more complete manuscript, as he is known to have done in the case of Francisco Álvares' account of his travels in Ethiopia,[46] or he might have been influenced by reading Góis' book. Thus, even if one prefers the Ramusio text of Cadamosto, there is still no proof that Prince Henry's interest in astronomy was directly connected with nautical science.

Evidence has been claimed for the Prince's competence in astronomy in the description in the catalogue of Ferdinand Columbus' library of a manuscript in Spanish (since lost) allegedly written by the Infante: *Libro en español de mano llamado secreto de los secretos de astrologia compuesto por el infante don Enrique*

---

42 Alvise da Cadamosto, 'Navigatione, Proemio' in Giovanni Battista Ramusio, *Navigationi et Viaggi*, Venice, Vol. I, 1550, f. 105 (r.).

43 Alvise da Cadamosto 'Navigation' in Fracanzano da Montalboddo, *Paesi nouamente ritrouati*, Vicenza, 1507, ed. Milan, 1508, facsimile edition Princeton University Press, 1916, sign.a.i. (r.).

44 Printed in *Viagens de Luís de Cadamosto e de Pedro de Sintra*, Academia Portuguesa de História, Lisbon, 1948, p. 4.

45 Duarte Leite, *op. cit.* Vol. I, pp. 131–134.

46 Armando Cortesão, 'A "Carta de Marear' de Pêro da Covilhã' in Idem *Esparsos*, Coimbra, 1975, Vol. III, p. 218.

de Portugal. The résumé, in the catalogue, of the contents of the work, examined and published by Guy Beaujouan, shows that it deals only with the techniques of astrological prediction and is clearly unconnected with anything related to nautical astronomy.[47] That the Prince could have been the author of the work was admitted by Beaujouan and Joaquim de Carvalho,[48] although this had been denied by Duarte Leite.[49] Magalhães Godinho, following the latter, suggested that the work might have been owned, rather than written by the Prince,[50] which to us seems more likely.

A further instance in which clear proof is lacking is that of the Prince's knowledge of ptolemaic cartography. Duarte Leite asserted that he believed that no copy of Ptolemy's *Geography* existed in Portugal in Prince Henry's life-time.[51] The contrary has been alleged by Damião Peres[52] on the basis of a statement attributed to one of the Prince's navigators Diogo Gomes:

> 'At one time the Infante D. Henrique, wishing to learn of distant regions in the western ocean [= Atlantic], whether there were islands or *terra firma* beyond the description of Ptolemy, sent out caravels to seek these lands.'[53]

For Leite nothing ensures that this statement was not a personal addition to Gomes' text by Martin Behaim who took down the navigator's account c. 1485 long after the Prince's death.[54] In 1960 Virginia Rau published the accounts of an Italian merchant Bartolomeo di Ser Vanni, resident in Portugal, showing that the Bishop of the Algarve, D. Álvaro Afonso, had bought from him a copy of the ptolemaic atlas between 1460 and 1461.[55] This is hardly proof that Prince Henry knew of Ptolemy's work, though the mid-16th century evidence provided by the Portuguese historian João de Barros, is perhaps a little more cogent than Leite was ready to allow. Barros wrote in 1552 that:

---

47 Guy Beaujouan, 'Fernand Colomb et le traité d'astrologie d'Henri le Navigateur' in *Romania*, Vol. LXXII (1961), pp. 96–105.

48 Joaquim de Carvalho, 'A propósito da atribuição do *Secreto de los Secretos de Astrologia* ao, Infante D. Henrique' in Idem, *Estudos sobre a Cultura Portuguesa do século XV*, Vol. I, Coimbra, 1949, pp. 283–361.

49 Duarte Leite, *op. cit.* Vol. I, p. 420 n. 2.

50 V. Magalhães Godinho, in Duarte Leite, *op. cit.* Vol. II, pp. 492–493.

51 Duarte Leite, *op. cit.* Vol. I, p. 159, and Idem. *Ácerca dos feitos do Crónica de Guiné*, Lisbon, 1941, pp. 102–103.

52 Damião Peres, 'A acção do Infante D. Henrique no desenvolvimento da ciência' in *Congresso Internacional dos Descobrimentos, 1960, Actas*, Lisbon, 1961, Vol. I, p. 124.

53 Diogo Gomes, 'De Prima Inventione Guineae', in J.M. Schmeller, 'Über Valentin Fernandez Aleman und seine Sammlung von Nachrichten über die Entdeckungen und Besitzungen der Portugiesen in Afrika und Asien bis zum Jahre 1508', in *Abhandlungen der Philosophisch-Philolog. Classe der Kóniglich Bayerischen Akademie der Wissenschaften*, Bd IV, Dritte Abth. Munich, 1847, p. 33.

54 Duarte Leite, *op. cit.* Vol. I, p. 148.

55 Virginia Rau, 'Bartolomeo di Ser Vanni, mercador-banqueiro florentino "estante" em Lisboa nos meados do século XV' in *Do Tempo e do História*, Vol. IV (197 1), p. 113.

'... the Prince knew from Ptolemy's maps and also from accounts of tribesmen (*alarves*) that the African coast continued steadily southwards below the equator, although he had no information concerning navigation along the coast.'[56]

Ptolemy's representation of a closed Indian Ocean cannot have brought much encouragement to any project the Prince might have had of reaching Asia by sailing round Africa. Duarte Leite denied that the Infante had ever intended that his navigators seek to reach India and explained away Pope Calixto III's Bull of 1456 in which the latter had granted to the Portuguese crown the whole coast of Africa 'as far as the Indians' (*usque ad indos*),[57] as referring to Ethiopia only.[58] In the Middle Ages, Ethiopia had often been known as 'Tertia India'[59] and Ethiopia for the Portuguese was held to extend as far as the Atlantic.[60]

That Prince Henry might have overcome the 'blocking effect' of Ptolemy's *Geography* through his reading of accounts of voyages round Africa in Antiquity, is suggested, but not proven, by the remarks of Damião de Góis in his little work on the religion of the Ethiopians, Góis wrote that,

'... in order to confirm and reap the harvest of his studies, Prince Henry commanded [...] the discovery of what he had hitherto only conjectured, that the Atlantic led into the Indian Ocean and that the Indian Ocean flowed into the Atlantic.'[61]

In his *Crónica do Príncipe D. João [11]* (1567), Góis wrote that the Prince had found evidence of the circumnavigability of the African continent from his readings of Strabo, Pliny, Pomponius Mela and Cornelius Nepos.[62] However it is relatively unlikely that the Prince had read Strabo, the text of which was unknown in the Middle Ages. Although the first manuscripts circulated from the first third of the 15th century, the first printed edition only appeared at Rome c. 1469.[63] Cornelius Nepos is quoted by Pomponius Mela and by Pliny on the circumnavigations of Hanno and Eudoxus.[64] That the Prince had direct access to and had read Pomponius Mela and Pliny is possible though doubtful.[65] At his death at Sagres in

---

56 João de Barros, *op. cit. ed. cit.* Déc. I, Liv. I, Cap. 2, p. 14.

57 Bull of Pope Calixto III, 13/3/1456 in *Monumenta Henricina*, Vol. XII, Coimbra, 1971, p. 288.

58 Duarte Leite, *op. cit.* Vol. I, pp. 96–122.

59 Jordan Catalani de Severac, 'Description des Merveilles de l'Asie' (XIVe siècle) in *Recueil de Voyages et de Mémoires*, publié par la Société de Géographie de Paris, Paris, 1839, Tome III, p. 55.

60 Duarte Pacheco Pereira, *Esmeraldo de Situ Orbis* (c. 1508), ed. cit. Bk. I, Chap. XXVII, p. 78.

61 Damião de Góis, *op. cit. loc. cit.*

62 Damião de Góis, *Crónica do Principe D. João [II]*, Lisbon, 1567, ed. Lisbon, 1977, pp. 20–22.

63 A. Diller, *The Textual Tradition of Strabo's Geography*, Amsterdam, 1975, p. 97 and p. 132.

64 Pomponius Mela, *De Chronographia*, III, 90–93; Pliny, *História Naturalis*, Bk 2, Chap. LXVI.

65 Manuscripts of Pomponius Mela, although known in Italy in the 15th century, do not appear to have circulated in Portugal in the Prince's life-time. Pomponius Mela's work was first translated in Portugal into Spanish by Joan Faras from the Rome 1492–93 edition. Cf C. M. Gormley, M.A. Rouse

1460, he left there no books on astronomy or science of any kind, only devotional ones.[66] The question of the possible influence on Prince Henry of Ptolemy's *Geography* or of accounts by writers of Antiquity on the circumnavigability of the African continent, must remain undecided.[67]

Two world maps are known to have reached Portugal from Italy in Prince Henry's life-time and which are said to have enlarged his notions of world geography. The Prince's brother D. Pedro is reported by António Galvão (1563) to have brought back from Venice in 1428, a world map which ' ... had all the parts of the world and the earth described.'[68] It also showed the strait of Magellan and the Cape of Good Hope! The map, according to Galvão ' ... helped and furthered the discoveries of Prince Henry'.[69] Galvão's description is too imprecise for us to more than hazard a conjecture of what this map really showed – perhaps a modified version of Ptolemy's world map with the *terra incognita* enclosing the lower part of the Indian Ocean cut out, leaving a future Cape of Good Hope and a future south-east Asian peninsula, such as can be seen in the later maps of Henricus Martellus. However that such a modification of Ptolemy's world map could have been attempted as early as 1428 is difficult to accept.

The other world map to reach Portugal from Italy in Prince Henry's life-time is the world map of Fra Mauro (1459), ordered to be made specially for the Portuguese King D. Afonso V. A copy of the original, since lost, is in the Marciana Library in Venice.[70] Citing a contemporary document, Luís de Albuquerque maintains that the map proved disappointing to the Portuguese authorities as it seemed already out of date in relation to the progress of their discoveries.[71] The information on the cartographical resources at the Prince's disposal is, however, too fragmentary for us to draw any solid conclusions and the Fra Mauro map, if it was not already out of date, arrived too late to exert any influence on the Prince.

---

and R.H. Rouse, 'The Medieval circulation of the *De Chronographia* of Pomponius Mela', in *Medieval Studies*, Vol. 46 (1984), pp. 266–320. Zurara, the Prince's biographer, only knew Pliny at second hand. See Duarte Leite, *Ácerca da 'Crónica dos feitos de Guiné'*, Lisbon, 1941, pp. 98–99. According to Joaquim de Carvalho (*Estudos sobre a Cultura Portuguesa do século XV*, Vol. I, Coimbra, 1949, pp. 69–70), readers of Pliny were rare in Portugal in the Prince's time.

66 A.J. Dias Dinis, *Estudos Henriquinos*, Vol. I, Coimbra, 1960, pp. 287–297 and pp. 470–476.

67 Duarte Leite (*op. cit.* Vol. II, pp. 313–318) dismissed out of hand any influence of ancient writers on the Prince in this connection.

68 António Galvão, *Tratado [...] dos diversos & desvayrados caminhos* Lisbon, 1563. Portuguese text and English translation in António Galvão, *The Discoveries of the World*, London, 1601, ed Hakluyt Society, 1st Series, No 30, London, 1862, reprint Burt Franklin, New York, n.d. pp. 66–67.

69 Idem. *loc. cit.*

70 R. Almagià & T.G. Leporace, *Il Mappamondo di Fra Mauro*, Rome, 1956, Facsimile reproduction and transcription of the legends.

71 Luís de Albuquerque, *Ciência e Experiência nos Descobrimentos Portugueses*, Lisbon, 1983, pp. 63–64. The document, dated 3/3/1462, recognizes that a certain Dr João Fernandes da Silveira had made

An unfounded tradition that Prince Henry was the inventor of the quadratic plane chart with parallels and meridians set at right angles, appears to have been put about by the French Jesuit Georges Fournier in 1643,[72] since no earlier source has so far been discovered. Fournier's words were repeated word for word by the Portuguese chief cosmographer Manuel Pimentel in 1699.[73] The earliest evidence of plane charts with latitudes is the chart of the Portuguese Pedro Reinel (c. 1504)[74] and that with longitudes, the sketch maps of Alessandro Zorzi based on others (lost) by Bartolomeo Colombus (c. 1503) in a manuscript dated to between then and 1538.[75] Nothing reliable is known of Prince Henry's possible influence on cartographical theory. Of his concern with traditional nautical charts, there remains the letter of the Regent D. Pedro, dated 20 October 1443, stating that Prince Henry had ordered the charting of the African coast beyond Cape Bojador after one of his navigators, Gil Eanes, had succeeded in rounding it.[76] The first mention of the use of instruments by Portuguese navigators for the practice of nautical astronomy comes after the Prince's death and concerns the navigator Diogo Gomes, who describes the use of a quadrant to measure the height of the pole star.[77] But as we have noted the dating of the testimony of Gomes has been the subject of much debate. Leite suspects the text to be an interpolation made by Martin Behaim.[78] Albuquerque holds the text to refer to Diogo Gomes' own experience and dates it to c. 1460–62.[79] No surely based conclusion has so far been reached on this point. Although it is well attested that Prince Henry was a patron of the University in

---

payments to the artists who worked on the Fra Mauro map, 'so that what they had already accomplished should not be wasted.' (*E esto por sse no perder o que ja em elle era fecto*), Document in João Martins da Silva Marques, *Descobrimentos Portugueses*, Vol. III (1461–1500) Lisbon, 1971, p. 20.

72 Georges Fournier, S.J., *Hydrographie contenant la théorie et la pratique de toutes les parties de la navigation*, Paris, 1643, 2nd edition, Paris, 1667, reproduced in facsimile, Grenoble, 1973, Chap. III, p. 505.

73 Manuel Pimentel, *Arte de Navegar*, 2nd edition, Lisbon, 1712, pp. 67–68. Modern edition published from the 2nd edition by A. Cortesão and Luís de Albuquerque, Agrupamento de Estudos de Cartografia Antiga, Vol. 7, Lisbon, 1969, pp. 135–142. The first edition of Pimentel's work appeared in 1699.

74 A. Cortesão & A. Teixeira da Mota, *Portugaliae Monumenta Cartographica*, Lisbon, 1960, Vol. I, Pl. 8.

75 R. Almagià, 'Intorno a quattro codici fiorentini e ad uno ferrarense dell'erudito Alessandro Zorzi', in *La Bibliofilia*, Vol. XXXVIII (1936), pp. 313–347.

76 Letter of the Regent D. Pedro, 22 October 1443, in *Monumenta Henricina*, Vol. VIII, Coimbra, 1967, p. 107.

77 Diogo Gomes, *op. cit. loc. cit.*

78 Duarte Leite, *op. cit.* Vol. I, pp. 376–81.

79 Luís de Albuquerque, *Os Guias Náuticos de Munique e de Evora*, Lisbon, 1965, pp. 28–31 and Idem. 'A Navegação Astronómica' in A. Cortesão, *História da Cartografia Portuguesa*, Coimbra, 1970, Vol. II, pp. 239–240.

Lisbon, and that he introduced new disciplines, one of them of *Astrolegia* (= astronomy/astrology),[80] there is no reason to believe, as has been claimed, that such teaching was calculated to benefit future navigators to enable them to practice nautical astronomy.[81]

If, in what has been said above, we have sought to dispel a number of illusions entertained about Prince Henry's settlement at Sagres and his alleged proficiency in the science of nautical astronomy, there does remain the very positive proof that his fame as a promoter of geographical discovery had already in the mid-fifteenth century, spread beyond the frontiers of Portugal to be recognized by the humanists of the Italian Renaissance. In an undated letter to Prince Henry shortly before his death, Poggio Bracciolini praised the Prince's achievements in the following terms:

> 'How illustrious indeed to have been the only one of such courage, of such resolution and of such planned purposefulness to have dared to do that which none so far had undertaken or attempted. You alone [have discovered] unknown seas in regions never seen, unknown races living outside the known world and savage peoples living at its farthest confines beyond the regular annual shifts of the sun's track, where none before had opened a way.'[82]

Following the customary style of the time, Poggio ascribes to Henry himself the achievements of his navigators. The reference to the 'regular annual shifts of the sun's track' shows a recognition that the Prince's navigators had gone beyond the Tropic of Cancer.

To conclude. During the Prince's life-time, not only had regular navigation to and from the islands of the Madeira archipelago and the Azores been made possible, with the beginnings of their colonization assured, but the exploration of the African coast beyond Cape Bojador systematically conducted as far as the present Sierra Leone.[83] In the face of the documents that have come down to us, all this had been achieved with traditional Mediterranean navigational techniques.[84] The nau-

---

80 *Monumenta Henricina*, Vol. IV, Coimbra, 1963, Doc. 3, 12 October 1431, p. 7.

81 A. Moreira de Sá, *O Infante D. Henrique e a Universidade*, Lisbon, 1960, p. 67. This point has been emphasized to us by Luís de Albuquerque in a private communication, as well as the fact that there is no proof that the Gil Eanes of Lagos, a student at the University in 1431, is the same person as the navigator who rounded Cape Bojador in 1434. Cf Idem. p. 97 & p. 117.

82 Poggio Bracciolini, Letter to Prince Henry of Portugal, n.d. in *Spicelegium Romanum*, Vol. X, Rome, 1844, p. 255.

83 A.J. Dias Dinis, O.F.M., 'Limite meridional dos Descobrimentos henriquinos' in *Itinerarium, Colectânea de Estudos*, Ano I, No 6 (1955), pp. 798–845. Cf also Duarte Leite, *Ácerca do Crónica dos feitos de Guiné*, Lisbon, 1941, pp. 243–69 and Damião Peres, *História dos Descobrimentos Portugueses* 3rd edition, Oporto, 1983, pp. 123–133. The above writers all refute the claims of Jaime Cortesão that Portuguese navigators had gone much further in the Prince's life-time. Cf Jaime Cortesão, *Os Descobrimentos Portugueses*, 4th edition, Livros Horizonte, Lisbon, 1984, Vol. II, Chap. X.

84 Damião Peres (*op. cit.* p. 187) quotes the authoritative opinions of two naval officers, Fontoura da Costa and Moreira Campos that regular voyages to and from the Azores could not have occurred without

tical astronomy that was to come later and be introduced by the Portuguese themselves remains their supreme contribution to maritime discovery,[85] to be taken up, used and copied in turn by Spanish,[86] French,[87] English[88] and German[89] navigators.[90]

---

a nautical astronomy to guide sailors. This may well be true but a historian cannot accept such a conjecture without supporting documents to prove it.

85 Luís de Albuquerque, 'A Navegação Astronómica' in A. Cortesão, *op. cit.* Vol. II, Chap. VII, pp. 225–461. An English edition of this work was published simultaneously.

86 Tables for the Rule of the Sun of Portuguese provenance were reproduced by the Spaniard Martín Fernándes de Enciso, *Suma de Geographia*, Seville, 1519.

87 The Rules for the Sun and the Pole Star appear in French in the 'Traité de la Sphère avec le Régime du Soleil et du Nord', attached to Jean Alphonse's *Cosmographie* (c. 1544), ed. Georges Musset, Paris, 1904, pp. 89–112 and pp. 117–127. Alphonse was of Portuguese origin and his French has many lusitanisms.

88 Roger Barlow in his *A Brief Summe of Geography* (1540–41), introduced to English navigators the Rule of the Sun and of the Pole Star, taken from Enciso's work. Cf *A Brief Summe of Geography by Roger Barlow*, ed. E.G.R. Taylor, Hakluyt Society, 2nd series, No. 69, London, 1932.

89 Cf Wolfgang Köberer, 'Ein nederdeutsches Navigations handbuch aus dem 16 jahrhundert', in *Mare Liberum*, Vol. I, Lisbon, 1990, pp. 153–170.

90 On the spread of Portuguese nautical science to the rest of Europe, cf Luís de Albuquerque, *Dúvidas e Certezas na História dos Descobrimentos*, 2a parte, Lisbon, 1991, Chap. V, pp. 93–11, and Idem, 'A Projecção da Náutica Portuguesa Quinhentista na Europa' in Idem. *Estudos de História*, Coimbra, Vol. 1, 1974, pp. 199–240.

IV

The Emergence of Nautical Astronomy in Portugal in the XVth century

Of the great oceans of the world, the Atlantic, because of its violence, was the last to be mastered by man. The task in its entirety had to wait for the Portuguese sailors of the Renaissance. Isidore of Seville (c. 570–636), a Christian writer of the late Roman Empire, had written of the Atlantic that it was 'incommensurable and uncrossable'.[1] Although Pliny (A.D. 23–79) refers vaguely to the Canary Islands,[2] all knowledge of them disappears in the Middle Ages until a Portuguese expedition under the command of the Italian Lanzarotto Malocello 'rediscovered' them in 1336.[3] Italian charts of the XIVth century begin progressively to show the Canaries, Madeira, Porto Santo and the Azores, but all aligned along a N/S axis without any appreciation of the relative distances between them or how far they lay from the European shore. The first written evidence of the Portuguese 'discovery' of the islands of Madeira and Porto Santo appears in 1419–20 and of the Azores in 1427, about the same time as they began to be colonised under the aegis of Prince Henry of Portugal, called the 'Navigator'.[4] The difficulties of returning to them on regular voyages was to motivate the Portuguese to develop methods of measurement using the Pole Star as a navigational aid and this led, not only to a greater accuracy in placing the islands on the charts, but also to a greater precision in the charting of the west African coastline which they were progressively exploring during the second half of the XVth century.

Claims that Portuguese nautical astronomy originated in Aragon and was transmitted from there to Portugal[5] or was introduced into Portugal from Germany by Regiomontanus and Martin Behaim[6] have long ago been shown to be baseless.[7] Its origins are indubitably Portuguese. The emergence of nautical astronomy in XVth century Portugal is closely connected with the use, by navigators, of the Mediterranean portulan chart and of the magnetic compass associated with it. For direction, mariners relied on magnetic rhumb lines on the chart which they converted to a course set by the compass. For distance they relied on a scale (*tronco das léguas*) on the chart to which they compared their estimation of distance run over a 24 hour period called a *singradura*.[8] The first use of astronomical observations was to check and improve the estimation of distance run.

The first recorded cases of observations by navigators in the Atlantic of angular distances between the horizon and a celestial body appear in the second half of the XVth century, the angular distances being given in terms of coded linear measurements. Thus in 1455, the Italian navigator Alvise da Cadamosto, at the mouth of the Gambia (Latitude 13° 30′ N) declared seeing the Pole Star 'low on the horizon at the height of a lance'.[9] This odd use of the term 'lance' to express an angular distance is also found in a treatise by Jacobus Angelus of Ulm, writing about 1490 in connection with a comet seen in 1402 with its tail to be 'one and half lances in length'.[10] Before him, Marco Polo, in a conversation between 1293 and 1310 with Peter of Abano, had spoken according to the latter, of the *polum antarcticum*, probably the 'Magellan clouds', being seen in the Indian Ocean '*elevatum quantitate lanceae militis*'.[11]

Other examples of coded linear measurements (for example, 'the height of a man') are given by the Portuguese mariner Pedro de Sintra (1462) in relation to the island of Shenge on the west African coast (Latitude 7° 55′ N)[12] and by the Venetian map-maker Fra Mauro, who in a legend on his world map of 1459 referring to the Indian Ocean, wrote of the 'Antarctic Pole' as being there observed at the height of a *brazo* – 'a (man's) arm'.[13] Coded linear measurements for angular distances appear to be of Arab origin, for in the work of Abd-al-Rahman al-Sufi, (10th century), we find *dzira* – 'an arm's length from elbow to forefinger' equal to 2° 20′; *kamat al-insan* – 'height of a man' and *rumh* – 'a lance', the latter being equivalent to 14°,[14] not very different from the 13° 30′ of the locality where Cadamosto observed the Pole Star at the height of a lance. No evidence is known to us showing how or where these coded Arab measurements reached the West.

The first reference to the use in the Atlantic in relation to the portulan chart, of an astronomical instrument (the quadrant), comes in a much debated text concerning the Portuguese navigator Diogo Gomes, who voyaged as far as the Cape Verde Islands in 1460–2. His oral recollections were taken down by Martin Behaim and copied by the German printer Valentim Fernandes, resident in Lisbon. Gomes recalls of his voyage:

> And I had a quadrant when I went to these places (that is, the neighbourhood of the Cape Verde Islands) and I wrote on the 'table' (limb) of the quadrant the height of the arctic pole and I found it (the quadrant) better than the chart. It is certain that one sees one's course on the chart, but once one has committed an error, one can never get back on the course one had set. (*Et ego habedam quadrantem quando ivi ad partes istas, et scripsi in tabula quadrantis altitudinem poli arctici, et ipsum meliorem inveni quam cartam. Certum est, quod in carta videtur via marinandi, sed semel errata nunquam redeunt ad primum propositum*).[15]

Although the last sentence is admittedly obscure, it is generally agreed that the quadrant was not used by Gomes to obtain directly the latitudes of the places he had visited. From the evidence of an archaic text annexed to the *Reportorio dos Tempos* published by Valentim Fernandes in the 1551 and 1565 editions of the work (1st ed. 1518),[16] it has been shown that the quadrant was first used merely to establish *relative* north–south distances between Lisbon and places on the west African coast, as well as distances separating such places from each other; the main purpose being to compare and correct estimated distances as marked according to the traditional method on the portulan chart.[17] From the evidence of the text of the *Reportorio*, the names of the places on the African coast were simply written directly onto the limb of the quadrant.[18] Only later, probably towards the turn of the XVth century and at least before c. 1504, the date of the first known chart with a meridian graduated in degrees of latitude,[19] was the quadrant used to obtain absolute latitudes instead of relative distances.

The delicate problem in this development was the choice of a module relating celestial angular distance to terrestrial linear distance. Portuguese navigators in the XVth century, including Bartolomeu Dias, used a module of $16\frac{2}{3}$ leagues (or 70 miles) to a degree (a 15 percent error). Later, in the XVIth century, this was replaced by a module of $17\frac{1}{2}$ leagues to a degree (a 7 percent error).[20] The first module comes from Medieval Arab sources which can be shown to be claiming the authority of Claudius Ptolemy. Medieval Latin translators carelessly replaced Arab miles with Roman ones, the latter being considerably shorter.[21] The origin of the second module is unknown and first appears in the two booklets in Portuguese on astronomical navigation, published in c. 1509 and c. 1516, with which we shall deal in a moment.[22] The Portuguese soldier-sailor Duarte Pacheco Pereira forms the unique exception among navigators, followed by no one, in that he adopts the figure of 18 leagues to a degree (only a 4 percent error).[23]

The earliest attempt in Europe to measure experimentally the value of the degree of the meridian seems to have been that of the Spaniard Antonio de Nebrija (1441–1522), who, some time after 1510, is said to have measured the distance between Burgo de Osma and Alcalá de Henares, both lying on the same meridian or nearly and separated from each other by one degree less eight minutes. He found it equal to 20 Castilian leagues.[24] If the Castilian league in question was the same as the marine league, this was not far from the correct value, but nothing is known of the circumstances and methods of Nebrija's measurement and the result does not seem to have been taken account of by sailors at the time.

In 1500, the Pole Star lay about 3° 25′ distant from the geographic pole. Portuguese navigators in the XVth century, using a value of 3° 30′ (a good approximation), employed, in order to obtain the latter's true altitude, two methods for correcting the apparent altitude of the Pole Star at different times during the night as it revolved round the geographic pole. The figure of a man, with his navel centred on the geographic pole and with his arms outstretched and feet together, was imagined in the sky. From observation of the successive positions of the Guards (two stars of the *Ursae minoris* constellation), in relation to the figure's head, arms and feet, the necessary amount of correction was added to, or subtracted from, the apparent altitude of the Pole Star. This simple method dealt only with four different positions of the Pole Star. When the principal Guard (Kochab) was 'in the head', the Pole Star was 3° below the pole. When the principal Guard was 'in the feet', the Pole Star was 3° above the pole. When the principal Guard was horizontally in line with the Pole Star, 'in the east arm' or 'in the west arm', no correction was introduced, though the actual error in the last two cases, requiring in fact a correction of ±1·5 degrees, was not held to be significant in practice.[25]

Later, the figure in the sky was replaced by a 'wheel' or *Roda do Norte*, 'the wheel of the North Star', with eight positions on its circumference created by drawing diagonals through the centre in imitation of a compass rose. On each of the eight points were marked the apparent latitude of Lisbon, when the Guards were in any one of these positions at given hours in the night. The latitude of Lisbon was taken to be 39° (actually 38° 44′ N). The navigator, before leaving Lisbon, observed the height of the Pole Star, noting the position of the Guards; then later during his voyage, observed the height of the Pole Star again when the Guards were in the same position. From the difference in the apparent height of the Pole Star, he could calculate the distance he had sailed in a north–south direction since leaving Lisbon. On the other hand if he were to observe the Pole Star at a different time from that at which he had done so in Lisbon, he could calculate the necessary correction.[26] The corrections for the eight positions were thus as follows:

For the north-west/south-east diagonal	±30′
For the north/south diagonal	±3°
For the north-east/south-west diagonal	±3° 30′
For the east/west diagonal	±1° 30′.

There is no clear evidence to show when the method of the 'figure in the sky' or the later one of 'the wheel', both known as the *Regimento do Norte* or 'Rule for the North Star', were first used by Portuguese navigators, but it is conjectured by Luís de Albuquerque that they appeared between 1455 and 1475.[27] The corrections for the Rule of the North Star, though not of great accuracy, were sufficient for sailors and had the advantage of being simple. They appear in both the two booklets of rules for applying nautical astronomy with which we shall deal presently.

As the Portuguese, in their explorations down the west coast of Africa, approached the equator (they crossed it between 1472 and 1474),[28] the Pole Star sank lower and lower to the horizon and eventually ceased to serve them as a guide. It was then that a crucial breakthrough took place, in which observation of the meridian altitude of the Sun was substituted for observation of the height of the Pole Star, thus enabling navigation in either of the two hemispheres. The quadrant proving less suited for observing the Sun, there emerged, at least by the 1480s, the nautical astrolabe of the Portuguese, first described in 1517 by the Italian Alexander Zorzi. This was a simple instrument consisting of a graduated ring and an alidade fitted with two pinnules set relatively close together so that, as Zorzi notes, the observation of the Sun's rays through the pinnules could be more easily stabilised against the rolling of the ship.[29]

The calculation of latitudes by observing meridian altitudes of the Sun, a technique indispensable for astrological prognostications, was known in Europe from at least the XIIth century,[30] but it was never used either for map-making,[31] or for navigation before the XVth. The first texts specifically written for sailors, which expound the rules of procedure for calculating latitudes from the observation of the meridian altitudes of the Sun, appear in Portugal, printed in Lisbon at the beginning of the XVIth century. These texts were:

(i) *Regimento do astrolabio e do quadrante*, (c. 1509), known as the *Regimento de Munique* since the only known extant copy is in the Bavarian State Library in Munich.

(ii) *Regimento da declinaçam do sol*, (c. 1516), known as the *Regimento de Evora*, as the only extant copy is in the Evora Public Library, Portugal.[32]

The first of these booklets expounds, with numerous examples, a simplified method, using fairly rough approximations; the second attains a much higher degree of precision. While these two booklets were only printed in the first two decades of the XVIth century, there is evidence from the previous century that Portuguese sailors were already using the data and the rules given in them, no doubt in manuscript form. Drawn up specifically for use by sailors, the booklets contain tables of figures from which the declination of the sun can be known at any given date. These tables, indispensable for the calculation of latitudes, were taken from the work of the Spanish Jewish astronomer-astrologer Abraham Zacuto.[33] Born in Salamanca about 1452, Zacuto was the author of a substantial book in Hebrew of some 336 pages, the *Hajibbur hagadol*, of which only nine pages contain the tables necessary for the calculation of latitudes. There is nothing in it relating to the rules of procedure for calculating latitudes,[34] such as are found in the booklets. Such rules however were commonly known to astronomers in the Middle Ages. The Jewish physician José Vizinho, born probably in Vizeu (Portugal), at a date unknown, describing himself as a disciple of Zacuto, made a partial translation from the Hebrew of the *Hajibbur hagadol* and had it printed at Leiria (Portugal) in 1496 with the title of *Almanach Perpetuum*.[35] It was from the tables contained in the *Almanach Perpetuum* that all the nautical tables used by the Portuguese navigators up until the middle of the XVIth century were calculated.

The calculation of latitudes from observation of the meridian altitude of the Sun, as practised by astronomers/astrologers, required a series of relatively complicated operations difficult for sailors to master.[36] The observer must accomplish the following six operations:

(i) Observe the meridian altitude of the Sun.

(ii) Find in the astronomical tables for the day, month and year of the observation, the longitude of the Sun. The term usually employed was the 'place of the sun' in one of the twelve signs of the Zodiac, each sign being divided into 30°.

(iii) Find, in the table provided for this, the declination of the Sun for its longitude on the day, month and year of the observation.

(iv) Carry out a number of arithmetical operations in a formula which depends on the observer's position north or south of the equator and on the direction, north or south of his shadow.

These four operations were to be completed by two further ones in the case of the direct use of Zacuto's *Almanach*:

(v) Since the *Almanach* had been prepared for the period 1473–1476, for each subsequent 4-year period ending in a leap year, one had to add in from the *Almanach*'s *tabula equationis solis* a corrective constant of 1′ 46″ designed to take account of an advance of the sun of 42′ 56″ over the said 4-year period. This advance was caused by the difference between the length of the tropical year in the then used Julian Calendar and the real length of the tropical year in the period concerned.[37]

(vi) As Zacuto's tables of declination only gave the values of the Sun's longitude in whole degrees, one had to make interpolations, which were laborious.

These interpolations, together with the adding in of the constant from the *tabula equationis solis*, rendered the direct use by navigators of Zacuto's *Almanach* exceedingly difficult. In order to simplify these operations for use by sailors, operations (ii) and (iii), in the booklets referred to, were made into one by enabling the determination of declination for each day of the year from a single table. The need to make interpolations (vi) was also eliminated.

In the *Regimento de Munique* (c. 1509), by sacrificing precision, simplification was carried to the point of neglecting the variation in declination over the 4-year period between leap years. The declinations were given on a 'single solar table', calculated for a given year, which scholars have been unable to identify with certainty, but which, according to Luís de Albuquerque, could be some year between 1483 and 1493.[38] Errors due to this simplification remained insignificant in relation to those arising from the readings of the instruments then used.[39] The *Regimento de Evora* (c. 1516), of greater precision, did, on the other hand, take account of the variation in declination during the four-year period for which it was prepared (i.e. 1517–1520). Navigators were thus relieved of the tedious work of having to do themselves the work of bringing Zacuto's tables up-to-date by the adding in of multiples of the constant from the *tabula equationis solis*. The *Regimento de Evora* turns out to be, as we shall see, an example of the model on which were based all the nautical tables used by Portuguese navigators from 1497 up to the middle of the XVIth century.

In trying to follow in detail how the Portuguese monarch's scientific advisers proceeded, we hope to show how the transfer of a technique from the domain of astrologers to that of mariners was achieved. Among these advisers, two were of singular importance: first José Vizinho, then Abraham Zacuto who came to Portugal in 1492; both played key roles. In the marginal notes in certain of his books, Columbus wrote that in 1485, (long before Zacuto's arrival in Portugal), the Portuguese King D. João II. had sent Vizinho 'to observe the height of the sun in the whole of Guinea (i.e. west Africa)',[40] and according to these same notes, that the Portuguese navigator Bartolomeu Dias had, in 1488, observed the latitude of the Cape of Good Hope (by taking the height of the Sun) with an astrolabe.[41] Unfortunately the evidence of Columbus' marginal notes in his books is extremely fragile, since the figures he quotes are always wide of the mark. Either they prove the incompetence of Vizinho and Dias in making their observations, or gross errors by Columbus in copying the figures. Italian scholars prefer the first solution; Portuguese ones the second. No solution to this vexed question appears anywhere in sight.

In the first note Columbus remarks that Vizinho had placed the island of Los Idolos in Guinea in Latitude 5° N and several minutes.[42] It is in fact situated between 9° 25′ N and 9° 32′ N; Vizinho had therefore apparently committed an error of 4° 25′.[43] In his second note, Columbus states that Dias had placed the Cape of Good Hope in Latitude 45° S,[44] whereas in reality it is 34° 22′ S. Whatever may have been the origin of Columbus' errors, confirmation that a systematic survey of the latitudes of places on the west African coast had indeed been undertaken by the Portuguese (but it is not known exactly when), comes from Duarte Pacheco Pereira in his work *Esmeraldo de Situ Orbis*, only written however about 1508. In it he writes that he had 'over a long period of years' in west Africa, 'observed the altitude of the sun and its declination to know in how many degrees each place lay in latitude from the equator toward the poles and that he had found that the circle of the equator passed over the promontory of Lopo Gonçalves' (actually 0° 36′ S).[45] Duarte Pacheco Pereira's survey was probably done, Luís de Albuquerque surmises, between 1485 and 1495.[46]

It can be conjectured that Vizinho's and Pacheco Pereira's task consisted in attempting to correlate the cartography of the traditional Mediterranean type nautical charts with that based on celestially determined coordinates in Ptolemy's *Geography*, copies of which were only to reach Portugal in the second half of the XVth century.[47] The *Geography* was, in the latter part of the XVth century, to provide the Portuguese with a theoretical framework for their projected voyages to India round the Cape of Good Hope.[48] Thus it was important for Pacheco Pereira to situate the equator (a theoretical marker) in relation to a feature of the African coast such as Cape Lopo Gonçalves inscribed (from direct experience) on the traditional nautical charts. Little is known of José Vizinho beyond the fact that he was the translator of Zacuto's book and that he had voyaged to Guinea. In 1483–4, according to the Portuguese historian João de Barros, he was one of a group of cosmographers called upon by the Portuguese King D. João II to examine the feasibility of Columbus' project of reaching Asia by sailing westward.[49] Barros also states that he, as one of the King's physicians, was together with other experts the inventor of 'the manner of navigating with the altitude of the sun, for which they had prepared tables of its declination, such as navigators use at present, now more accurate than at the start'.[50] In the expression 'now more accurate than at the start' (*já mais apuramente do que começou*), can perhaps be discerned a possible distinction between the rough method of approximation of the *Regimento de Munique* (c. 1509), and the more sophisticated one found in the *Regimento de Evora* (c. 1516). Zacuto, it is claimed, was the author of the model followed in the latter, while Vizinho (aided probably by Rodrigo) was the author of the former.[51] The tables in the *Regimento de Munique* were, if we can trust the evidence of Columbus, used for Bartolomeu Dias' voyage (1488), but there is no evidence to show that they had been used during the voyages of Diogo Cão (1484–86),[52] who had reached as far as 22° 10′ S on the coast of present day Namibia.[53]

From 1493, the second of the two important figures, Abraham Zacuto plays the major role. He is the author of the first four-year tables drawn up for use by navigators.[54] Expelled from Spain in 1492 with other Jews, Zacuto was brought to Lisbon by King D. João II, where he resided for four years during the reigns of King D. João II and of his successor King Manuel. In relation to the latter he was termed 'King's Mathematician',[55] and 'King's Astronomer'.[56] Forced to leave Portugal in 1497, he fled to Tunis and later to Damascus where he died in 1522.[57] The little information available on Zacuto's role in the development of nautical science in Portugal appears in Gaspar Correa's chronicle *Lendas da India*, a work whose historical reliability is the subject of contradictory evaluations by Portuguese scholars.[58] Correa was a man of mediocre education and understood little of nautical astronomy. He does, however, explain quite correctly how Zacuto had prepared a set of rules (*Regimento*) consisting of four-year tables of the Sun's declination, but he mistakenly makes Zacuto the inventor of the 'astrolabe' (*sic*) and of the rules for the Pole Star.[59] Zacuto's *Regimento*, referred to by Correa, contained the first 4-year tables drawn up for a period *later* than those for the period 1473–6 given in his major work printed in 1496. They were drawn up for the period 1497–1500 and were prepared by him for Vasco da Gama's voyage (1497) and were also later used for the voyage of Pedro Alvares Cabral (1500).[60] The Spaniard Martín Fernándes de Enciso obtained these tables secretly and published them in his work *Suma de Geographia*, Seville, 1519,[61] and it is from his publication that we know of them. The tables were later translated into English from Enciso's book by Roger Barlow in his manuscript work of 1540–1, *A Brief Summe of Geography*.[62]

If the oldest 4-year tables for navigators stand to be those prepared by Zacuto for Vasco da Gama's voyage, the oldest *printed* 4-year tables are those contained in the *Regimento de Evora* (c. 1516), calculated for the later period of 1517–20. It was thus on Zacuto's model that Portuguese as well as European navigators were to base all their tables until they were eventually replaced by new ones brought out in 1537 by the Portuguese mathematician Pedro Nunes for the period 1537–40.[63]

The precision of Portuguese nautical astronomy in practice was noted in 1505 by the Italian traveller, Lunardo da Chá Masser, who voyaged to the Orient in a Portuguese ship; he wrote:

> The Portuguese navigate by the altitude of the sun or by that of the Pole Star with the aid of an astrolabe. And they are utterly confident while out of sight of land for three months and they know exactly where they are and at the end of a given number of days, I found myself at the intended place of arrival, which is indeed a remarkable thing.[64]

In the southern hemisphere the Portuguese navigator Pero Anes had, by 1508, worked out a rule for navigating by observing the altitude of certain stars of the Southern Cross, as had been done in the northern hemisphere for the Pole Star.[65]

Portuguese nautical science was to spread to the rest of Europe mainly through works written by Spaniards. After Enciso, copied by Barlow, Pedro de Medina's *Arte de Navegar* (1545) went into seven editions in French and Martín Cortés' *Breve Compendio de la Sphera y de la Arte de Navegar* (1551) also went into seven editions in English. Cortés expounds a graphic method of calculating the declination of the Sun, a method first proposed in nautical circles by the Portuguese navigator Francisco Rodrigues (c. 1513).[66]

After a major breakthrough in the XVth century in the application to navigation by Vizinho and Zacuto of astronomical techniques used by astrologers, Iberian progress in nautical astronomy remained limited to improvements in precision in the XVIth and XVIIth centuries. No new major innovations occurred that were usable by navigators. Luís de Albuquerque has written that nautical astronomy only appeared in Portugal when the need for it was felt.[67] The conditions for its emergence were there from at least the XIVth century and existed throughout the XVth century; as Guy Beaujouan has noted, a common fund of Judeo-Luso-Castilian astronomical knowledge ran along what might be described as an axis linking Salamanca to Lisbon.[68] Mariners certainly did not take the initiative in drawing on this fund of knowledge and astronomers/astrologers are hardly likely to have sought to persuade sailors of its advantage to them. Who then was instrumental in getting the two parties to work together? On this crucial point evidence is at present lacking. Our conjecture is that the role of King D. João II is likely to have been central. Only royal command could have made an astrologer/physician like José Vizinho submit to the discomfort of a voyage to Guinea in a caravel in 1485 in order to observe the Sun with a astrolabe.

ACKNOWLEDGEMENT

The reader will observe that this synthesis draws heavily on the very deep and thorough work of the late Professor Luís de Albuquerque whom I was privileged to know and work with over a number of years. What I have written would not have been possible had I not benefited constantly from his immense kindness and generosity in forwarding to me not only copies of his very considerable writings, but also numerous photocopies of rare out-of-print works difficult to find outside Portugal.

REFERENCES AND NOTES

1 Isidore of Seville (1969). *De natura rerum*, edited with French translation by Jacques Fontaine, Bordeaux, p. 307.

2 Pliny, *Natural History*, Bks V and VI.

3 Charles Verlinden (1958). Lanzarotto Malocello et la découverte portugaise des Canaries, *Revue belge de philologie et d'histoire*, Tome XXXVI, pp. 1173–1209.

4 Damião Peres (1943). *História dos Descobrimentos Portugueses*, Oporto, 3rd ed. Oporto (1983), pp. 58, 68.

5 José Mª Millas Vallicrosa (1987). La cultura cosmográfica en la corona de Aragón durante el reinado de los Reyes Católicos, Idem. *Nuevos Estudios sobre la Historia de la Ciencia Española*, Madrid, Vol. II, p. 301.

6 A. Humboldt (1836). *Examen critique de la géographie du Nouveau Continent et des progrès de l'astronomie nautique aux XVe et XVIe siècles*, Paris, Tome I, pp. 274–5.

7 Guy Beaujouan (1966). Science livresque et art nautique, Michel Mollat & Paul Adam, (eds) *Les aspects internationaux de la Découverte Océanique*, Paris, pp. 64–6 (for Aragon). J. Bensaude (1912/17), *L'astronomie nautique au Portugal*, Berne, facsimile reprint N. Israel, Amsterdam (1967), pp. 5–61. H. Wagner (1918), Die Entwicklung der Wissenschaftlichen Nautik im Beginn des Zeitalters der Entdeckungen nach neuern Anschauungen, in *Annalen der Hydrographie und Maritimen Meteorologie*, Vol. XXXVI, p. 282.

[8] On the Mediterranean Portulan chart cf. W. G. L. Randles (1988). From the Mediterranean Portulan Chart to the Marine World Chart of the Great Discoveries: the crisis of cartography in the sixteenth century, *Imago Mundi*, Vol. 40, pp. 115–118.

[9] Alvise da Cadamosto (1948). *Viagens*, ed. Academia Portuguesa de História, Lisbon, p. 58 (Italian text with Portuguese translation).

[10] Cf. Jane L. Jervis (1985). *Cometary Theory in Fifteenth Century Europe*, Dordrecht-Boston-Lancaster, p. 38.

[11] Peter of Abano (1565), *Conciliator*, Venice, Diff. 67, f. 101 (v.) (written about 1303). Cf. also Guy Beaujouan, (1975); *Rapports Entre Théorie et Pratique au Moyen Age*, J. E. Murdoch & E. D. Sylla, (eds), *The Cultural Context of Medieval Learning*, Dordrecht (Holland), p. 466, note 77.

[12] Pedro de Sintra, in Alvise da Cadamosto, *op. cit. ed. cit.* p. 49.

[13] Placido Zurla (1806). *Il Mappamondo di Fra Mauro*, Venice, p. 49 (*el polo antartico levado uno brazo*).

[14] Cf. H. C. F. C. Schjellerup (1874) [trans]. *Description des étoiles fixes composée au milieu du dixième siècle de notre ère par l'astronome persan Abd-al-Rahman al-Sufi*, St Petersburg, p. 27.

[15] Diogo Gomes, (1847). *De Prima Inventione Guineae*, in J. M. Schmeller, Über Valentim Fernandes Alemã und seine Sammlung von Nachrichten über die Entdeckungen und Bezitzungen der Portugiesen in Afrika und Asien bis zum Jahre 1508, *Abhandlungen der Philosoph. – Philolog. Classe der Königlich Bayerischen Akademie der Wissenschaften*, Bd.IV, Dritte Abth. Munich, p. 53.

[16] Valentim Fernandes (first ed. 1518). *Reportorio dos Tempos*, Lisbon, facsimile reprint of the 1563 edition by J. Bensaude, Geneva, (1919), p. 141.

[17] Luís de Albuquerque (1965). *Os Guias náuticos de Munique e de Evora*, Lisbon, p. 30.

[18] Luciano Pereira da Silva (1945). A Arte de navegar dos Portugueses desde o Infante a D. João de Castro, Idem, *Obras completas*, Lisbon, Vol. II, p. 294. António Barbosa, (1948); *Novos subsídios para a história da ciência náutica portuguesa da época dos Descobrimentos*, Oporto, p. 152. Luís de Albuquerque, (1975); A navegação astronómica, *Estudos de História*, Coimbra, Vol. III, p. 39. Guy Beaujouan, Reflexions sur les rapports entre théorie et pratique au Moyen Age, *op. cit.*, pp. 465–467.

[19] The chart of Pedro Reinel (c. 1504), reproduced in A. Cortesão & A. Teixeira da Mota, (1960). *Portugaliae Monumenta Cartographica*, Lisbon, Vol. I, Pl. III.

[20] A. Fontoura da Costa, (1939). *A Marinharia dos Descobrimentos*, Lisbon, pp. 210–216. A. Teixeira da Mota, (1960); Bartolomeu Dias e o valor do grau terrestre, *Congresso Internacional de História dos Descobrimentos*, Lisbon, *Actas*, Lisbon, (1961), Vol. II, pp. 199–209. Fontoura da Costa took the degree to be equivalent to 111 kms and the Portuguese league as equal to 5920 metres. (Cf. A. Fontoura da Costa, *op. cit.* p. 216.)

[21] C. A. Nallino (1944). Il valor metrico del grado di meridiano secondo i geografi arabi, *Raccolta di scritti editi ed inediti*, Vol. 4, Rome, pp. 412–416.

[22] A. Fontoura da Costa (1938). La lieue marine des portugais aux XVe et XVIe siècle, *Comptes rendus du Congrès International de Géographie*, Amsterdam, Vol. II, p. 7.

[23] Duarte Pacheco Pereira, *Esmeraldo de Situ Orbis* (c. 1508), ed. A. Epifánio da Silva Dias (1905); Lisbon, reprint Lisbon (1975), Liv. I, Cap. 2, p. 23. A. Fontoura da Costa (1939). *A Marinharia dos Descobrimentos*, Lisbon, p. 216.

[24] Armando Cotarelo Valledor (1947). *Nebrija Cientifico*, Madrid, pp. 33–34. Mario Méndez Bejarano (1923). *Diccionario de Escritores, Maestros y Oradores, naturales de Sevilla*, Seville, Tome II, p. 35.

[25] Manuel dos Reis (1961). O Regimento do Norte na astronomia náutica portuguesa da época dos Descobrimentos, *Boletim da Academia das Ciências de Lisboa*, Nova Série, Vol. XXXIII, pp. 108–127.

[26] Luis de Albuquerque (1965). *Os Guias náuticos de Munique e de Evora*, Lisbon, pp. 22–26.

[27] Idem. p. 27.

[28] A. Fontoura da Costa (1938). Descobrimentos marítimos Africanos dos Portugueses com D. Henrique, D. Afonso V e D. João II, *I Congresso da História da Expansão Portuguesa no Mundo*, 1ª Secção, Lisbon, p. 26.

[29] Alesandro Zorzi. Informatiõ hauuto Io Alexandro da Portogalesi .1517. in Veneza, Ms Magl. XIII, 80, fls. 131–140 (v.). Biblioteca Nacional, Florence. Printed in Francisco Leite de Faria & A. Teixeira da Mota, *Novidades Náuticas e Ultramarinas numa Informação dada em Veneza em 1517*, Centro de Estudos da Cartografia Antiga, Vol. XCIX, Lisbon, (1977), p. 20; also in *Memórias da Academia das Ciências de Lisboa*, Tome XX (1977), p. 20.

[30] Cf. Emmanuel Poulle (1971). Les conditions de la navigation astronomique au XVe siècle, *Revista da Universidade de Coimbra*, Vol. 24, p. 35.

[31] Cf. J. K. Wright (1932). Notes on the knowledge of latitudes and longitudes in the Middle Ages, *Isis*, Vol. V, pp. 97–98.

[32] Both works have been reprinted in facsimilé editions by J. Bensaude in the series *Histoire de la Science Nautique Portugaise*, Vols I & II, Munich, (1914); by Luís de Albuquerque, *Os Guias náuticos de Munique e de Evora*, Lisbon, (1965), (transcription of the Munich text with variants and the new chapters in the Evora text); in a facsimilé reprint of the Munich and Evora texts with introduction by Luís de Albuquerque taken from his 1965 work, but lacking Anexos B, C and D and his bibliography = *Guia náutico de Munique e Guia náutico de Evora*, Comissão Nacional para as Comemorações dos Descobrimentos Portugueses & Instituto Nacional de Pilotagem dos Portos, Lisbon, (1991).

[33] J. Bensaude, (1912/17). *L'astronomie nautique au Portugal à l'époque des Grandes Découvertes*, Berne, facsimilé reprint N. Israel, Amsterdam, (1967), p. 26; Luís de Albuquerque, *op. cit.* pp. 76–82. For Emmanuel Poulle, the fact that the longitudes of the sun in the *Regimento de Munique* are given in whole degrees makes it impossible to identify the source. Poulle however sees the source of the *Regimento de Evora* as being the Alphonsine tables, 'without it being necessary to bring in the work of Zacuto' (Emmanuel Poulle, *art. cit.* pp. 43–44). To this point Luís de Albuquerque makes the rejoinder that Valentim Fernandes, who published the same tables as those in the *Regimento de Evora* in his *Reportório dos Tempos* (1518), declares that they had been prepared by Gaspar Nicolas from Zacuto's *Almanach Perpetuum*; Luís de Albuquerque (1977), Abraham Zacuto et l'*Almanach Perpetuum*, Idem. Estudos de História, Coimbra, Vol. V, pp. 61, 66.) Cf. also Valentim Fernandes, *Reportório dos Tempos*, facsimilé ed. by J. Bensaude (1919). *Histoire de la Science Nautique*, Vol. VII, Geneva, p. 52.

[34] J. Bensaude, *L'astronomie nautique au Portugal...*, p. 26.

[35] Francisco Cantera Burgos (1935). *Abraham Zacuto*, Madrid, p. 57.

[36] Cf. on this point A. Fontoura da Costa (1939). *A Marinharia dos Descobrimentos*, Lisbon, p. 91.

[37] Cf. António Barbosa (1928). O *Almanach Perpetuum* de Abraham Zacuto e as Tábuas Náuticas Portuguesas, *O Instituto* vol. 75, pp. 541–562.

[38] Luís de Albuquerque (1977). Abraham Zacuto et l'*Almanach Perpetuum*, Idem. *Estudos de História*, Vol. V, Coimbra, p. 80.

[39] Idem, *op. cit.* p. 76 and J. Bensaude, *L'astronomie nautique au Portugal*, p. 118.

[40] '*...rex Portugaliae misit in Guinea anno Domini 1485 magister Ihosepius, fixicus eius astrologus,* [*ad com*]*piendum altitudinem solis in totta Guinea...*' Note by Columbus in his copy of Pope Pius II's *Historia rerum ubique gestarum*, Venice (1477), reproduced by George E. Nunn, (1924). *The Geographical Conceptions of Columbus*, New York, reprint New York, (1977), pp. 6–7.

[41] '*...hoc anno de .88...Bartolomeus Didacus...navigaverit...usque...cabo de Boa Esperança... quique in eo loco invenit se distare per astrolabium ultra linea equinoctiali gradus 45...*' Note in Columbus' copy of Pius II, *op. cit.* reproduced by George E. Nunn, in *op. cit.* pp. 7–8.

[42] '*...alliis in die xi marcii invenit se distare ad equinoctiali gradus V minute in insula vocatur de los Idolos...*' Cf. Nunn, *op. cit.* pp. 6–7.

[43] A. Magnaghi (1928). I presunti errori de C. Colombo e la determinazione delle latitudini, *Bollettino de la Reale Società Geografica Italiana*, Vol. IV, pp. 462–463. Magnaghi tries rather unsuccessfully to defend Columbus.

[44] See note 40.

[45] Duarte Pacheco Pereira, *Esmeraldo de Situ Orbis* (c. 1508), ed. A. Epifánio da Silva Dias, Lisbon, (1905), reprint Lisbon, (1975), Liv. II, Cap. 11, p. 127.

[46] Luís de Albuquerque (1965). *Os guias náuticos de Munique e de Evora*, Lisbon, pp. 63–64.

[47] W. G. L. Randles (1986). O redescobrimento da *Geografia* de Ptolemeu na Itália do

Renascimento e o seu impacte em Espanha e Portugal durante os Descobrimentos, *A Abertura do Mundo, Estudos de História dos Descobrimentos Europeus*, eds. Francisco Contente Domingues & Luís Filipe Barreto, Vol. I Lisbon, pp. 145–49.

[48] Idem. (1989). La configuration cartographique du continent africain avant et après le voyage de Bartolomeu Dias : hypothèses et enseignements, *Congresso Internacional Bartolomeu Dias e a sua época*, Oporto, Vol. II, pp. 111–119.

[49] João de Barros (1932). *Asia*, Déc. I, Liv. III, Cap. 11, ed. Coimbra, p. 113.

[50] Idem. Déc. I, Liv. IV, Cap. 2, *ed. cit.* p. 127. '*...esta maneira de navegar per altura do sol, de que fizeram suas tavoadas para declinaçam delle: como se ora usa entre os navegantes, já mais apuradamente do que começou...*' Among the other experts were a certain Master Rodrigo of whom nothing is known, and Martin Behaim, of whom Ravenstein, at the beginning of this century, showed that his contribution to Portuguese nautical science was negligible. Cf. E. G. Ravenstein (1908). *Martin Behaim and his Globe*, London.

[51] Luís de Albuquerque (1987). Introduction, facsimile reprint of Abraham Zacuto, *Almanach Perpetuum*, (Leiria, 1496), Imprensa Nacional – Casa da Moeda, Lisbon, p. 47.

[52] A. Fontoura da Costa claims, citing no evidence, that they were so used. Cf. A. Fontoura da Costa, (1939). *A. Marinharia dos Descobrimentos*, Lisbon, p. 95.

[53] W. G. L. Randles (1988). *Bartolomeu Dias and the Discovery of the South-east Passage Linking the Atlantic to the Indian Ocean (1488)*, Centro de Estudos de História e Cartografia Antiga, Vol. 188, Lisbon, p. 22.

[54] Luís de Albuquerque (1977). Abraham Zacuto et l'*Almanach Perpetuum*, Idem. *Estudos de História*, Vol. V, Coimbra, p. 80.

[55] Sousa Viterbo (1898). *Trabalhos náuticos dos Portugueses. Séculos XVI e XVII*, Lisbon, facsimile reprint Imprensa Nacional – Casa da Moeda, Lisbon, (1988), Pt I, p. 326.

[56] Abraham Zacuto (1496). *Almanach Perpetuum*, Leiria, facsimile reprint Lisobn, (1987), *explicit*.

[57] Francisco Cantera Burgos, *op. cit.* pp. 37–39.

[58] A. Fontoura da Costa calls him a writer of 'rare probity' (A. Fontoura da Costa, *A Marinharia dos Descobrimentos*, *ed. cit.* p. 96); Duarte Leite (*História dos Descobrimentos*, Lisbon (1959), Vol. I, p. 143) and Luís de Albuquerque (Abraham Zacuto et l'*Almanach Perpetuum*, Idem. *Estudos de História*, Vol. V, Coimbra, (1977), p. 62) are both perhaps too severe in their criticism.

[59] Gaspar Correa (1898). *Lendas da India*, Tome I, Lisbon, pp. 163–169. The work was written in Goa between 1529 and 1561, but was not printed until the 19th century.

[60] Luciano Pereira da Silva (1945). A Arte de Navegar dos Portugueses desde o Infante a D. João de Castro, Idem. *Obras Completas*, Lisbon, Vol. II, pp. 315–317.

[61] Martin Fernándes de Enciso (1519). *Suma de Geographia*, Seville, sign. a.viii (v.)–b.vii (r.).

[62] Roger Barlow (1932). *A Brief Summe of Geography* ed. with notes by E. G. R. Taylor, Hakluyt Society, 2nd series, Vol. LXIX, London.

[63] Pedro Nunes (1537). *Tratado da Sphera*, Lisbon, Idem. *Obras*, Vol. I, Lisbon, (1940), pp. 233–37.

[64] Lunardo da Chá Masser (c. 1505). Relazione, *Memórias da Academia Real das Sciencias de Lisboa*, Nova Série, Tome VII, Parte I, Lisbon, (1895), pp. 78–79. Observation of the Pole Star with an astrolabe is extremely difficult and much easier with a quadrant or a *balestilha* (Jacob's staff). The first Portuguese reference to the latter occurs in 1514. Cf. Luís de Albuquerque, *Os guias náuticos de Munique e de Evora*, pp. 106–107.

[65] João de Lisboa (c. 1514). *Livro de Marinharia*, ed. Brito Rebelo, Lisbon (1903), p. 22, 37–41; Luís de Albuquerque (1975). A Navegacão astronómica, Idem. *Estudos de História*, Vol. III, Coimbra, pp. 234–255.

[66] Martín Cortes (1551). *Breve Compendio de la Sphera y de la Arte de Navegar*, Seville, Segunda Parte, Cap. VII; Cf. Luís de Albuquerque, *Ibid.* pp. 218–226.

[67] Luís de Albuquerque (1991). *Dúvidas e Certezas na História dos Descobrimentos*, 2ª parte, Lisbon, p. 52.

[68] Guy Beaujouan (1969). *L'astronomie dans la péninsule ibérique à la fin du Moyen Age*, Agrupamento de Estudos de Cartografia Antiga, Vol. XXIV, Coimbra, p. 17.

V

The Recovery of Ptolemy's *Geography* in Renaissance Italy and its impact in Spain and Portugal in the period of the Discoveries*

In 1400 a copy of the manuscript of Ptolemy's *Geography* was obtained from Constantinople by the Florentine patron of letters Palla Strozzi who persuaded the Greek scholar Manuel Chrysoloras to translate it into Latin; the work being finally completed by Jacopo Angelus in about 1406.[1] Copies are known to have circulated in Italy and France in the first quarter of the 15th century.[2] In Southern Germany, the principles of the *Geography* were known and applied to map-making from 1425 onwards.[3] Much controversy has centered on the question of the authenticity of the maps (as opposed to the text) and their date of origin. Let it be sufficient here to note that it is generally admitted that the world map, on a pseudo-conical projection, is the work of Agathodaimon of Alexandria, who may have been a contemporary of Ptolemy.[4]

Ptolemy's *Geography* was only progressively understood and its message absorbed in the 15th century. First the maps, then the relation between land and water and lastly the mathematical principles were successively digested, first by university scholars, later (and only partially) by sailors. The third projection was never understood by sailors, and the first and second, though recognized by them as useful, were never applied in navigation in the 16th century.[5]

---

* First published in Portuguese in Francisco Contente Domingues & Luís Filipe Barreto, eds, *Estudos de História dos Descobrimentos Europeus*, (Lisbon, 1986), Vol. I, pp. 145–151.

[1] See Vespasiano da Bisticci (d.1498), *Vite di uomini illustri del secolo XV*, ed. Ludovico Frati, 3rd edn (Bologna, 1893), Vol. 3, pp. 9–10; Joseph Fischer S.J., 'De Cl. Ptolemaei Vita Operibus Geographia praesertim eivsque fatis, Pars Prior Commentatio', in Idem. ed., *Clavdii Ptolemaei Geographiae Codex Urbinas Graecvs 82, Tomus Prodromus*, (Lyons and Leipzig, 1932), pp. 171–191 and G.R. Crone, *Maps and their Makers*, (London, 1978), pp. 35.

[2] Joseph Fischer, *op. cit.* p. 191 and G.R. Crone, *op. cit.* p. 36.

[3] See Dana B. Durand, *The Vienna Klosterneuburg Map Corpus of the fifteenth century*, (Leiden, 1952), *passim*.

[4] See G.R. Crone, *op. cit.* p. 4.

[5] On Ptolemy's projections see S.Y. Edgerton, *The Renaissance Rediscovery of Linear Perspective*, (New York and London, 1975), chapters VII and VIII.

In Spain, 15th century manuscript copies of Ptolemy's *Geography* are held in the University libraries at Valencia and Salamanca.[6] For Valencia there are clear indications that Ptolemy's *Geography* was read, studied and understood in the last two decades of the 15th century by Jacob Perez de Valencia (died 1490 or 1491), who refers to Ptolemy in his Commentary on the Psalms, published at Valencia in 1484,[7] At Salamanca Antonio de Nebrija (1442–1522) published his little Cosmography there between 1497 and 1504 in which he shows a clear knowledge of Ptolemy's mapping principles.[8] Of these two authors the first was a prelate, the second a university scholar and neither was connected with maritime circles.

In Portugal, clear evidence that Ptolemy's *Geography* was known in the 15th century, appears in the *Oration of Obedience* of King D. João II, pronounced by Vasco Fernandes de Lucena before the Pope in 1485.[9]

These proofs of knowledge of Ptolemy's *Geography* in the Iberian Peninsula in the last two decades of the 15th century are late; the Discoveries were then well under way. It is during the earlier period and especially during the lifetime of the Infante D. Henrique that the question is crucial.

Duarte Leite, with his customary voltairian verve, roundly denied that Ptolemy was known in Portugal in the Infante's lifetime,[10] refusing even the testimony of João de Barros, writing admittedly much later,[11] that the Infante had learned from Ptolemy that the west African coast stretched southward beyond the equator, '*a qual terra* [África] *segundo mostrauã as távoas de Ptholemeu, & assi pela informaçãm que tinha dos alarues, sabia ser continua huma a outra, té se meter debaixo*

[6] The University Library of Valencia holds a manuscript copy, dated to not later than 1458, of Jacopo Angelus' Latin version of Ptolemy's *Geography*. See: *Cosmografía de Claudio Ptolomeo*, ed. with Spanish trans. by Victor Navarro Brotóns, Enrique Rodrigues Galdeano & Angel Aguirre, (Valencia, 1983), 2 vols. Two 15th-century manuscript copies of Ptolemy's *Geography* are in the Library of the University of Salamanca; see Guy Beaujouan, *Manuscrits scientifiques médiévaux de l'Université de Salamanque et de ses 'Colegios Mayores'*, (Bordeaux, 1962), p. 159 and pp. 161–162.

[7] Jacob Perez de Valencia, [*Commentaria in Psalmos*], (Valencia, 1484), sign.ii.ii^{v}–sign.ii.vi^{v}.

[8] Antonio de Nebrija, *In Cosmographiae libros introductorium*, (Salamanca, s.d. [1497/1504]). Mod. edn of Latin text of the (Paris, 1533) edition with Spanish trans. in *La Ciencia de la Tierra, Cosmografia y Cosmógrafos Salmantinos del Renascimiento*, eds Cirilo Flórez Miguel, Pablo García Castillo & Roberto Albares Albares; translator, José Sanz Ramos; published by the Caja de Ahorros y Monte de Piedad de Salamanca, Colección Salamanca en el Descubrimiento de America, (Salamanca, 1990), pp. 235–282.

[9] *Valasci ferdinãdi utriusque iuris cõsulti Illustrissimi regis Portugallie oratoris ad Innocentium viii. pontificem maximum de obedientia Oratio*, [Rome, 1492?], facsimile edn of Latin text with transcription and Portuguese trans. by A. Fontoura da Costa, *Às Portas da Índia em 1484*, (Lisbon, 1936), p. 80.

[10] Duarte Leite, *Ácerca da 'Crónica dos feitos de Guiné'*, (Lisbon, 1941), pp. 102–103 and idem. 'A sabedoria do Infante' in *História dos Descobrimentos*, Vol. I, (Lisbon, 1959), p. 122.

[11] The first *Década* of Barros' *Ásia*, had been completed by 1539; see I.S. Révah, 'João de Barros', in *Revista do Livro*, Rio de Janeiro, No 9, (1958), p. 68.

da linha equinocial ...' (he knew from what Ptolemy's maps showed and from information from local tribes that this land [i.e. the African coast] was continuous stretching below the equator).[12]

In 1960 Virgínia Rau was able to throw new light on the question. The accounts of the Florentine merchant, resident in Lisbon, Bartolomeo di Iacopo di Ser Vanni, show that in the years 1460–61 (the Infante died in 1460), the Bishop of the Algarve, D. Alvaro Afonso, had bought from him a copy of the *tavole di Tolomeo, prepared by Piero del Massaio dipintore* (the tables [i.e. maps] of Ptolemy, prepared by the painter Piero del Massaio).[13] The Piero del Massaio in question appears to be the same as the Pietro del Massajo, or Pietro Massari, known to have been engaged in producing copies of Ptolemy's *Geography* in Florence c. 1458–72.[14] This still does not settle whether the Infante knew of Ptolemy, though in our opinion, Barros' testimony should be allowed to carry more weight than that accorded it by Duarte Leite.

The lesson drawn from Ptolemy by the Infante that the African coast stretched south of the equator may have been useful in countering the traditional belief in the torrid zone, where, so it was held, man could not live. But a reading of Ptolemy could also lead to the discouraging recognition of the fact that African coast stretched far to the southward as a barrier hindering circumnavigation of it. Worse, Ptolemy showed the Indian Ocean as a closed sea, inaccessible from the Atlantic. The recovery of Ptolemy's *Geography* in the West, cannot be said, as far as the world map in its pure form is concerned, to have been an incentive to the Portuguese in their attempt to reach India by circumnavigating the dark continent.

There is however evidence that the idea of the India Ocean as a closed lake was abandoned in Italy at least from the middle of the 15th century. Marco Polo, on his return voyage from China to India by sea had shown that this was not so. Fra Mauro, on his world map of 1459 states that the Indian Ocean is not a *stagnon* (a closed sea) and a copy of his work had been sent to Portugal in the Infante's lifetime.[15]

Without more evidence, it is impossible to speculate any further on the probable influence of Ptolemy on the Infante's projects.

---

12 João de Barros, *Ásia*, Déc. I, Liv. I, Cap. II, (Lisbon, 1552), edn (Lisbon, 1932), reprint (Lisbon, 1988) p. 14. It is true that the reference to 'information from local tribes' renders Barros' testimony somewhat shaky, since in 1460 the date of the Infante's death, the Portuguese were still a long way from the equator and no Africans they could have met could have told them of the regions to the south of it.

13 Virgínia Rau, 'Bartolomeo di Iacopo di Ser Vanni, mercador-banqueiro florentino "estante" em Lisboa nos meados do século XV', in *Do Tempo e da História*, Vol. IV, (1971), p. 113.

14 See H. O[mont], introduction to *Géographie de Ptolémée*, traduction latine de Jacopo d'Angiolo de Florence, reproduction réduite des cartes et plans du manuscrit latin 4802 de la Bibliothèque Nationale [de Paris], (Paris, s.d.), pp. 1–3 and G.R. Crone, *op. cit.* pp. 36–37.

15 Placido Zurla, *Il Mappamondo di Fra Mauro*, (Venice, 1806), p. 52 and pp. 84–85.

Though the Ptolemaic principle of an enclosed Indian Ocean was obviously abandoned in the second half of the 15th century, as far as a sea connection between the Atlantic and the Indian Ocean was concerned, the idea (Ptolemaic) persisted in maps and texts of the 16th century that the coastline of the American continent, of which stretches had been here and there discovered, was continuous and uninterrrupted from Labrador to the southern limits of present-day Argentina, the coast then curving round to join the mythical austral continent, which in turn linked up with the south-east Asian peninsula. This idea, well prior to Magellan's voyage, starts to appear on Juan de la Cosa's chart of c. 1508,[16] and more distinctly on Lopo Homem's Mappamundi of 1519.[17] It is expressed in words by Duarte Pacheco Pereira in his *Esmeraldo de Situ Orbis* of c. 1508:

> 'Since this land beyond the Ocean is so large and on this side we have Europe, Africa and Asia, it is clear that the Ocean is placed between these two lands with land on either side of it, and we can therefore affirm that the Ocean does not surround the earth as the philosophers declared, but rather that the earth must surround the seas, since it lies in the concavity of the earth and in its centre. Therefore I conclude that the Ocean is only a very large lake set in the concavity of the earth ...'[18]

Pacheco's 'Ptolemaic idea' which he contrasts with that of the philosophers (Aristotelian philosophers, as we shall see later) was expounded before him, by the German Hieronymus Münzer (Jerónimo Monetário in Portuguese) in a letter to the Portuguese King D. João II of 14th July 1493, a Portuguese translation of which was printed in the *Regimento de Munique*, a book of nautical tables (c. 1509). Münzer enjoins D. João II,

> 'not to be led astray by Alfraganus and others without experience, who said that only a fourth part of the earth was uncovered by water and that the other three parts lay under the sea'.[19]

The experience of the Portuguese navigations, declared Münzer, had left no doubt that, 'the earth dies not lie under the sea, but on the contrary, the sea lies in the

[16] Reproduced by Kenneth Nebenzahl, *Atlas of Columbus and the Great Discoveries*, (Chicago, New York, San Franscisco, 1990), Plate 10.

[17] In Bibliothèque Nationale, Paris. Reproduced in Armando Cortesão & A. Teixeira da Mota, *Portugaliae Monumenta Cartographica*, (Coimbra, 1960), Vol. I.

[18] Duarte Pacheco Pereira, *Esmeraldo de Situ Orbis*, Liv. I, Cap. II, edn (Lisbon, 1905), reprint (Lisbon, 1975), pp. 21–22. English translation by G.H.T. Kimble, in Duarte Pacheco Pereira, *Esmeraldo de Situ Orbis*, Hakluyt Society, 2nd series, Vol. LXXIX, (1936), Kraus reprint, (Nendeln, 1967), pp. 10–13.

[19] Hieronymus Münzer, Letter, reproduced in Luís de Albuquerque, *Os Guias Náuticos de Munique e Évora*, (Lisbon, 1965), pp. 185–187. The text is also printed in Henry Vignaud, *Histoire critique de la Grande Entreprise de Christophe Colomb*, Vol. II, (Paris, 1911), pp. 620–622. Vignaud gives a partial version in Latin from the original manuscript, together with the Portuguese translation.

earth and is encircled by it.[20] Münzer, who had visited Portugal, is here taking sides with Ptolemy against scholastic Aristotelianism (the reference to Alfraganus (al-Farghani, d. after 861) is unfounded since the latter rather perpetuated Ptolemy's ideas than scholastic ones.[21] Thus Ptolemy, supported by the maritime experience of the Portuguese, is put forward by Münzer, to refute the long standing scholastic belief (developed from Aristotle's thought) that the earth and the water were separate spheres of different sizes and with centres that did not coincide.[22]

It is evident that until the Portuguese had explored south of the equator and had shown that the southern hemisphere was not entirely covered by water, Ptolemy and Aristotle were, as they had remained all through the Middle Ages, but rival authorities and allegiance to one or the other could only be a matter of 'opinion' or of institutional or disciplinary affiliation. The navigations of the Portuguese had introduced a much more cogent criterion to enable a theory choice to be made: – experience – visual experience – *a experiencia he madre das cousas* (experience is the mother of knowledge) as Duarte Pacheco Pereira was to express it.[23]

Perhaps the part of Ptolemy's *Geography* which sailors had the greatest difficulty in grasping and applying in their navigation techniques was his representation of the curvature of the earth and of the convergence of the meridians. The portulan chart, developed in the Middle Ages, had served the Portuguese well enough in all their major voyages of discovery. When, in the 16th century, parallels and meridians were introduced on the charts, meridians were represented as parallel to one another and meridians and parallels were shown cutting each other at right angles.

The *Livro de Marinharia* of the Portuguese pilot João de Lisboa, a manuscript of c. 1550, contains a number of texts not attributable to João de Lisboa himself. One of these can be dated to not later than 1519, since it is reproduced in Spanish trans-

---

20 Hieronymus Münzer, *loc. cit.*

21 Cf Alfragan, *Differentie scientie astrorum*, ed. F.J. Carmody, (Berkeley, 1943).

22 On this scholastic tradition see Giuseppe Boffito, 'Intorno alla "Quaestio de Terra et Aqua" attribuita a Dante', in *Memorie della Real Accademia delle Scienze di Torino*, Ser. 2, Vol. 51, (1902), pp. 73–159, Vol. 52, (1903), pp. 257–342 (covers the whole medieval period); Sigmund Günther, 'Aeltere und neuere Hypothesen über die chronische Versetzung des Erdschwerpunktes durch Wassermassen', in *Studien zur Geschichte der Mathematischen und Physikalischen Geographie*, (Halle, 1879), Vol. III, pp. 129–215; Lucien Gallois, *Les Géographes allemands de la Renaissance*, (Paris, 1890), reprint (Amsterdam, 1963), chapter VIII; Pierre Duhem, *Les origines de la statique*, (Paris, 1906), Tome II, pp. 93–97; Pierre Duhem, *Le système du monde*, Tome IX, (Paris, 1958), chapters XVI and XVII; W.G.L. Randles, *De la terre plate au globe terrestre*, (Paris, 1980), chapter II, No 2; Edward Grant, 'In Defense of the Earth's centrality and immobility: scholastic reactions to Copernicanism in the seventeenth century' in *Transactions of the American Philosophical Society*, Vol. 74, (1984), pp. 22–32; cf the section on the 'Terraqueous Sphere' in which Grant adds to and nuances the position held in the aforecited work.

23 Duarte Pacheco Pereira, *op. cit.*, Livro IV, Cap. I, *ed. cit.*, p. 152, (p. 165 of the Eng. trans. *cit.*).

lation by Martín Fernández de Enciso in his *Suma de Geographia*, Seville, 1519.[24] In a style anything but academic and more typical of the language used in instructions for pilots, the author of the text states that,

> '... asy se poderia e deviam de fazer as cartas em fegura de quadrãte pera se conformasem com o corpo esperico que he redomdo / porem como as cousas de todo o esperico vão jumtas fazemse em plano por lomgetude por que os que maream nã sam estrolliqos e se algum o he por açidente e por que em plano comprendem milhor pratiqua com aquillo que seus entendimentos allcãção da teoriqa segundo abelidade de cada hum por iso eu vemdo que devia por obra a otilidade comum e nã a particolar acordey de a fazer em plano pera que o comum a entendese milhor e pera os particolares que mais alcamçã baste que he a ymaginatiua emtemdam que he fegura de quadramte cõsyderãdo que imdo desda ycanunçiall comtra os pollos em cimquo graos demenue hum grao em a redomdeza do esperico atee 40 graos da ycanunçiall e de 40 atee 60 vay demenuimdo atee que chegados os sesenta se demenue em cimquo de latetude dous de lomgetude e dally se vay acreçemtãdo a demenuyçam atee a fim ... '[25]

Putting this text into English offers considerable difficulties; we propose the following free rendering.

> 'Charts can and should be made to show the rounded quarter of the globe (*quadrante*) in order to follow the nature of the sphere, which is round. However since the meridians of every sphere converge (*como as cousas de todo o esperico vão jumtas*), they are [here] shown parallel to one another (*em plano*), as far as longitude is concerned, because those who navigate are not astronomers (*estroliqos*) and if it happens that one of them is, it is only by accident, and by showing the meridians parallel to one another (*em plano*), they understand better for practical purposes with something they can grasp in terms of theory, each according to his capacities. For this reason then, in order to provide for common users and not for the specialized, I have decided to make the chart a plane chart (*em plano*) so that the ordinary [sailor] may better understand it. And for the specialized, who can grasp more, let them imagine to themselves that the chart is in the form of a quarter of the globe and that as one moves from the equator to the poles, every five degrees, the roundness of the sphere diminishes one degree, until [one gets to] 40 degrees from the equator and from 40 degrees to 60 degrees there is a reduction of two degrees [of longitide] for every five of latitude until 60 degrees is reached and from there on the rate of reduction increases up [to the pole]'.

It is obvious that while the author of this text realizes the necessity of representing the convergence of the meridians, he is quite unable to propose a basic modification of the portulan chart to take this into account. The only temporary solution to the problem was to instruct pilots to take specific corrective measures on specific routes. The earliest example of such specific instructions is noted by the Portuguese soldier-sailor-cosmographer D. João de Castro in connection with the route

---

24 Martín Fernández de Enciso, *Suma de Geographia*, (Seville, 1519), sign. b.ix^v–b.x^r.

25 João de Lisboa, *Livro de Marinharia*, ed. J. Brito Rebelo, (Lisbon, 1903), pp. 197–198.

between Brazil and the Cape of Good Hope. The pilots' estimation on their charts for the run of a ship in normal circumstances in twenty four hours (*singradura*) was 40 leagues, but on the route from Brazil (Cape Frio, Lat. S. 22° 51') to the Cape of Good Hope (Cape Agulhas, Lat. S. 34° 50'), they artifically 'speeded up' the ship, allowing from 70 to 80 leagues to a *singradura*, since otherwise they tended to come upon the Cape of Good Hope before they were normally expecting it, or found that they had already run past it when their calculated estimation on the charts told them that it should shortly be in sight.[26]

Using charts with meridians parallel to one another, (or which implied that they were parallel to one another), the pilots could not represent on them the fact that the South Atlantic was much 'narrower' than the Atlantic in middle latitudes. They therefore used the above strategem to compensate for the fact.

Only in mid-16th century, with Pedro Nunes' demonstration of the existence of the loxodromic curve with a method for correcting deviation along it together with his instructions for re-setting the ship's course on a great circle, then later followed by Mercator's invention of his projection, was the problem of convergent meridians overcome for sailors. Yet more than a century was to elapse before Ptolemy's map-making principles were to become familiar and regularly applied for use by navigators in Portugal and elsewhere.

[26] D. João de Castro, 'Roteiro de Lisboa a Goa (1538)', in Idem. *Obras Completas*, (Coimbra, 1968), Vol.I, pp. 206–207.

VI

From the Mediterranean portulan chart to the marine world chart of the Great Discoveries: the crisis in cartography in the sixteenth century

The origins of the Mediterranean portulan chart are shrouded in mystery and remain the object of much controversy.[1] The earliest surviving example, the so-called Pisan chart, actually of Genoese origin, is dated to c. 1290.[2] Most agree that the framework of the portulan chart is not based on any projection.[3] It is probable, though not yet clearly established, that the coastal outlines preceded and existed independently of the rhumb network, which became added later with the appearance of the magnetic compass.[4] A late description of the construction of the portulan chart by the Spaniard Martin Cortés, (1551) describes it as built up from two perpendicular axes N/S and E/W intersecting at the centre of the chart. Through this centre are drawn from 12 to 32 rhumb lines representing the points of the compass rose.[5] Cortés however only tells how to draw the network of rhumbs to make a copy of an existing chart, but not how to integrate newly discovered coastlines into the network. A scale in leagues or miles is drawn in a corner. No indication what-

---

1 The opposed positions of the two principal protagonists in the controversy, Hermann Wagner, 'Das Rätsel der Kompasskarten ...' (1895) and Heinrich Winter, 'Hermann Wagners wahre Position in der Kontroverse um die Kompasskarten ... ' (1948), are reprinted in Köberer, Wolfgang, (ed.) *Das rechte Fundament der Seefahrt* (Hamburg, 1982). Wagner maintained that there was a continuity relating the portulan charts to ancient Greek maps. Winter rejected this, claiming that they were solely related to the invention of the magnetic compass. *Cf* also the 'continuist' position of Richard Uhden in 'Die antiken Grundlagen der Mittelalterlichen Seekarten', *Imago Mundi* (1935), 1–19.

2 Bibliothèque Nationale, Paris, Rés Ge B 1118.

3 Wagner, Hermann, 'The origins of the Medieval Italian nautical charts' in *Report of the 6th International Geographical Congress* (London, 1895), 695–703.

4 K. Kretschmer concluded that it appeared that the rhumb network could have preceded the coastal outlines, but admitted that he could not be absolutely sure. *See*: Kretschmer, K, *Die Italienischen Portolane des Mittelalters* (Berlin, 1909); Facsimile reprint by George Olms (Hildersheim, 1962), 39–40. We find it hard to follow H. Winter who insists that all was related to the magnetic compass, *See*: Winter, H., 'The origin of the sea chart,' *Imago Mundi* 13 (1956), 39–52.

5 Cortés, Martín, *Breve Compendio de la Sphera y de la Arte de Navegar* (Seville, 1551), ff. lxi (v.)–lxv (v).

ever of the use of coordinates of latitude or longitude can be inferred from the organization of the portulan chart.[6]

To know the position of his ship on a chart, a mariner needed to know two things, distance and direction. Direction, he extracted from the chart by seeking the nearest rhumb parallel to his intended course.[7] He then set his course on this rhumb according to his magnetic compass. Distance he obtained after having estimated the run of the ship in leagues over a twenty four hour period called a *singladura* or *singradura*. He then took, with a pair of dividers, this distance, as given by the scale, and transferred it to the rhumb of his course. By placing one point of his dividers at his place of departure, the other then gave the position of the ship along the rhumb line.[8] This method of navigation worked well enough in the Mediterranean where no ship would be out of sight of land for more than a few days and where the curvature of the earth was of negligeable importance (the magnetic compass treats the earth as if it were flat). It is generally agreed that celestial navigation was never regularly practised in the Mediterranean.[9] It is true that because of magnetic declination, the axis of the Mediterranean was tilted on the charts towards the ENE from 7 degrees to 8 degrees in its western part and from 9 degrees to 10 degrees in its eastern part, but, this did not interfere with the navigator's use of the chart.[10]

Once Europeans began the exploration of the Atlantic and more especially their progression along the West African coast, the portulan system of navigation began to show its inadequacy because of accumulated errors of estimated distance on the long north-south voyages undertaken.

During the second half of the fifteenth century, the Portuguese developed two forms of celestial navigation as a remedy for this problem: first the calculation of latitude, about 1460, from observation of the height above the horizon of the pole star, then, later between 1481 and 1485, from the observation of the height of the sun at midday. We shall here only resume the main points of this development and pass over the finer details and considerable controversy connected with the circumstances in which it was achieved by the Portuguese.

The evidence is based essentially on two texts: the first concerning the use of the pole star, is to be found in the recollections of a Portuguese navigator Diogo Gomes

---

6 Wagner, H., *art. cit.*

7 Cf the description in a relatively late text, Faleiro, Francisco, *Tratado del Esphera y del Arte del Marear* (Seville, 1535). Facsimile edition by J. Bensaúde (Munich, 1915) 59.

8 Cf the description in a relatively late text: Lavanha, João Baptista, 'Tratado da Arte de Navegar', (1588), MS quoted in Cortesão, A., *Cartografia e Cartógrafos Portugueses* (Lisbon, 1935), vol. II, 335.

9 Cf A. Teixeira da Mota, 'A Arte de navegar no Mediterrâneo nos séculos XIII–XVII e a criação da navegação astronómica no Atlântico e no Indico', in *Anais do Clube militar naval* (Lisbon, 1957), 5–6 and António Barbosa, *Novos subsídios para a história da ciência náutica portuguesa da época dos descobrimentos* (Oporto, 1948), 3–38 (1st pub. 1938).

10 Wagner, Hermann, *art. cit.*, p. 700.

who explored the Guinea coast between 1460 and 1462; the second concerning the use of the sun, is to be found in a manuscript note by Columbus in the margin of his personal copy of Aeneas Silvius Piccolomini's *Historia rerum ubique gestarum*, Venice, 1477, and it is dated 1485.

The first text reads: 'And I had a quadrant when I went to these places [i.e. Guinea] and I wrote on the table (*tabula*) [= limb][11] of the quadrant the height of the arctic pole and I found it [the quadrant] better than the chart. It is certain that one sees one's course on the chart, but once one has committed an error, one can never get back on one's bearing.'[12]

The meaning of the last sentence is admittedly obscure. Two hypotheses have been suggested to explain how the quadrant was used, either to check estimations of distance run by comparing changing heights of the pole star, or to check places previously marked on the chart by procedures of pure estimation.[13] Debate has centered on whether the quadrant was actually used on board ship or only ashore,[14] probably the former.

---

[11] On the interpretation of the word *tabula*, *cf* Barbosa, António, *op. cit.*, 146–152; Taylor, E.G.R., 'The navigating manual of Columbus', *Bollettino Civico*, 1 (Genoa: Istituto Colombino, 1953), 36–37; Beaujouan, Guy, 'Science livresque et art nautique au XVe siècle', in Mollat, Michel et Adam, P., *Les Aspects internationaux de la Découverte océanique aux XVe et XVIe siècles, (1960)*, (Paris, 1966), 29–31.

[12] Diogo Gomes, *De Prima Inventione Guineae*, in Schmeller, J.M., 'Über Valentĩ Fernandez Alemã und seine Sammlung von Nachrichten über die Entdeckungen und Besitzungen der portugiesen in Afrika und Asien bis zum Jahre 1508', in *Abhandlungen der Philosophisch-Philolog. Classe der Königlich Bayerischen Akademie der Wissenschaften*, Bd. IV, (Dritte Abth. Munich, 1847), 33. '*Et ego habebam quadrantem, quando ivi ad partes istas, et scripsi in tabula quadrantis altitudinem poli arctici, et ipsum meliorem inveni quam cartam. Certum est quod in carta videbitur via marinandi, sed semel errata nunquam redeunt ad primum propositum.*'

The considerable controversy centered on this text has been reviewed by Vitorino Magalhães Godinho, 'Evolução e estado actual de algumas questiões', in Leite, Duarte, *História dos Descobrimentos*, organização, notas e estudo final de V. Magalhães Godinho (Lisbon, 1960), Vol. II, 502–4; by Beaujouan, Guy, *art. cit.*, 69–73, and by Albuquerque, Luís de, *Os Guias Náuticos de Munique e Évora* (Lisbon, 1965), 28–31.

The last sentence of my translation is somewhat different from that of Guy Beaujouan, *art. cit.* p. 69: ('Il est certain que sur la carte on voit la route maritime, mais une fois qu'elle est fausse, on ne revient jamais au but primitivement fixé'). My translation is inspired from the remarks of Professor Léon Bourdon of the Sorbonne and from the suggested translation of Hermann Wagner, 'Die Entwicklung der Wissenschaftlichen Nautik im Beginn des Zeitalters der Entdeckungen nach neuern Anschauungen', in *Annalen der Hydrographie und Meteorologie*, XXXXVI (1918), Heft III/ IV, 112: ('Wenn sich in der Karte einmal Fehler (sc. auf Grund blosser Kurs und Distanzmessungen) eingeschlichen hat, so kommt man nicht an das Ziel.)

[13] Albuquerque, Luís, de *op. cit.*, 29–30 & Idem, *Ciência e Experiência nos Descobrimentos Portugueses* (Lisbon, 1983), 72.

[14] Guy Beaujouan, (*art. cit.* 70 & 89) maintains that the quadrant was only used on land. A. Teixeira da Mota claims it was used on board ship. *Cf* A. Teixeira da Mota, note at the end (p. 87) of Beaujouan's paper.

These crude attempts to introduce celestially determined references into a 'distance and direction' cartography, became progressively refined, as can be seen from the evidence of the second text concerning the method for observing the height of the sun. In Columbus' words this reads: 'The King of Portugal, in the year of our Lord 1485 sent to Guinea Master Joseph [Vizinho] his physician and astrologer to measure the height of the sun in the whole of Guinea, which he did thoroughly and reported [the results] to the most Serene King, I being present.'[15]

The late Admiral Teixeira da Mota noted that the distinct effects of this planned surveying of the West African coastline[16] from celestial data can be readily seen by comparing the oldest known Portuguese nautical chart dated *c.* 1471–1482,[17] with the so-called 'Cantino' chart of 1502.[18] The West African coast on the Cantino chart is correctly placed as to latitude (that is in relation to the equator and the tropics, for the Cantino chart has no actual scale of latitudes). This does not occur on the earlier 1471–1482 chart.[19]

---

15 Christopher Columbus, *Postilla* in his personal copy of Aeneas Silvius Piccolomini, *Historia rerum ubique gestarum* (Venice, 1477), reproduced with French translation in Pierre d'Ailly, *Imago Mundi*, edited and translated by Edmond Buron (Paris, 1930), Tome III, 751–752. That his name was Joseph Vizinho, can be inferred from a reference to him as the translator of the canons of Abraham Zacuto's *Almanach Perpetuum (Leiria, 1496)*, facsimile edition by J. Bensaúde, Berne, 1915, last folio. *Cf also* João de Barros, *Décadas da Asia*, (1552), Déc. I, Liv. IV, Cap. II, who refers to him as one of the King's physicians.

16 The Portuguese soldier-sailor-cosmographer Duarte Pacheco Pereira, declared (c. 1508) that during many years, he had observed the degrees of latitude in equatorial Guinea. *Cf* Pereira, Duarte Pacheco, *Esmeraldo de Situ Orbis*, (c. 1508), Bk II, Chap. XI, English translation by G.H.T. Kimble (Hakluyt Society, 2nd series, 79, London, 1937, reprint Kraus 1967), 136.

17 Reproduced in Cortesão, A. and Teixeira da Mota, A., *Portugaliae Monumenta Cartographica* (Lisbon, 1960), 1, Plate 2.

18 *Idem.* Plates 4 and 5.

19 Teixeira da Mota, A., *A Evolução da Ciência náutica durante os séculos XV–XVI na cartografia portuguesa da época*, Agrupamento de Estudos de Cartografia, Antiga, Secção de Lisboa, III (Lisbon, 1961), 13–14; *Idem. A África no Planisfério Português anónimo 'Cantino' (1502)*, Centro de Estudos de Cartografia Antiga, Junta de Investigações do Ultramar, Secção de Lisboa, CII (Lisbon, 1977), 5, & Idem. 'Africa in the anonymous Portuguese "Cantino" Planisphere (1502)', in Cornelis Koeman, (ed.) 'Land und Seekarten im Mittelalter und in der frühen Neuzeit', in *Wolfenbütteler Forschungen*, Bd 7 (Munich, 1980), 126.

It should be noted however that in the same note by Columbus, Vizinho's alleged placing of the island of *Los Idolos*, near Konakry (present day Guinea) was more than 4 degrees in error and that the Castle of El Mina was also more than 4 degrees in error. *Cf* Magnaghi, A., 'I presunti errori che vengono attributi a Colombo nella determinazione delle latitudini, in *Bollettino Soc. Geog. Ital.* (1928), 462–65. Fontoura da Costa, A., *A Marinharia dos Descobrimentos*, 2nd ed. (Lisbon, 1939), 37, n. 49, claims that the errors were Columbus' when writing the note, not Vizinho's. Duarte Pacheco Pereira, writing albeit somewhat later *c.* 1508, gives the latitudes of many places on the West African coast. He is rarely more than 40 minutes out, a fact which should absolve the Portuguese from accusations that their techniques of measurement were grossly inaccurate. *Cf* Duarte Pacheco Pereira, *Esmeraldo de Situ Orbis*, (*c.* 1508), English translation *cit.*, 183–84.

The first known chart to carry a separate and distinct scale of latitudes in degrees appears a few years later. It is that of Pedro Reinel of *c.* 1504.[20]

The result of this gradual introduction of celestially determined coordinates into the 15th century portulan charts was that the West African coast became correctly drawn according to celestially determined latitude, while places in the Mediterranean remained (and were to remain until the late seventeenth century) several degrees too high in latitude where the effects of magnetic declination had placed them on the Medieval charts.[21]

The unholy marriage of the old portulan chart and what was to become the new plane chart, was first noted by the Portuguese pilot João de Lisboa in 1514: 'Because of magnetic declination [the importance of which was just being realised], we find', he said, 'that on the charts, all the coasts are falsely [situated] with errors of up to 11¼ degrees or even twice as much. In order to correct this, we should navigate according to truth [i.e. true geographical north] though when [sailing] near the coast, as long as the chart has not been corrected, you should navigate according to past custom, for with the false, you should navigate falsely, and with the true, truly.'[22] By this he meant that one should not relate celestially determined latitudes to a chart which had been drawn from data based only on distance and direction, though one could navigate well enough, and especially near the coast with such a chart if one followed the methods proper to it. This point has been well brought out by Professor Luis de Albuquerque.[23]

A reverse procedure to that advocated by João de Lisboa, (and it was as common as it was scientifically unsound), was to introduce on a plane chart, in a region where a change in magnetic declination was markedly noticeable, a separate meridian (sometimes oblique) to which the navigator transferred his attention when sailing there. Such was the case of the oblique meridian in the neighbourhood of New Foundland, first found in the above mentioned chart of the Portuguese Pedro Reinel (*c.* 1504), or the two graduated meridians not reflecting each other on the western and eastern margins, accompanied by two equators overlapping each other

---

[20] Cortesão, A. and Teixeira da Mota. A., *Portugaliae Monumenta Cartographica*, 1. (Lisbon, 1960) Plate 8.

[21] This anomaly was noted by the Portuguese mathematician Pedro Nunes in 1537, Pedro Nunes, *Tratado, ... em defensam da carta do marear* (Lisbon, 1537), in *Obras*, Lisbon, 1940, Vol. 1, 189–190. Rome he said was in 41½ degrees, but on the portulan chart at 46 degrees. Rhodes was at latitude 36 degrees but on the charts at 42 degrees. *Cf* Albuquerque, Luís de, *Ciência e Experiência nos Descobrimentos Portugueses* (Lisbon, 1983), 77.

[22] Albuquerque, Luís de, *Tratado da Agulha de Marear de João de Lisboa, reconstituição do seu texto seguida de uma versão francesa com anotações*, Centro de Estudos de Cartografia Antiga, Junta de Investigações do Ultramar, Serie Separatas, CL (Coimbra, 1982), 140.

[23] Albuquerque, Luís de, *Ciência e Experiência* . . . 73–75. and *Idem.* "Considerações sobre a Carta-Portulano", in *Revista da Universidade de Coimbra*, 31, (1984), 21–22.

on the King-Hamy chart of 1502–1504, or the auxiliary meridians in the Western Atlantic on Diego Gutierrez's chart of 1550.[24]

Pilots, being conservative minded by nature, did not quickly adopt celestial methods of navigation. An astronomer, Master Joan Faras, who accompanied Pedro Alvares Cabral on his voyage to India in 1500, during which he discovered Brazil, wrote from a place on the latter's eastern coastline at 16° 21' Lat S.[25] that the pilot's estimations of distance run were much greater than his. 'One will only know who is right when we reach the Cape of Good Hope, they with the chart, or I with the chart and the [mariner's] astrolabe'.[26]

It was on this route from the east coast of Brazil to the Cape of Good Hope, later to be used by all ships sailing from Europe to the Orient in the 16th century, that the curvature of the earth and the convergence of the meridians were to become a distinct problem in cartography, since the portulan chart with its scale of degrees of latitude tending to transform it into a plane chart, took no account of this phenomenon, but we shall return to this point later.

The Spaniard Martín Fernández de Enciso, in his *Suma de Geographia* (1519) wrote that, ' ... charts should be made in the form of quadrants (by which he meant 'quarters' of a sphere) so as to conform to the spherical body [of the earth], which is round. But as the gores (*costas*) of every sphere draw together (*van juntas*) we draw them on a plane surface (*en plano*) as regards longitude, because sailors are not astronomers (*astrologos*) and if one happens to be, it is only by chance, and because on a plane surface, sailors understand better with something that their minds can grasp in the way of theory, according to the capacities of each.'[27]

---

24 The chart of Pedro Reinel is reproduced in A. Cortesão and A. Teixeira da Mota, *Portugaliae Monumenta Cartographica*, 1 (Lisbon, 1960), Plate 8; *Cf* Albuquerque, Luís de, *Ciência e Experiência* ... 70. The King-Hamy chart, in the Huntington Library, San Marino, California, is reproduced in Nordenskiold, A. E., *Periplus* (Stockholm, 1897), Plate XLV. Diego Gutierrez's chart of 1550 is reproduced by D. Gernez, 'Les cartes avec échelle de latitudes auxiliaire pour la rêgion de Terre Neuve', in *Communications de l'Académie de Marine de Belgique*, VI (Antwerp, 1952), 104. Gernez noted 28 charts of this type. Alonso de Santa Cruz (*c.* 1555) described how cartographers artificially 'jacked up' the whole American continent 3 degrees to compensate for magnetic declination. *Cf* Alonso de Santa Cruz, *Libro de las Longitudines*, edited by Antonio Blazquez y Delgado Aguilera (Seville, 1921), 38–39. The Portuguese Francisco Faleiro and the Spaniard Martín Cortés declared themselves firmly against auxiliary meridians (*Cf* Francisco Faleiro cited by J. Pulido Rubio, *El Piloto Mayor de la Casa de Contratación de Sevilla* . . . (Seville, 1950), 104; and Martín Cortés, *Breve Compendio de la Sphera* . . . (Seville, 1551), f. lxxiii (v.).

25 Greenlee, William B., *The voyage of Pedro Álvares Cabral* (Hakluyt Society, 2nd series, 81, London, 1937, Reprint Kraus, 1967), 36.

26 Mestre Joan Faras to King Manuel, Vera Cruz, Brazil, May 1st 1500; Portuguese text in Joaquim Barradas de Carvalho, *La traduction espagnole du 'De Situ Orbis' de Pomponius Mela par Maître Joan Faras et les notes marginales de Duarte Pacheco Pereira* (Junta de Investigações do Ultramar, Centro de Estudos de Cartografia Antiga, Lisbon, 1974), 107–108 with facsimile. English translation, somewhat different to mine in Greenlee, William B., *op. cit.* 37.

27 Martín Fernández de Enciso, *Suma de Geographia* (Seville, 1519), sign. b.ix (v.)–b.x. (r.).

Thus did the plane chart, with parallels and meridians drawn at right angles, become accepted as the standard instrument of navigation in the 16th century. The first chart with a scale of longitude shown along the equator is that of the Portuguese Jorge Reinel (*c.* 1519),[28] but this new addition was of negligeable value for the sailor since no reliable methods of measuring longitude at sea were as yet available.

In 1537, the Portuguese mathematician Pedro Nunes declared that he considered the plane chart the only realistic cartographical instrument for navigation and he wrote a treatise in defense of it, in which at the same time he drew attention to its defects. Noting in his *Tratado ... em defensam do carta de Marear* that the meridians were all equidistant from one another, that, is parallel to each other, he stated that ' ... one cannot draw any sort of planisphere with curved lines that might be better suited to our way of navigating than [what we already have] in the [plane] chart. [...] What does it matter', he concluded, 'if on the chart, an island or the mainland is shown more stretched out [in an East/West direction] than it is in reality, as long as it is shown having as many degrees [of longitude] as it should have from East to West, and as long as I [the navigator] am aware, in making the necessary calculations, that the degrees concerned correspond to smaller distances than the chart actually shows, since it is a 'quadratic' (*quadrado*) chart [by which he meant that the longitudes on all the parallels were in equatorial degrees]. We get a better use out of the [plane] chart because the directions [i.e. parallels and meridians] are all straight lines parallel to one another, than we suffer from its defects for it being a quadratic [plane] chart'.[29] In speaking of the 'necessary calculations' Nunes would appear to be referring to tables showing the reductions in terrestrial distance between given meridians for each parallel lying from the equator up to the poles. A table of this kind can be found in the works of Johann Werner (1514) and of Peter Apian (1524).[30] Pilots in the 16th century do not seem to have been able to use such tables. They did make corrective calculations in their use of the plane chart, but not of the kind Nunes the academic cartographer envisaged.

The Portuguese sailor-soldier-scientist D. João de Castro noted, while on a

---

[28] Reproduced in Cortesão, A. and Teixeira da Mota, A., *Portugaliae Monumenta Cartographica*, I (Lisbon, 1960), Plate 24.

[29] Pedro Nunes, *Tratado . . . em defensam do Carta de Marear* (Lisbon, 1537), in Pedro Nunes, *Obras*, I (Lisbon, 1940), 176–77.

[30] Werner, Johann, *Libellus de quatuor terrarum orbis in plano figurationibus . . .* , (Nuremberg, 1514), sign. g. vi (r.) and Apian, Peter, *Cosmographicus liber* (Landshut, 1524), col. 37. Durand, Dana B., *The Vienna Klosterneuburg map corpus of the 15th century* (Leiden, 1952), 170–173), states that such a table can already be found in the *Practica Astrolabii* of Andalò di Negro (*c.* 1270–*c.* 1348). Professor Emmanuel Poulle of the Ecole des Chartes, Paris however assures me that it is a later interpolation. The existence of such tables would not therefore seem to predate the rediscovery of Ptolemy's *Geography* at the beginning of the fifteenth century.

voyage to India in 1538, that during the crossing of the South Atlantic from Brazil to the tip of Africa, once the ships had reached Cape Frio (Lat. S. 22° 51', Long. W. 43° 03') and had set their course for the Cape of Good Hope, (Lat. S. 34° 50', Long. E. 20°), the pilots, in their daily estimations of distance run for this route, artificially increased them to make the ship seem to sail faster than it would be expected to according to the chart – 70 to 80 leagues in each twenty four hour period instead of the usual 40 leagues under similar conditions of wind. 'The two coasts of Brazil and the Cape of Good Hope', said Castro, who never himself mentions converging meridians, 'are shown on the charts further apart than God set them on the globe or in the world'.[31] As an example of what could happen if these precautions were not taken, Castro cites the case in 1505 of a ship whose pilots, believing themselves to be at the mouth of the Congo river, then met a ship whose crew told them that they were in fact off Cape Correntes on the coast of Mozambique.[32] Their ship had overrun the Cape, when in fact they believed that they were still west of it.

Pedro Nunes had proposed, as a solution to the problem of representing converging meridians, that charts should be drawn in the form of an atlas of a series of large scale regional maps (*quarteirões*) on which the meridians would be spaced out according to the intervals on the median parallel just as Ptolemy had done for his separate provincial maps. The margin of error at the top and bottom of the chart would be minimal, said Nunes. Within a band of 18 degrees of latitude north and south of the equator, the meridians could be spaced out according to the equatorial degree, thus making charts in those latitudes quadratic plane charts.[33] However there is no evidence that this solution was ever put into practice.

Another proposal, closer to the practical capacities of mariners, was that of the Spaniard Andrés Garcia de Céspedes in 1606. He suggested putting more than one scale or 'trunk of leagues' (*tronco das leguas*) on the map, one near the equator and another in high latitudes with a smaller scale.[34] No known example of such a chart has however come to our knowledge. Mariners seem to have continued to use the quadratic plane chart throughout the 16th century and well into the 17th.

---

[31] D. João de Castro, 'Roteiro de Lisboa a Goa', (1538) in *Obras completas*, 1 (Coimbra, 1968), 207. Castro erroneously attributed the cause of the phenomenon, for which the pilots introduced their correction, to magnetic declination. Andrés Garcia de Céspedes pointed out his error in his *Regimento de Navegacion* (Madrid, 1606), f. 131 (n).

[32] D. João de Castro, *loc. cit.*

[33] Pedro Nunes, *op. cit.*, 197–98.

[34] Andrés Garcia de Céspedes, *op. cit.* f. 102 (r.)–(v.) From the evidence of Gaspar Jorge Couto in his work *Regimento náutico*, (1608), it appears that it was the Portuguese cartographer João Baptista Lavanha, who was the first in 1588 to propose the introduction of more than one scale of leagues. (*Cf* A. Fontoura da Costa, *A Marinharia dos Descobrimentos*, 2nd edition, Lisbon, 1939, p. 253.

Mercator's projection, which appeared in 1569, did not, on the whole, begin to be used by sailors until late in the 17th century.

Thus, during the period in question, the inadequacies of the Mediterranean portulan chart were only partially compensated for. If the introduction of celestially determined latitudes had necessarily transformed the portulan chart into a quadratic plane chart, the remaining inadequacies of the latter were only palliated with the aid of *ad hoc* correctives introduced in certain areas of the chart, such as auxiliary scales of latitudes, or 'speeding up' the ship on the Brazil-Cape of Good Hope route. Yet taken all in all, although the crisis of cartography was in no way surmounted in the 16th century, the plane chart with all its faults, seems to have served mariners for better or for worse enabling them to navigate on known routes across the whole globe.[35]

---

35 Grateful thanks are here expressed to Professor Léon Bourdon of the Sorbonne, without whose help and guidance, the meaning of many parts of the Portuguese, Latin and Spanish texts would have eluded the writer.

VII

BARTOLOMEU DIAS AND THE DISCOVERY OF THE SOUTH-EAST PASSAGE LINKING THE ATLANTIC TO THE INDIAN OCEAN (1488)

The voyage of Bartolomeu Dias, which established the southern limit of the African continent, thus proving that a navigable passage from the Atlantic to the Indian Ocean existed, began with his departure from Lisbon in August 1487 and ended with his return in December 1488 (1).

It is therefore fitting that, in Sagres, where there began under Prince Henry, called the Navigator, the original drive to explore the whole terrestrial globe, we here celebrate the quincentenary of Bartolomeu Dias' singular achievement.

Dias' voyage was part of a concerted project to reach the Indian subcontinent from Lisbon, by sea round Africa. The guiding mind behind this project was that of the Portuguese King D. João II (1455-1495). Already as Prince, before coming to the throne, D. João had, at the age of 19, received in 1474 from his father King D. Afonso V the entire responsibility for maritime enterprise (2). Coming to the throne in 1481, he devoted his full energies to his project.

Up to the moment when D. João II took charge in 1474 of the organization of the discoveries, the African coast had, under D. Afonso V, been explored to a point just beyond the equator: Cape St. Catherine in 1° 53' Lat. S. (3).

Between 1474 and 1488, over a period of fourteen years, two navigators, Diogo Cão and Bartolomeu Dias were to explore and carefully chart the African coast from Cape St. Catherine to the Cape of Good Hope and beyond, as far as the *Rio do Infante,* placed by Duarte Pacheco Pereira (c. 1508) in 32° 40' Lat. S. (4) and identified by E. Axelson as the Keiskama river (5).

---

(1) A. Fontoura da Costa, *Às Portas da Índia em 1484,* Lisbon, 1935, pp. 31-37. Of this rare and very important study, only 350 copies were printed in book form. The original appeared in *Anais do Clube Militar Naval,* Lisbon, 1935.

(2) Damião Peres, *História dos Descobrimentos Portugueses,* 3rd edition, Lisbon, 1983, pp. 118 and 193. For the original document providing the evidence, Cf. J. Bensaude, *L'Astronomie nautique au Portugal,* Berne, 1912-1917, Reprint Amsterdam 1967, pp. 271-4 and A. Brásio, *Monumenta Missionaria Africana,* Lisbon, 1954, Vol. IV, pp. 12-15.

(3) A. Fontoura da Costa, *op. cit.,* p. 5.

(4) Duarte Pacheco Pereira, *Esmeraldo de Situ Orbis,* (c. 1508), Lisbon, 1905, facsimile reprint Lisbon, 1975, p. 149. English translation by G.H.T. Kimble, Hakluyt Society, 2nd series, Vol. LXXIX, London, 1937, Reprint Kraus, 1967, p. 161.

(5) E. Axelson, *Portuguese in South East Africa, 1488-1600,* Cape Town, 1973, p. 15 and Idem, *South-East Africa, 1488-1530,* London, 1940, Appendix II, pp. 169-171.

Before the voyages of Cão and Dias, what were the ideas current in Portugal concerning the configuration of the African continent? To answer this question, we must ask what was the knowledge and extent of influence in Portugal of Ptolemy's *Geography*. It is well known that a copy had been brought at the beginning of the 15th century from Constantinople to Florence, where numerous copies were made (6). It is known that a copy existed in Portugal in the second half of the 15th century, having been bought by the Bishop of the Algarve, D. Alvaro Afonso in 1460-61 from the Florentine merchant Bartolomeo di Iacopo di Ser Vanni (7).

Although Ptolemy represented the African continent as extending well below the equator to 16º Lat. S., his world map showed the East African coast, below the *Promontorium Prassum* in 15º Lat. S., as turning eastwards to link up with the south-east Asian peninsula, thus transforming the Indian Ocean into an enclosed lake and implicitly blocking any hope of a navigable passage from the Atlantic.

But Ptolemy's representation of an enclosed Indian Ocean does not seem to have been followed in late Medieval *mappaemundi*. Fra Mauro's World Map of 1459, a copy of which was sent to Portugal in Prince Henry's lifetime is a case in point. A legend on it specifically states that the Indian Ocean is not a *stagnon* (a closed sea) (8).

Arab geographers of the Middle Ages, to whom Ptolemy's *Geography* was familiar, in contrast to scholars in the West, had not followed Ptolemy in his representation of the Indian Ocean (9). Fra Mauro may have felt their influence.

We do not know what kind of maps were current in Portugal in the reign of D. João II (1481-1495), but we do know that they existed. In Fernão Lopes de Castanheda's *História do Descobrimento e Conquista da India,* (1551) and in Francisco Alvares, *A Verdadeira Informação das Terras do Preste João* (1540) we have a reference to a chart *(carta de marear)* given to Pero

---

(6) Vespasiano da Bisticci, *Vite di Uomini illustri del secolo XV,* modern edition, Bologna, 1893, Vol. III, pp. 9-10 and G.R. Crone, *Maps and their Makers,* London, 1978, pp. 36-37.

(7) Virgínia Rau, 'Bartolomeo di Iacopo di Ser Vanni, mercador-banqueiro florentino 'estante' em Lisboa nos meados do século xv', in *Do Tempo e da História,* Vol. IV, Lisbon, 1971, p. 113.

(8) Placido Zurla, *Il Mappamundo di Fra Mauro,* Venice, 1806, p. 52 & pp. 84-85. Fra Mauro's map does not seem to have impressed the Portuguese, who had had it copied at some expense. Cf. the *Carta da Quitação* to Dr. João Fernandes da Silveira, 3/2/1462, in João Martins da Silva Marques, *Descobrimentos Portugueses,* Vol. III, (1461-1500), Lisbon, 1971, p. 20, ('E deu e despendeo trimta ducados tres quartos aos pyntores que pyntarom o papa [sic] mundo em Veneza. E esto por sse no perder o que ja em elle era fecto'). Cf. also the interpretation of this text by Luís de Albuquerque, *Ciência e experiência nos Descobrimentos Portugueses,* Lisbon, 1983, pp. 63-64.

(9) Cf. Charles Issawi, 'Arab Geography and the circumnavigation of Africa', in *Osiris,* Vol. 10, (1952), pp. 117-128.

da Covilhã and Afonso de Paiva before their departure by land in May 1487 to reconnoitre the East for D. João II, thus before Dias set sail in August (10).

In Ramusio's edition of Alvares' work, there appears the statement that the chart had been copied from a *mappamundi* by the Bishop of Viseu [Diogo Ortiz] de Calçadilha, Master Rodrigo and Master Moses. The three were scientific advisers to D. João II (11). On the existence of a sea passage from the Atlantic to the Indian Ocean, they claimed that they had had found some record of the fact *(dicevano haverne trovata non so che memoria)* (12).

If, as is probable, D. João II had indeed accepted on their chart such an apparent modification of Ptolemy's representation, whereby a strait was made to pierce the latter's *Terra Incognita* running along the southern edge of the Indian Ocean, there yet remained two major questions to be answered: how far south did the African continent extend and, no less important, how far east did it extend? In an undeservedly little known study: 'A viagem de Bartolomeu Dias e as concepções geopolíticas de D. João II' the late Admiral Teixeira da Mota made a marked contribution in showing how hypotheses were formulated to answer these questions. He insisted in particular on the fact that the Portuguese underestimated the width in longitude of the African continent, narrowing it by about 15 degrees or by more than 1500 kilometers (13). But D. João II's hypotheses were to prove too optimistic and they led to a cruel disillusion, which retarded his project of circumnavigation.

Diogo Cão and Bartolomeu Dias were the first Iberian navigators to mark the extent of their discoveries on each of their voyages by setting up stone pillars called *padrões* (sing. *padrão*), carrying inscriptions with their date of erection. Cão set up four and Dias three (14).

---

(10) Fernão Lopes de Castanheda, *História do Descobrimento e Conquista da Índia*, Lisbon, 1551, Livro I, Cap. I (Copy in the British Library) and Francisco Alvares, *A Verdadeira Informaçam das Terras do Preste João*, Lisbon, 1540, Cap. CIV, (p. 279 of the Lisbon, 1943 edition, Agência Geral das Colónias).

(11) Master Rodrigo and Master Moses are mentioned by João de Barros, *Décadas da Ásia*, Déc. I, Liv. IV, Cap. 2. On Diogo Ortiz, cf. Guy Beaujouan, *L'Astronomie dans la péninsule ibérique à la fin du Moyen Age*, Agrupamento de Estudos de Cartografia Antiga, Vol. XXIV, Coimbra, 1969, p. 15.

(12) G.B. Ramusio, *Navigationi et Viaggi*, Venise, 1550, Vol. I, f. 254 (v). This passage is not found in the Portuguese text published in Lisbon in 1540, but there are good reasons to believe it to be authentic. Cf. Ramusio's introductory *Discorso*, f. 204 (r.) in which he explains that both his version and the version published in Lisbon were summarized from a larger original.

(13) A. Teixeira da Mota, 'A Viagem de Bartolomeu Dias e as concepções geopolíticas de D. João II', in *Boletim da Sociedade de Geografia de Lisboa*, Oct./Dec. (1958), pp. 297-322. Cf. also Idem, 'Africa in the anonymous Portuguese 'Cantino' Planisphere (1502), in Cornelis Koeman, (ed.), 'Land- und Seekarten im Mittelalter und in der frühen Neuzeit', in *Wolfenbütteler Forschungen*, Vol. 7, Munich, 1980, pp. 129-136.

(14) A. Fontoura da Costa, *op. cit.*, pp. 22-26 and p. 40. Cf. also the very thorough

To fully understand the achievement of Dias we must rapidly look at that of Cão, his predecessor.

Diogo Cão made two voyages down the African coast, the first in 1482-84 in the course of which he discovered the mouth of the Congo river. After placing a *padrão* (the *padrão de S. Jorge)* on the south side of the river's mouth on 23 April 1483, he sailed on southward as far as Cape Lobo (13° 26' Lat. S.), where on 28 August 1483, he set up the *padrão de S. Agostinho.* At this point, he seems to have believed that he had reached the southern limit of the African continent and that here the coast turned eastwards as is suggested on the chart of Cristoforo Soligo (c. 1490), an Italian chart copied from a Portuguese original (15).

Such were the elated hopes in Portugal of now holding the Indian Ocean within easy reach, that D. João II had his ambassador in Rome, Vasco Fernandes de Lucena, declare in triumph in an Oration of Obedience before the Pope in 1485, that the Portuguese had, the previous year come within a few days voyage close to the *Promontorium Prassum* (16).

Cão's next voyage, 1485-86, was however to dash all these high expectations. The coast continued relentlessly southwards. Cão placed two more *padrões,* one at *Monte Negro* (now Cabo Negro, 15° 42' Lat. S.) and the last one at *Cabo Padrão* (now Cape Cross in 21° 47' Lat. S.). Just beyond, at the *Serra Parda,* (today Ponta dos Farilhões in 22° 10' Lat. S.), (17) he gave up and turned back to Portugal, to be thereafter quietly forgotten for his failure to have made the breakthrough to the *Promontorium Prassum* (18).

The Martellus map of c. 1489 carries a legend in front of the *Monte Negro* which reads: 'This mountain, called *Monte Negro,* was reached by the fleet of [John] the second, King of Portugal; the fleet was commanded by Diegus Canus, who, in memory of this set up a marble column, with the emblem of the Cross, and the said fleet proceeded onward as far as the *Serra Parda,* which is distant 1000 miles from the *Monte Negro* and from there it went no further *(et hic moritur)* (19).

---

study which has not dated by E.G. Ravenstein, 'The voyages of Diogo Cão and Bartolomeu Dias, 1482-88', in *Geographical Journal,* 1900, (off print of 31 pp.).

(15) Cristoforo Soligo, Venise, (c. 1490), Codice Egerton 73, No. 33, British Library. Cf. also A. Fontoura da Costa, *op. cit.,* p. 45 and Damião Peres, *op. cit.,* p. 215.

(16) *Valasci ferdinãdi utriusque iuris cõsulti Illustrissimi regis Portugaliae oratoris, ad Innocentiũ .VIII., pontificẽ maximũ de obedientia Oratio,* Rome, 1485, reproduced in facsimile by A. Fontoura da Costa, *op. cit.*

(17) A. Fontoura da Costa, *op. cit.,* pp. 19-20.

(18) Damião Peres, *op. cit.,* pp. 217-218.

(19) Legend on the map of Henricus Martellus, Florence, c. 1489, British Museum, Add. Ms 15760, ff. 68 (v.)-68 (r.). The translation is based on the suggested French translation by Professor Léon Bourdon of the Sorbonne. It has been claimed that the expression *et hic moritur* implies that Cão died at this place and did not return to Portugal. It is now accepted that the text cannot mean this, since it has been shown that Cão did return to Portugal. Cf. note by A. Teixeira da Mota at the end of the paper by Augusto César da Silva Castro Junior, 'Diogo Cão e a legenda de Henrique Martelo' in *Actas do Congresso*

FIG. 1 — Enlargement of the southern part of Africa in Pietro Coppo's *Atlas,* (c. 1520), (Biblioteca Comunale dell'Archiginnasio, Bologna, Italy, Cod. A 117). The importance of the legend off the coast of south-west Africa as evidence that Diogo Cão did return to Portugal and did not die at the *Monte Negro* was referred to by Admiral Teixeira da Mota in a remark printed at the end of the paper given by Augusto César da Silva Castro Júnior, «Diogo Cão e a legenda de Henrique Martello», in *Actas do Congresso Internacional de História dos Descobrimentos,* Lisbon, 1961, Vol. II, p. 108. Cf. also Roberto Almagià, «The Atlas of Pietro Coppo», in *Imago Mundi,* Vol. VII, (1950), pp. 48-50 and Américo da Costa Ramalho, «Sobre a data da Morte de Diogo Cão» in *Estudos sobre a época do Renascimento,* Coimbra, 1969, pp. 2-8.

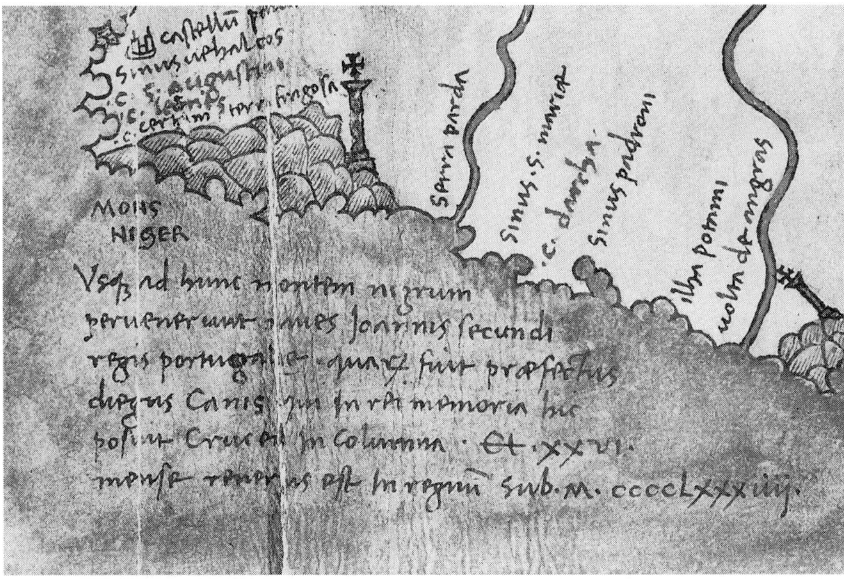

FIG. 2 — Enlarged reproduction of the legend on the map of Africa in Pietro Coppo's *Atlas* (c. 1520), (Biblioteca Comunale dell'Archiginnasio, Bologna, Cod. A 117). The legend proves that Diogo Cão returned in person to Portugal and did not die at the *Monte Negro* as some scholars have claimed. The Latin text reads as follows:

> «usque ad hunc Montem Nigrum pervenerunt naves
> Johannis secundi regis Portugalliae quạrum fuit
> praefectus diegus canis, qui in rei memoria hic
> posuit crucem in columna. Et XXVI mense reversus
> est in regnum MCCCCLXXXIIII (= 1484).»

The English translation reads:

> «The ships of John II, King of Portugal reached
> as far as this Monte Negro. They were under the command
> of Diogo Cão, who in memory of this accomplishment
> set up a cross on a column. And in 26 months he
> returned to the kingdom [of Portugal] in 1484.»

We turn now to Bartolomeu Dias. Dias left Portugal in late July or early August of 1487 (20), and continued down the African coast past Cão's last *padrão,* as far as the *Angra das Voltas* (21) identified as the modern Luderitz Bay in 26° 35′ Lat. S. (22). Here, according to Barros, Dias was held up by a storm which prevented him from advancing. He therefore put out to sea and ran before it with sails at half mast for thirteen days. When it ceased he turned east again to look for the coast and not finding it, turned north where he then struck it at a place called the *Angra dos Vaqueiros,* the modern Fish Bay (23), close to the present Mossel Bay.

He thus had, without realizing it and without seeing it, passed the southernmost limit of the African continent and reached a point along its coast which now seemed oriented in an eastward direction.

Gago Coutinho, Teixeira da Mota and Axelson all doubt the reality of Barros' 'storm'. They prefer an explanation in terms of southerly prevailing winds obliging Dias to steer on a tack out to sea (24). Dias would also have had to struggle against the Agulhas current. (25). That Dias rounded the Cape well to the south of it is more than probable, but how far south is a question that we shall come back to.

From the *Angra dos Vaqueiros,* Dias saw, with great relief that, in the words of Duarte Pacheco Pereira, '... the coast here turned northwards and northeastwards... giving great hope of the discovery of India' (26). He continued along it as far as a place he named the *Penedo das Fontes* (27) in 33° 44′ Lat. S. and 26° 38′ Long. E., now known as False Island or Kwaai-

---

Internacional da História dos Descobrimentos, Lisbon, 1961, Vol. II, pp. 107-109. Cf. also Damião Peres, *op. cit.,* pp. 216-218. Teixeira da Mota cites a legend on the Atlas of Pietro Coppo c. 1520 which states that Cão did indeed return to Portugal. The figure of 1000 miles cannot but be a gross exaggeration.

(20) A. Fontoura da Costa, *op. cit.,* p. 31.

(21) João de Barros, *Décadas da Asia,* Lisbon, 1552, Déc. I, Liv. III, Cap. IV, and Duarte Pacheco Pereira, *Esmeraldo de Situ Orbis,* (c. 1508), Liv. III, Cap. 5, (p. 140 of the edition of A. Epifânio da Silva Dias, Lisbon, 1905, facsimile edition, Lisbon, 1975). English translation by G.H.T. Kimble, *ed. cit.,* p. 151.

(22) E. Axelson, *Portuguese in South Africa, 1488-1600,* Cape Town, 1973, p. 13, Barros puts the *Angra das Voltas* in 29° Lat. S. which is considered an error by Axelson.

(23) E. Axelson, *Portuguese in South East Africa, 1488-1530,* London, 1940, p. 17.

(24) Gago Coutinho, *A Náutica dos Descobrimentos,* Lisbon, 1952, Vol. I, pp. 230-231 and p. 242; A. Teixeira da Mota, *Bartolomeu Dias descobridor do Càbo de Boa Esperança,* Lisboa, 1955, p. 17; E. Axelson, *Portuguese in South-East Africa, 1488-1600,* Cape Town, 1973, p. 14.

(25) E.G. Ravenstein, *op. cit.,* p. 22.

(26) Duarte Pacheco Pereira, *Esmeraldo de Situ Orbis,* (c. 1508), Liv. IV, Cap. 7, ed. Epifânio da Silva Dias, Lisbon, 1905, facsimile edition, Lisbon, 1975, p. 142, English trans. G.H.T. Kimble, *ed. cit.,* p. 154.

(27) João de Barros, *Décadas da Ásia,* Dec. I, Liv. III, Cap. IV.; Duarte Pacheco Pereira, *op. cit.,* Liv. III, Cap. 9, *ed. cit.* p. 147, English trans. cit. p. 160. Barros and Pacheco confuse the *Penedo das Fontes* with the *Ilha da Cruz* in Algoa Bay. Cf. E. Axelson, *op. cit.,* p. 15.

hoek. Here he set up another *padrão*, the *Padrão de S. Gregório,* probably on March 12, 1488. The remains of the *padrão* were found by Axelson in 1938 (28).

The Martellus map of 1489 shows the limit of Dias' voyage with a legend that reads: 'In the course of their last voyage in 1489 [sic] the Portuguese reached the *ilha de fonte*' by which is meant the *Penedo das Fontes.*

At the *Penedo das Fontes,* the crew demanded that Dias turn back, claiming that they had discovered enough coastline and that they could at least take home the important news that the coast ran eastwards, from which it could be deduced that behind them lay a great cape and that the best counsel would be to go back and discover it (29). They agreed to continue eastwards for another two or three days, which was done, and this led to the discovery of a river which they named the *Rio do Infante,* later put by Duarte Pacheco Pereira in 32° 40' Lat. S. (30) probably the present Keiskama river (31), and this was the furthest point reached by Dias.

Returning westwards, Dias and his men were given a view, in Barros' words, of that great and notable Cape, hidden for so many hundreds of years and which when it is seen, not only reveals itself, but opens up an entire new world of countries.' (32) At the tip of the Cape of Good Hope, Dias set up the *Padrão de S. Felipe,* probably on 6 June, 1488 (33). His last *padrão,* the *Padrão de Santiago,* he set up on a point west of Luderitz Bay. The place was identified by Axelson and he recovered some of the fragments of the *padrão* (34).

Dias, on his return to Portugal, made a report of his voyage to D. João II, which Christopher Columbus, being present, described in a marginal note in his copy of Pierre d'Ailly's *Imago Mundi.* Dias, in Columbus' words, reported how he had overrun the Cape and how he had reached a latitude of 45° (the Cape being in reality in latitude 34° 22′ S.). The navigator 'described his voyage league by league on a chart laid before the King.' (35).

Various hypotheses have been proposed to explain the figure of 45° Lat. S., which some see as apparently referring to the Cape itself, and others to a

---

(28) E. Axelson, *Portuguese in South-East Africa, 1488-1530,* London, 1940, p. 20 and Appendix III, pp. 172-181.

(29) João de Barros, *op. cit. loc. cit.*

(30) Duarte Pacheco Pereira, *Esmeraldo de Situ Orbis,* ed. cit. Liv. III, Cap. IX, p. 148-49; English trans., pp. 160-61.

(31) E. Axelson, *Portuguese in South-East Africa, 1488-1600,* Cape Town, 1973, p. 15.

(32) João de Barros, *op. cit. loc. cit.*

(33) E. Axelson, *op. cit.*, p. 18.

(34) Idem, *op. cit.*, p. 19.

(35) Christopher Colombus, *Postilla* in Pierre d'Ailly, *Imago Mundi,* reproduced in A. Fontoura da Costa, *op. cit.*, p. 52. Another *postilla* by Columbus in his copy of Aeneas Silvius Piccolomini's *Historia rerum ubique gestarum,* Venice, 1477, reproduced by Fontoura da Costa, *op. cit.*, p. 52 confirms the figure of 45°.

point further south. Ravenstein accepted it as a latitude at sea beyond the Cape (36), Bensaude and Fontoura da Costa held it to be a gross error imputable to Columbus (37). Jaime Cortesão claimed that the figure was deliberately invented by D. João II to mislead Columbus into believing that the Cape route was much longer than it really was (38). This argument is part of Jaime Cortesão's general thesis regarding the Portuguese crown's policy of secrecy *(política de sigilo)* and it has not won wide acceptance. Damião Peres held the figure of 45º to be acceptable, a consequence of the 'storm' which drove Dias south and he believed that the 1500 kilometers (450 leagues) corresponding to the distance given by Columbus between Cão's last *padrão* and the latitude of 45º could have been sailed in the thirteen days mentioned by Barros (39), but this we find it hard to accept. The figure of 45º remains for us an enigma.

How were the results of Dias' voyage interpreted by D. João II? Had the most southern tip of Africa really been found, and if so how far did the continent still extend eastwards before allowing navigators to proceed northwards? Did the cape that Dias had found really represent the major turning point of the continent's coastline? Was this the much sought after *Promontorium Prassum,* or did the *Promontorium Prassum* lie still further on, pointing East and yet to be reached? (40) The profile of the Martellus map of 1489, itself ressembling closely the profile of Martin Behaim's globe of 1492 would seem to lend support to such an hypothesis. On the Martellus map, the extremity of the second eastward pointing cape is masked by the legend describing Dias' voyage and how the Portuguese had reached the *Penedo das Fontes.*

That, after Dias' return, it was still thought in Portugal that the principal cape was the *Promontorium Prassum* and that it was yet to be discovered, might possibly be inferred from an oration delivered in Beja by João Teixeira in March 1489, in which he declared that: '... day by day, we strive to reach those promontories the *Promontorium Prassum* and the *Promontorium Raptum* and those sands through which passes the Nile, and so from the latter come to the approaches of the Indian Ocean.' (41)

---

(36) E.G. Ravenstein, *op. cit.,* p. 17.

(37) J. Bensaude, *L'Astronomie nautique au Portugal,* Berne, 1912/17, reprint, Amsterdam, 1967, p. 109, and A. Fontoura da Costa, *A Marinharia dos Descobrimentos,* 2nd ed., Lisbon, 1939, p. 37, n. 48.

(38) J. Cortesão, *A. Política de Sigilo,* Lisbon, 1960, pp. 50-51.

(39) Damião Peres, *Historia dos Descobrimentos,* 3rd edition, Lisbon, 1983, pp. 236-237.

(40) The historian João de Barros declared in 1552 that the *Promontorium Prassum* had indeed been judged by D. João II to be 'the extremity of the continent' *(fim daquela terra),* Cf. João de Barros, *Décadas da Asia,* Lisbon, 1552, Dec. I, Liv. III, Cap. IV.

(41) João Teixeira, *Oratio habita insigni viro Ioanne Teixeira...* (1489), Coimbra, 1562, sign. A. ix(r.).

Yet, for Valentim Fernandes in the introduction to his edition of Marco Polo's *Travels* published in Lisbon in 1502, there seemed no doubt in his mind of the decisive nature of Dias' accomplishment. 'Dias', he said, 'had been chosen like Joshua to enter into that New World, which we can indeed call the Promised Land.' (42)

The historian João de Barros was later, in 1552, to set out the new configuration of the African continent contrasting it with the representation given it by Ptolemy. Beyond the region of Agisymba, the southern limit known to Ptolemy, wrote Barros, lies another land unknown in his day, of which the coastline is now well explored. It starts at the *Prassum Promontorium* in 15 1/2 degrees Lat. S., called by the inhabitants Mozambique. Its western limit, unknown to Ptolemy lies in 5° Lat. S. in the land of the *Pangelungos*, that is, near the mouth of the Congo. Between these two limits, to the east and to the west, lies the great and famous Cape of Good Hope, for thousands of years unknown to the world. (43)

The adding to Ptolemy's *oikoumene* of this enormous triangular wedge was, very largely, the remarkable accomplishment of Bartolomeu Dias. It remained for Vasco da Gama to complete the exploration of the stretch of coastline between the *Rio do Infante* and Sofala, but it was Dias who, for the first time in history, successfully proved the existence of a southeast passage linking the Atlantic to the Indian Ocean.

---

(42) Valentim Fernandes, *O Livro de Marco Polo*, Lisbon, 1502, re-published by Francisco Maria Esteves Pereira, Lisbon, 1922, sign. A.ij (v.).

(43) João de Barros, *Décadas da Asia*, Déc. I, Liv. VIII, Cap. IV.

VIII

La configuration cartographique du continent africain avant et après le voyage de Bartolomeu Dias: hypothèses et enseignements

La configuration du continent africain dans la cartographie antérieure au voyage de Bartolomeu Dias pose le problème de l'éventuelle synthèse tentée par les dirigeants portugais, ou par ceux dont ils avaient sollicité l'expertise, entre la représentation de l'Afrique dans la *Géographie* de Ptolémée et la représentation du continent noir sur les cartes nautiques, dressés par les navigateurs jusqu'aux limites atteintes dans leurs voyages.

Nous savons, grâce à Virgínia Rau, et contrairement à ce qu'avaient affirmé Duarte Leite et Magalhães Godinho, que la *Géographie* (ou du moins les cartes qui l'accompagnaient) fut connue au Portugal, sinon l'année de la mort de l'Infant D. Henrique, du moins l'année suivante [1].

Selon Ptolémée, les derniers lieux habités sur la côte occidentale de l'Afrique correspondent au *Hypodromus Æthiopium*, à 5° 15′ de lat. N, et, sur la côte orientale au *Promontorium Prassum*, à 15° de lat. S [2]. Au-delà, la côte occidentale s'infléchit indéfiniment vers l'ouest et la côte orientale vers l'est pour rejoindre l'Asie du sud-est.

---

[1] Virgínia Rau, «Bartolomeo di Iacopo di Ser Vanni Mercador-Banqueiro 'estante' em Lisboa nos meados do século XV», in *Do Tempo e da História*, vol. IV (1971), pp. 97-117. L'évêque de l'Algarve, D. Álvaro Afonso, a acheté, entre 1460 et 1461, un «livro delle tavole di Tolomeo» au marchand italien Ser Vanni. Pour Duarte Leite, cf. *História dos Descobrimentos*, Lisbonne, 1959-1961, vol. I, p. 159, et pour Magalhães Godinho, cf. *ibid.*, vol. II, p. 507.

[2] Ptolémée, *Geography*, trad. d'Edward Luther Stevenson, New York, 1932.

Bien que très évidente dans les manuscrits et les éditions imprimées de sa *Géographie*, la conviction de Ptolémée que l'Océan Indien constitue un lac fermé, sans communication avec l'Atlantique, fut, sans doute sous l'influence de la cartographie arabe [3], rapidement abandonnée en Europe.

Chez Idrisi (1154) [4], l'Océan Indien communique avec l'Atlantique, et il en est de même sur la carte de Pietro Vesconte (*c.* 1320), qui s'en est visiblement inspiré [5]. Fra Mauro, dans sa mappemonde dressée pour le roi de Portugal Afonso V en 1459, suit la même tradition, héritée des Arabes; une légende, sur la carte, dit expressément que l'Océan Indien n'est pas un *stagnon* (une mer fermée) [6].

Il est douteux que la mappemonde de Fra Mauro ait beaucoup apporté aux Portugais car s'il indique, dans l'île de Diab (= Madagascar) qu'il représente séparée du continent africain par un étroit canal (canal de Mozambique), plusieurs villes de la côte orientale, Braua (= Brava), Maabase (= Mombasa), Chelue (= Kilva), elles sont toutes situées dans un désordre chaotique [7].

Les Portugais ne semblent d'ailleurs pas avoir accordé grande importance aux informations qu'apportait cette carte, car le roi donna l'ordre de régler les frais des dessinateurs sans attendre l'achèvement de l'œuvre: «E deu [le Dr. João Fernandes da Silveira] e despendeo trimta ducados tres quartos aos pyntores que pintarom o papa (*sic*) mundi em Veneza. E esto por sse no perder o que ja em elle era fecto» [8].

Venons-en à l'analyse de la configuration supposée de l'Afrique autour de la date importante de 1486.

---

3 Charles Issawi, «Arab Geography and the circumnavigation of Africa», in *Osiris*, vol. X (1952), pp. 117-128.

4 Idrisi, *Mappemonde*, Ms Arabe, Ms Pococke 375, fols. 4 (*r*)-4 (*v*), Bodleian Library, Oxford.

5 Pietro Vesconte, Ms Vatican, Ms Pal. Lat. 1362 A, folios 1 (*v*)-2.

6 *Il Mappemondo di Fra Mauro*, éd. T.G. Leporace et R. Almagià, Rome, 1956, Tavola XI, p. 27. («Adoncha sença alguna dubitation se può affermar che questa parte austral e de garbin sia navigabile e che quel mar indiano sia oceano e non stagnon . . .»).

7 *Ibid.*, éd. cit., Tavole III et IX. Cf. Aussi G.R. Crone, «Fra Mauro's representation of the Indan Ocean and the eastern islands», in *Studi Colombiani*, Gênes, vol. III (1951), pp. 57-64.

8 Cf. Carta da Quitação do Dr. João Fernandes da Silveira, 3-3-1462, in João Martins da Silva Marques, *Descobrimentos Portuguese*, vol. III (1461-1500), Lisbonne, 1971, p. 20. Cf. également l'interprétation de ce texte par Luís de Albuquerque, *Ciência e Experiência nos Descobrimentos Portugueses*, Lisbonne, 1983, pp. 63-64.

Deux ans auparavant, Diogo Cão avait révélé que la côte occidentale de l'Afrique s'étendait jusqu'au Cap Lobo (actuellement Cabo de Santa Maria, à 13° 26′ de lat. S). L'examen de la carte italienne de Soligo (*c.* 1489, copie d'une carte portugaise), ainsi que l'analyse du «Discours d'Obédience» de Vasco Fernandes de Lucena devant le Pape de Rome, en 1484, permirent à Fontoura da Costa, à Damião Peres et à Teixeira da Mota de soutenir que Diogo Cão avait cru être parvenu à l'extrémité du continent, et qu'il était persuadé que l'Océan Indien ne se trouvait, pour reprendre les termes de Lucena, qu'«à quelques jours de route» [9].

Vers la fin de l'année 1485 ou au début de 1486, en tout cas avant le retour de Diogo Cão de son second voyage, le navigateur João Afonso de Aveiro rentre à Lisbonne avec la nouvelle qu'il avait découvert, au fond du Golfe de Guinée, le royaume du Bénin, où on lui avait parlé d'un potentat appelé l'Ogané [10], qui se trouvait 250 lieues plus à l'est. Comme celui-ci, à l'instar du Prêtre Jean (le Négus d'Ethiopie), restait toujours caché derrière des rideaux de soie, le roi D. João II pensa qu'il ne pouvait s'agir que de la même personne. «C'est pourquoi», écrit João de Barros, D. João II «consultant avec les cosmographes de ce royaume la carte de l'Afrique de Ptolémée, et tenant compte de l'ordre dans lequel les découvreurs avaient placé les *padrões* le long de la côte, ainsi que de la distance de 250 lieues vers l'est, où les gens du Bénin disaient se trouver les terres du prince Ogané, il [D. João II] estimait que ce prince devait être le Prêtre Jean [. . .], et il lui semblait aussi que, si ses navires continuaient à suivre la côte qu'ils étaient en train de découvrir, ils ne pourraient manquer d'aborder au *Prassum Promontorium*, qui se trouvait à l'extrémité de ce continent» [11].

Voici donc, à cette date-clé de 1486, comment apparaît aux dirigeants portugais la configuration du continent africain: d'abord una façade Atlantique, s'étendant de l'équateur au Cap Lobo, puis une autre façade donnant sur l'Océan Indien et s'étendant jusqu'au *Promontorium Prassum*, situé par Ptolémée à la latitude de 15° S.

---

[9] A. Fontoura da Costa, *Às portas da Índia en 1484*, Lisbonne, 1936, p. 80. Damião Peres, *Descobrimentos Portugueses*, Lisbonne, 3e éd., 1983, pp. 206-215. A. Teixeira da Mota, «A Viagem de Bartolomeu Dias e as concepções geopolíticas de D. João II», in *Boletim da Sociedade de Geografia*, Oct.-Déc. 1958, pp. 297-322. C'est surtout Teixeira da Mota qui a le plus amplement développé cette thèse.

[10] Identifié par Raymond Mauny à l'Oni d'Ife, du peuple Yoruba.

[11] João de Barros, *Ásia*, Déc. I, Liv. III, Cap. IV, éd. Coimbra, 1932, p. 84.

Comme l'a montré Teixeira da Mota, on représentait les deux façades beaucoup plus rapprochées qu'elles ne l'étaient en réalité, aussi l'étendue de l'Afrique en longitude se trouvait-elle réduite d'environ 15°, soit 1.600 kilomètres [12]. Si l'on compare la carte d'Henricus Martellus (*c.* 1490) à celle dite de Cantino, cette vision des choses est manifeste. L'explication tient en partie à la confusion de l'Ogané avec le Prêtre Jean et à l'idée que l'on se faisait de l'extension du royaume de celui-ci vers l'ouest.

Dans l'hypothèse évoquée des deux façades, nul ne savait, en 1486, comment et où elles allaient se rejoindre au sud. Rien ne laissait prévoir qu'entre les deux extrêmes connus, l'un à l'est, d'après Ptolémée, l'autre à l'ouest, d'après le premier voyage de Diogo Cão, le continent africain pourrait éventuellement se développer en une saillie plongeant encore plus vers le sud, bien au-delà de la ligne joignant directement les limites des deux façades. Le *Promontorium Prassum* devant constituer l'extrémité du continent vers l'est, qui eût pu penser qu'il existerait encore un autre promontoire en direction du sud.

Le deuxième voyage de Diogo Cão, rentré à Lisbonne en 1486 après le retour de João Afonso de Aveiro, n'a pu apporter qu'une profonde déception. Loin de marquer le point où la côte africaine se tournait vers l'Océan Indien, le Cap Lobo n'en était qu'un accident. Elle se prolongeait encore sans relâche vers le sud. Cão, découragé, s'était arrêté à la *Serra Parda*, identifiée comme la Ponta dos Farilhões, que l'on situe aujourd'hui á la lat. de 22° 10′ S.

Neanmoins, l'obstination de D. João II à chercher le passage du sud-est vers l'Orient ne fléchissait pas.

Bartolomeu Dias, parti l'année suivante, en août 1487, contourne le Cap de Bonne Espérance et s'arrête au Rio do Infante, sur la côte orientale de l'Afrique. Il n'aperçoit, ne découvre l'extrémité du continent que lors de son voyage de retour.

Quels enseignements tira-t-on de son voyage? Le Cap de Bonne Espérance était-il bien le *Promontorium Prassum* si recherché? Ou s'agissait-il d'un cap entièrement «nouveau», le *Promontorium Prassum* se situant encore plus loin, au-delà du Rio do Infante?

---

12 A. Teixeira da Mota, *art. cit.*, p. 311, et également du même, «Africa in the Anonymous Portuguese 'Cantino' Planisphere (1502)», in Cornelis Koeman (éd.), «Land-und Seekarten im Mittelalter und in der frühen Neuzeit», *Wolfenbütteler Forschungen*, Bd. 7, 1980, pp. 123-136.

Plusieurs témoignages nous sont parvenus à ce sujet. Dans un discours prononcé à Beja le 1er mars 1489, l'année suivante au retour de Dias, João Teixeira déclarait: «Chaque jour, nous nous efforçons de nous rapprocher de ces promontoires, que ce soit le *Promontorium Prassum* ou le *Promontorium Raptum*, et aussi des sables du Nil, par lesquels on parvient au bord de l'Océan Indien, et de là au *Sinus Barbaricus*, source d'infinies richesses» [13].

Faut-il en déduire que Teixeira ne tenait pas le voyage de Dias pour un exploit majeur, qu'il ne considérait pas le Cap de Bonne Espérance comme le cap important marquant l'extrémité du continent et que, pour lui, le cap-clé, le *Promontorium Prassum* restait à découvrir? Ou faut-il voir, au contraire, dans ces formules stylisées, l'expression d'un humaniste s'adressant á d'autres humanistes habitués à la géographie ptoléméenne et, de ce fait, relativement indifférents à l'univers mental des cartes nautiques des pilotes? On remarquera une nette ressemblance entre les paroles de Teixeira: «Chaque jour nous nous efforçons de nous rapprocher de ces promontoires, que ce soit le *Promontorium Prassum* ou le *Promontorium Raptum*», et celles qu'avait prononcées Vasco Fernandes de Lucena cinq ans auparavant: «[. . .] la navigation portugaise n'est éloignée [du Golfe Arabique] que de peu de jours de route. En effet [. . .] les Portugais abordèrent l'année dernière près du *Promontorium Prassum*, où commence le Golfe Arabique [. . .]». Teixeira aurait-il calqué son discours sur celui de son prédécesseur, prononcé à Rome?

Les deux hypothèses ne sont pas mutuellement contradictoires. Comment Teixeira eût-il pu parler de l'importance, plus tard reconnue, du Cap de Bonne Espérance, sans avoir pu le comparer au *Promontorium Prassum*, que l'on n'avait pas encore aperçu, alors que, dans son esprit et dans celui de ses interlocuteurs, c'était ce dernier qui demeurait la référence?

De la découverte du Cap de Bonne Espérance, Colomb, présent lorsque Bartolomeu Dias fit son rapport devant D. João II, observait dans une apostille: «Noter qu'en cette année '88, au mois de décembre, Bartolomeu Dias [. . .] aborda à Lisbonne. Il fit rapport au dit Roi [D. João II] [. . .] qu'il avait atteint un promontoire auquel il donna le nom de Cap de Bonne Espérance, que nous avons estimé être en

---

13 João Teixeira, *Oratio habita ab insigni viro Ioanne Teyxeira*, Coimbra, 1562. sign. A viii (*r*).

Agesimba»[14]. Rien de plus. La région d'*Agesimba* rappelle encore le contexte de la *Géographie* de Ptolémée, dans lequel tout se situe.

Ce n'est qu'*après* le voyage de Vasco da Gama et la découverte de la non-existence du *Promontorium Prassum* (plus tard identifié au Cap Delgado), que l'on attribua toute son importance au Cap de Bonne Espérance.

En 1502, Valentim Fernandes peut écrire: «Ce cap si noble, ce renom d'espérance, donné non par l'intelligence humaine, mais par la divine providence, prophétisant que Bartolomeu Dias [. . .] serait digne, comme Josué, d'entrer dans ce monde nouveau, que nous sommes en droit de nommer la terre promise»[15].

Mais en 1488 Dias était-il si sûr, et à Lisbonne était-on prêt à croire qu'il était entré dans la «terre promise» autrement dit l'Océan Indien? La carte d'Henricus Martellus, de *c.* 1490, révèle bien les doutes qui pouvaient subsister à cet égard.

Duarte Pacheco Pereira, dans son *Esmeraldo de Situ Orbis* (*c.* 1508), écrivait que Bartolomeu Dias «. . . voyant que cette côte et ce rivage [la côte africaine au-delà du Cap de Bonne Espérance] tournait ensuite au nord-nord-est, se dirigeait vers l'Ethiopie sous l'Egypte et de là vers le golfe d'Arabie, où l'on pensait et espérait découvrir l'Inde, lui donna pour cette raison le nom de «cap de Bonne Espérance . . .»[16].

Mais l'infléchissement de la côte vers le nord-nord-est ne prouvait pas que l'on avait atteint le tournant majeur du continent. La tendance de la côte, évoqué par Pereira, ne déterminait rien, elle n'apportait qu'un espoir.

Ce n'est en effet qu'à la suite du voyage de Vasco da Gama, lorsque Duarte Pacheco Pereira écrit que: «C'est à ce promontoire [le Cap de Bonne Espérance] que l'Afrique arrive au bout de la Mer Océane, là qu'elle est séparé de l'Asie . . .»[17], que l'on saisit la signification réelle du Cap de Bonne Espérance.

Il est très important de souligner la distinction entre, d'une part, le pas décisif de Dias franchissant le vrai cap majeur du continent

---

14 Colomb, Apostille dans son édition de Pierre d'Ailly, *Imago Mundi*, éd. E. Buron, Paris, 1930, t. I, pp. 207-208.

15 Valentim Fernandes, «Epistola sobre a tralladaçam», in *O Livro de Marco Polo*, Lisbonne, 1502, éd. F.M. Esteves Pereira, Lisbonne, 1922, sign. A iv (*v*).

16 Duarte Pacheco Pereira, *Esmeraldo de Situ Orbis* (*c.* 1508), éd. Epifânio da Silva Dias, Lisbonne, 1905, rééd. Lisbonne, 1975, Liv. III. Cap. 7, p. 142.

17 *Idem.* Liv. III, Cap. 7, éd. cit., p. 143.

africain et, d'autre part, la prise de conscience de ce fait, aussi bien chez les marins que chez les humanistes. Il y aura, entre la réalité et la perception de cette réalité, un décalage de neuf ans, séparant le retour de Dias du retour de Gama.

Ne minimisons pas l'énorme défi auquel s'était mesuré Dias en contournant le Cap. L'expérience avait laissé chez ses hommes un souvenir marquant, comme le rappelait João de Barros parlant des préparatifs du voyage de Gama. Il évoque «... la fureur des mers de ce grand Cap de Bonne Espérance qui, dans l'esprit des marins, commençait à susciter une nouvelle légende, de périls, comme autrefois celle du Cap Bojador»[18].

Gaspar Correia (*c.* 1495 - *c.* 1565), qui attribue le voyage de Dias à João Infante, personnage qui l'avait accompagné, confirme les paroles de João de Barros: «Il [= João Infante / Dias] avait trouvé que les mers étaient si violentes que l'on ne pouvait pas naviguer [autour du Cap de Bonne Espérance] avec des caravelles» («achava os mares tão grandes que os não podia navegar com as caravelas») [...] mais qu'avec de grands navires qui résisteraient à la mer [...] jusqu'à ce qu'il découvre l'extrémité de la côte, sans aucun doute il avait un espoir certain qu'au-delà il trouverait de grandes terres» («mas com navios grandes que sofressem o mar, [...] até lhe descobrir o cabo [da costa] sem duvida tenha certa esperança que allem delle, acharia grandes terras»)[19].

Castanheda fait lui aussi allusion «aux très grandes tempêtes» («muyto grandes tormentas») qu'avait rencontrées Dias autour du Cap[20].

Le départ de Dias avait eu lieu en août 1487. Trois mois plus tôt, le 7 mai, D. João II avait expédié par voie de terre deux émissaires: Pero da Covilhã et Afonso de Paiva, auxquels il avait recommandé de recueillir des renseignements sur les conditions de navigation dans l'Océan

---

18 João de Barros, *Ásia*, Lisbonne, 1552, Déc. I, Liv. IV, Cap. I, éd. Coimbra, 1932, p. 122.

19 Gaspar Correia, *Lendas da Índia*, Lisbonne, 1858 t. I, p. 8.

20 Fernão Lopes de Castanheda, *História do Descobrimento & Conquista da India, pelos portugueses*, Lisbonne, 1551, Liv. I, Cap. I, p. 2 (exemplaire de la British Library). Dans la seconde édition de l'ouvrage, en 1554, l'expression «très grandes tempêtes» a disparu, peut-être parce que la réputation du Cap de faire subir des tempêtes aux navigateurs concernait la seule expérience de Dias, et n'apparaissait pas comme une caractéristique permanente. Barros note que, lors du voyage de Vasco da Gama, on a franchi le Cap «avec moins de tempêtes et de danger que n'en attendaient les marins» («com menos tormenta e perigo de que os marinheiros esperavam»). (Cf. João de Barros, *Ásia*, Déc. I, Liv. IV, Cap. III, éd. Coimbra, 1932, p. 129).

Indien. On leur donna «une carte marine copiée d'une mappemonde», très probablement une carte sur laquelle figurait une synthèse de la cartographie ptoléméenne et des cartes nautiques portugaises. On enjoignait à Covilhã de «voir si, dans les mers du Prêtre Jean [= le Négus d'Ethiopie], il y avait quelque information selon laquelle on pourrait passer dans les mers de l'Atlantique»[21]. Avec quelques marchands maures, Covilhã parvint jusqu'à Sofala, où «des marins et quelques Arabes lui firent entendre qu'il était possible de naviguer en direction de l'ouest, le long de toute cette côte dont on ne connaissait pas la fin»[22]. «Covilhã décida d'informer le Roi de tout ce qu'il avait vu [...], d'où il concluait que, après avoir navigué en Guinée et poursuivi leur route avec la terre en vue, les caravelles du roi, demandant la côte de l'île de la Lune [= Madagascar] et de Sofala, pourraient facilement pénétrer dans ces mers orientales et venir aborder à la côte de Calicut, parce que partout il y avait la mer»[23].

Castanheda répète à peu près les mêmes termes les propos de Francisco Álvares, et ajoute que Covilhã renvoya au Portugal, par l'intermédiaire d'un voyageur juif, des informations concernant l'Océan Indien, les accompagnant de la carte qu'on lui avait remise, sur laquelle il avait marqué les noms des lieux qu'il avait visités[24].

Parmi les preuves avancées par Covilhã de la possibilité de longer la côte africaine de la Guinée jusqu'à Sofala se trouvait, selon Castanheda, l'affirmation «que les gens de Sofala étaient aussi noirs que ceux de Guinée et que, de ce fait, il lui sembla qu'il s'agissait d'une seule et même côte»[25]. C'était là une hypothèse plausible, mais non concluante. Autre preuve avancée, selon Francisco Álvares: la certitude des maures de Sofala «qu'il était possible de naviguer, en direction

---

21 Francisco Álvares, «Viaggio della Ethiopia», in G.B. Ramusio, *Navigationi et Viaggi*, Venise, 1550, vol, I, fol. 254 (*r*). Sur la mission de Covilhã, cf. Conde de Ficalho, *Viagens de Pedro da Covilhã*, Lisbonne, 1898, pp. 54-66.

22 Francisco Álvares, *loc. cit.*

23 Francisco Álvares, *loc. cit.*

24 Fernão Lopes de Castanheda, *História do Descobrimento & Conquista da Índia pelos portugueses*, Lisbonne, 1551, Liv. I, Cap. I, pp. 2-4. Ces précisions ont disparu de la seconde édition de l'ouvrage de Castanheda, mais il n'y a pas lieu de considérer les indications de la première édition comme suspectes, comme le souligne Armando Cortesão, «A Carta de Marear de Pero da Covilhã», in *Esparsos*, Lisbonne, 1975, vol. III, pp. 220-223. Par ailleurs, nous ne suivrons pas l'auteur dans son obsession de vouloir soutenir qu'il y eut des voyages secrets entre celui de Dias et celui de Gama.

25 Fernão Lopes de Castanheda, *op. cit.*, *loc. cit.*

de l'ouest, le long de toute cette dite côte, dont on ne connaissait pas la fin»[26]. On trouve ici la conviction des arabes que l'Atlantique communiquait avec l'Océan Indien, et qu'il n'existait pas de *terra incognita* ptoléménne qui empêchât le passage parce que, selon l'expression d'Álvares, «partout il y avait la mer».

Mais cette conviction des Arabes correspondait plutôt à un postulat cosmographique (la continuité de toutes les mers) qu'à quelque expérience empirique, car aucun Arabe, que l'on sache, ne s'était jamais aventuré au-delà du Cap Correntes. Il apparaît donc que Covilhã n'a pu recueillir aucune information concrète à propos de la côte au-delà de Sofala, et que l'exploit de Dias n'en était que plus remarquable.

La patiente obstination à soutenir, année après année, les découvertes progressives de Cão et de Dias, alliée à l'initiative hardie de la mission de Paiva et de Covilhã, démontre l'étonnante envergure et la très grande cohérence du projet «Inde» de D. João II.

Bien que la réussite capitale de Dias n'ait été clairement perçue qu'après la découverte, par Vasco da Gama, de l'évolution sans surprises de la côte africaine séparant le Rio do Infante de Sofala, côte dépourvue de toute saillie gênante se prolongeant vers l'est, il n'en demeure pas moins que ce fut ce courageux navigateur qui réalisa la percée qui permit à l'Europe d'accéder par mer au monde oriental.

---

26 Francisco Álvares, éd. cit., *loc. cit.*

* Je tiens à exprimer ici ma gratitude au Professeur Léon Bourdon, de la Sorbonne, car sans son aide et ses conseils précieux, je n'aurais sans doute pas saisi le sens de certaines parties des textes portugais et italiens.

IX

The evaluation of Columbus' 'India' project by Portuguese and Spanish cosmographers in the light of the geographical science of the period

> ' ... standosi [Colombo] in Portogallo cominciò) a congietturare che, siccome quei Portoghesi camminavano tanto lontano al mezzo dì, medisimamente si potrebbe camminare alla volta dell' Occidente, e che di ragione si potrebbe trovare terra in quel cammino'. Fernando Colombo, *Le Historie della Vita e dei fatti di Cristoforo Colombo*, Venice, 1571 (ed. Rinaldo Caddeo, Milan, 1930) I, Chap. V, p. 39.
>
> 'While in Portugal, it began to occur to Columbus that, just as the Portuguese were sailing so far to the southward, one might in the same way, sail westward and that logically one should find land in that direction'.

Without resources of his own, Columbus was, in order to realize his project of sailing westwards to Asia, obliged to seek the aid of state powers, first that of Portugal, then later that of Spain.[1] Before finally obtaining the support of the Catholic sovereigns of Spain, Columbus had his project examined on three successive occasions, only to see it rejected, first in Portugal by a committee of cosmographers appointed in 1483–84[2] by the King D. João II[3] and then in Spain by a *Junta* of experts whose opinion was sought twice by the Catholic sovereigns in 1486–87 and in 1491.[4]

It is our intention here to analyse the validity of the arguments presented against the project and to show that far from being absurd, they reflected a respectable theory based on the writings of Aristotle. Columbus' own arguments, which were refuted in the name of the above theory, were, as we shall show, also based on Aristotle.

---

1 The financing of the voyage is analysed by Rinaldo Caddeo, in appendix F to his edition of Fernando Columbus' *Le Historie della Vita e dei fatti di Cristoforo Colombo* (Milan, 1930), II, pp. 346–65. Caddeo concludes that the greater part of the financing of the voyage was provided by Italian merchants resident in Spain.

2 Peres, Damião, *História dos Descobrimentos Portugueses*, 3rd ed. (Oporto, 1983), p. 256.

3 Their evaluation of Columbus' project is described by João de Barros, *Ásia*, Déc. I, Lisbon, 1552, Liv. III, Cap. XI (modern edition, Coimbra, 1932), p. 112.

4 Harrisse, Henri, *Christophe Colomb* (Paris, 1882) I, pp. 361, 363.

Several cosmographical parameters determined Columbus' project: (a) the distance across the Ocean separating Asia from Europe, a distance which depended on the east-west extent of the *oikumene,* or the Eurasian land-mass lying between Spain and China, (b) the circumference of the globe and the number of miles or nautical leagues in a degree and (c) that which it was possible to infer, regarding these distances, from the theory of the relations between the size of the surfaces of water and earth such as the followers of Aristotle held them to be in the Middle Ages.

Out of Aristotle's thought the scholastic thinkers of the Middle Ages had developed two doctrines, which in our context, can be seen to contradict each other. The first of these doctrines held that the *oikumene* was of large extent, and that therefore the distance between Europe and Asia was very small. The second doctrine held exactly the opposite to be the case.

Roger Bacon (*c.* 1219–1292), basing himself on a passage in Aristotle's *De Caelo*, held in his work *Opus Majus* (composed between 1266 and 1268), the first of these two doctrines. Where Aristotle had tentatively remarked that 'those who imagine that the region around the Pillars of Hercules joins the regions of India [...] are not suggesting anything utterly incredible,'[5] Bacon made him say that between Spain and India there was a *mare parvum* (a short sea distance).[6] Bacon seems to have been encouraged to allege this by the remark of Seneca, who, in his *Quaestiones Naturales*, had said that the sea between Spain and India could be crossed in 'a few days, if one had a good wind to drive the ship.'[7] Bacon further quoted Pliny and the prophet Esdras in support of his claim for a large *oikumene*. Esdras had declared that six parts of the earth formed dry land and a seventh part water,[8] and Pliny had written that one could readily sail from the Arabian gulf to the pillars of Hercules,[9] a voyage which Bacon seems to have interpreted as across the Atlantic.[10]

Bacon's words were closely repeated by the French cardinal Pierre d'Ailly (1350–1420) in his work *Imago Mundi*, Louvain (?), 1483 (?),[11] a work that Columbus copiously annotated, adopting Bacon's interpretation of Aristotle, Pliny and Esdras.[12]

---

5 Aristotle, *De Caelo*, II, 4, 298a (Translated by W.K.C. Guthrie, London, 1971), p. 253.

6 Bacon, Roger, *Opus Majus*, ed. J.H. Bridges, London 1897 (Reprint Minerva, Frankfurt/Main, 1964) I, pp. 290–292.

7 Seneca, *Naturales Quaestiones* (translated by Thomas H. Corcoran, Loeb edition, London, 1971) Bk I, pref. 13.

8 Esdras, III, 6:42.

9 Pliny, *Natural History* (translated by H. Rackham, Loeb edition I, London, 1972) Bk II, Chap. LXVII, pp. 303 and 305.

10 Bacon, *op. cit.* p. 291.

11 Pierre d'Ailly, *Imago Mundi*, Louvain, (?) 1483 (?) (ed. E. Buron, Paris, 1930), I, Chap. VIII, p. 211.

12 Columbus' annotations are reproduced in Buron's edition of Pierre d'Ailly, *ed. cit.*, I, p. 208.

Bacon and d'Ailly are among the few writers who held that Spain and India were not very far apart. Most of the scholastic writers of the fourteenth and fifteenth centuries seized on other passages in Aristotle's *De Caelo*, taking up the latter's doctrine of the four elements: earth, water, air and fire – arranged in concentric spheres round a common centre, the centre of the universe.[13] As we shall see, it was by logical deduction from this aspect of Aristotle's thought that they came to believe that the distance between India and Spain was indeed very large.

Aristotle, and his Greek commentators after him, had, in his doctrine of the elements, left unsolved the problem of how dry land could emerge (as it obviously did) from the concentric sphere of water, to enable men and animals to live and breathe air. Aristotle had allowed his rigorous logic relating to the elements to lead him into this absurd position, declaring specifically regarding their concentricity that an element could only be in its natural place if it were 'lodged' by the element in the sphere coming immediately after it. Thus his logic required that the earth be entirely covered by a uniform layer of water.[14]

Only in the Middle Ages did scholars begin to try to deal with this problem, proposing various solutions, most of them requiring the intervention of a divine miracle to maintain the dry land out of the water.[15]

The Spanish converted Jew, Paul de Burgos (*c.* 1350–1435)[16] in his *Additiones* to the *Postillae super totam Bibliam* of the Frenchman Nicolas de Lyra (1270–1349), held that on the first day of God's Creation, the elements were placed in a perfectly concentric order, but that on the third day, when according to Genesis[17] (Fig. 1) God commanded the waters to gather together so that the dry land should appear, he shifted the sphere of the water so that its centre no longer coincided with the centre of the earth and so that a part of its surface could emerge.[18]

---

13 Aristotle, *De Caelo*, Bk II, 4, 287b *ed. cit.* p. 161.

14 Idem. *ed. cit.* Bk IV, Chap. III, 310b; *cf* also Duhem, Pierre, *Le Système du Monde*, Paris, 1958, Vol. 9, pp. 88–91.

15 On this considerable and complex question which much occupied Medieval scholars, *cf* the thorough studies by Giuseppe Boffito, 'Intorno alla "Quaestio de Terra et Aqua" attributa a Dante', in *Memorie della Reale Accademia delle Scienze di Torino*, Second Series, 52 (1903), pp. 257–342; Duhem, Pierre, *Les origines de la Statique* (Paris, 1906), I, pp. 93–97 and Idem, *Le Système du Monde*, IX (Paris, 1958), Chapters XVI and XVII.

16 On Paul de Burgos, *cf* Martínez Añibarro y Rives, Manuel, *Intento de un diccionario biográfico y bibliográfico de autores de la provincia de Burgos* (Madrid, 1889–90), pp. 470–479, and Labrosse, Henri, in *Etudes Franciscaines*, Vol. 19 (1908), pp. 370–71. Labrosse notes that there were 27 editions of Nicolas de Lyra's work with Paul de Burgos' 'Additiones' printed before 1500. On Paul de Burgos' life, *cf* Serrano, Luciano, *Los conversos don Pablo de Santa Maria y don Alonso de Cartagena* (Madrid, 1942).

17 Genesis, 1, 9.

18 Burgos, Paul de, 'Additiones' in *Postillae Nicolai de Lyra super totam bibliam cum additionibus* (Nuremberg, 1481) (no pagination).

Fig. 1. Engraving of the spheres of earth and water before and after the *congregatio aquae* on the Third Day of the Creation, according to Paul de Burgos, *Additiones* [written in 1429] in Nicolas de Lyra, *Postillae Nicolai de Lyra super totam bibliam cum additionibus*, Nuremberg, 1481 (no pagination). 'In this figure let A be the centre of the earth, which is also the centre of the universe, and let B, D, G, E, be the sphere of the earth centred on A and let M, N, O, be the sphere of the waters centred on A before their being gathered together. For the waters, according to their primordial nature surrounded the earth in an equal manner, given that every part of the waters tends in like manner towards the centre of the earth, which is the centre of the universe. Now God [on the Third Day] commanded that this sphere of water be gathered together in one place in such a way that the dry land might appear. Thus let the second sphere of the water, which is that which was gathered together after this commandment, be where is now the sphere C, D, E, which is equal in size to the sphere M, N. O. Let the centre of the sphere of the water be F which is excentric, as has been said, since it is separated and removed from the centre A which is the centre of the universe.'

The consequence of Paul de Burgos' doctrine was that the *oikumene* could not be of greater extent than half of the circumference of the sphere of the water (and was very probably much less), and thus, contrary to the position held by Roger Bacon and Pierre d'Ailly, the distance between India and Spain could only be very considerable.

It was this doctrine of Paul de Burgos that the cosmographers of the Catholic sovereigns defended at their *Junta* at Salamanca in 1486–87 and again at their *Junta* at Santa Fé, near Grenada in 1491, in order to deny the feasibility of Columbus' project. Columbus on each occasion repeatedly defended the doctrines of Roger Bacon and Pierre d'Ailly.

Las Casas describes in his *Historia de las Indias* how the cosmographers at Salamanca had declared, ' ... that of this lower sphere of water and earth, only a very small part lay uncovered, for all the rest was under water, and for this reason one could not navigate except along the shores and coasts [of the *oikumene*] as the Portuguese did along [the coast of] Guinea.'[19] Fernando Columbus who repeats, in his *Historie*, almost word for word Las Casas' description, adds something which Las Casas does not: 'When scholars might be agreed that one could reach [by land] the limit of the Orient [i.e. the eastern extremity of the *oikumene*], they would be ready to admit that one could travel [by sea], from the [western] limit of Spain to the western limit [of the Atlantic] (*dal fine della Spagna fino all'ultimo Occidente*).[20] In other words, the cosmographers very reasonably argued, Columbus project could be envisaged only when the exact east-west extent of the *oikumene* were known.

In response to Columbus' quotation of Seneca that a voyage across the Atlantic could be achieved in a few days with a fair wind, the cosmographers retorted with a quotation from another of Seneca's works, the *Suasoriae* that 'many had doubted that the Ocean [i.e. the Atlantic] could be navigated, as it was reputed to be infinite.'[21]

In rebuttal of the pseudo-aristotelian doctrine of Paul de Burgos, as expounded by the members of the Salamanca *Junta*, Columbus had, in the words of the Portuguese historian João de Barros already in 1483–84, replied to the cosmographers of the King of Portugal D. João II that since the Portuguese had discovered ' ... the islands of the Azores and a large part of the African continent, unknown to the Iberian peoples [...], by analogy, there might well exist other islands and lands

---

19 Las Casas, Bartolomé de, *Historia de las Indias*, Lib. I, Cap. XXIX (éd. B.A.E., Tome XCV, Madrid, 1957), p. 111.

20 Columbus, Fernando, *Le Historie della Vita e dei fatti di Cristoforo Colombo* (Venice, 1571, ed. Rinaldo Caddeo, Milan, 1930), 1, Cap. XII, pp. 106–107.

21 Las Casas, *op. cit. loc. cit.* Seneca, *Suasoriae*, I (translated by M. Winterbottom, Loeb edition, London, 1974), p. 485.

to the west; for Nature could not have been so disorderly in the make-up of the whole orb, as to have intended that, for living beings to live and multiply, a greater share be assigned to the element of the water than to that of the earth uncovered [by the water]'.[22]

The members of the second Spanish *Junta* held at Santa Fé, continued to oppose Columbus' project with the same arguments drawn from the theory of Paul de Burgos, arguments which they erroneously attributed to Nicholas de Lyra, since the commentaries on the verses of the Bible by both writers were printed together.

An Italian, Alexandro Geraldini (1455–1525) present at the deliberations of the *Junta*, and who wrote between 1520 and 1521, related how the Spanish cosmographers, quoting Nicholas de Lyra (in reality Paul of Burgos), had '... said that the whole of man's habitat (*totam terrae humanae compaginem*) [in other words the *oikumene*] which extends across the sea from the Fortunate Islands [the Canary islands] as far as the East, had no edge bent down toward the lower part of the sphere (*nulla latere habere per inferiorem partem sphaera obtorta*)'.[23] Geraldini adds that he had remarked in public to the experts of the *Junta* that the experience of the Portuguese had disproved Paul de Burgos' theory, for their navigators had reached far into the southern hemisphere and had 'observed new stars under the sky of the Antipodes'.[24] This was doubtless an allusion to the voyage of Bartolomeu Dias in 1488 during which he had discovered the Cape of Good Hope. Columbus himself had been present when Dias made his report to the King D. João II.[25]

The immense extension of the African continent into the southern hemisphere had obviously given support to the theory of an *oikumene* of large size and would have given encouragement to those who maintained that the distance between Asia and Europe was quite small.

Columbus, in his letter of 7 July 1503, refers to Portuguese voyages far southward along the African coast beyond the equator, to justify his preference for Marinus of Tyre's east-west extension of the *oikumene* (225 degrees as against Ptolemy's 180 degrees), for 'Marinus', wrote Columbus, 'makes Africa extend to

---

22 João de Barros, *op. cit.* Déc. I, Liv. III, Cap. XI *ed. cit.* p. 112. The three cosmographers were José [Vizinho], Master Rodrigo and Diogo Ortiz de Calçadilha, a Castillian who had taken political refuge in Portugal in 1475 from Salamanca, where he had held the chair of Astrology at the University since 1469. *Cf* Beaujouan, Guy, 'L'astronomie nautique dans la péninsule ibérique à la fin du Moyen Age', *Agrupamento de Estudos de Cartografia Antiga, Secção de Coimbra*, XXIV (Coimbra, 1969) XXIV, p. 15.

23 Geraldini, Alexandro, *Itinerarium ad regiones sub aequinoctiali plaga constitutas* (Rome, 1631) pp. 204–205.

24 Idem. *loc. cit.*

25 *Cf* Columbus' marginal note in his copy of Pierre d'Ailly's *Imago Mundi*, reproduced by Nunn George E., *The Geographical Conceptions of Columbus* (American Geographical Society Research Series No. 14, New York, 1924, reprint Octagon Books 1977), pp. 7–8. (Latin text and English translation).

24 degrees Lat S., a latitude which the Portuguese have confirmed'.[26] The Portuguese had actually gone further, for the Cape of Good Hope is in 34°50' Lat S.

If we accept that, from the medieval point of view, the *oikumene* was of circular shape, it can be seen that the smaller the size of the *oikumene* in comparison to the surface of the Ocean, the more feasible the project of reaching India by a circumnavigation of the African continent seemed over that of a voyage across the Atlantic. But this only held true until the discovery of the Cape of Good Hope and its deep extension into the southern hemisphere. On the other hand, if the *oikumene* were shown to be of large size in relation to the surface of the Ocean, the distance between Spain and Asia would necessarily be small, and this could but encourage those who, like Columbus, believed that the shortest route to Asia was that which led west, rather than round Africa.

Now in Portugal, well before the Spanish *Juntas*, how had, in the course of the fifteenth century, the position of D. João II's cosmographers evolved on the subject of the relationship between the surfaces of earth and water, and as a result, on the supposed size of the *oikumene* and of the extent of the Atlantic? Here we must bring in a third model concerning this relationship, that proposed by Claudius Ptolemy (*c.* 90–168 A.D.) in his *Geography* of which the Florentine patron of the arts Palla Strozzi had had a Greek manuscript brought from Constantinople in 1400, to have it translated in 1406 into Latin by Manuel Chrysoloras and Jacopo Angiolo della Scarperia.[27]

No work published during the Renaissance was to so revolutionize the science of mathematical geography. Known in Portugal from at least 1484,[28] and in all probability in Spain[29] at the time of the *Juntas*, Ptolemy's work was apparently ignored by the Spanish cosmographers, who do not seem to have been aware of the incom-

---

26 Columbus' letter of 7/7/1503 ('Marino en Ethiopia escrive aliende la linea equinocial mas de 24 grados, i ahora que los Portugueses la navegan le fallan cierto') in *Select Documents illustrating the Four Voyages of Columbus*, ed. and trans. by Cecil Jane (Hakluyt Society, Second Series, No. LXX, Vol. II, London, 1933; Reprint Kraus, 1967), pp. 83–84. George Nunn claims that Marinus of Tyre had no influence on Columbus' ideas until his fourth voyage. *Cf* Nunn, George E., 'The Three Maplets attributed to Bartholomew Columbus', in *Imago Mundi*, IX (1952), p. 18 and Idem, 'Marinus of Tyre's Place in Columbus' concepts' in *Imago Mundi*, 11 (1937), p. 35. Nunn gives no satisfactory proof to support his claim.

27 Vespasiano da Bisticci, *Vite di Uomini illustri* (ed. Bologna, 1893) 3, pp. 9–10 and pp. 216–217.

28 Fernandes de Lucena, Vasco, *Valasci ferdinandi utriusque iuris consulti Illustrissimi regis Portugalliae oratoris ad Innocentium VIII, pontificem maximum de obedientia Oratio* (Rome, 1485); reproduced in facsimilé in Fontoura da Costa, A. *Às Portas do India em 1484* (Lisbon, 1936), p. 80.

29 Guy Beaujouan mentions the existence, among the scientific manuscripts at the University of Salamanca, of two manuscripts of Ptolemy's *Geography* of which one was copied in 1456. *Cf* Beaujouan, Guy, *Manuscrits scientifiques médiévaux de l'Université de Salamanque* (Bordeaux, 1962), p. 159 and pp. 161–162.

patibility between its basic theoretical tenets and the pseudo-aristotelian doctrine of Paul de Burgos, which they so obediently followed.

In contrast to the pseudo-aristotelian theory of late medieval scholastic science, according to which the earth-*oikumene* emerged like an island out of the water, Ptolemy declared that 'the continuous surface of the earth and of the seas forms one sphere, whose centre is the same as that of the celestial bodies.'[30] Columbus seems to have followed this view himself, for according to Las Casas, ' ... since all the water and the earth in the world made up one sphere [...] Columbus believed that it was possible to circumnavigate it from east to west'.[31]

Ptolemy, in his *Geography*, took the position that instead of the earth being immersed in the water, it had hollows in it, in which the oceans lay in the form of lakes isolated from one another. Such a view could only reinforce the idea of an *oikumene* of large size and thus reduce the extent of the Ocean.

A recognition of the clash between the two theories of the relationship between earth and water, the pseudo-aristotelian on the one hand, and the ptolemaic on the other, appears clearly in the work of a Spaniard, Jacob Perez de Valencia. In his Commentary on the Psalms, published in Valencia in 1484, he comes out resolutely in favour of the ptolemaic theory, as opposed to the pseudo-aristotelian. ' ... for some imagine', he wrote, 'that the earth lies in the water like a light ball, or like an apple in a basin full of water, of which only the summit appears above the water [...] (Fig. 2). This view is obviously irrational'.[32] '[In reality]', he continued, 'the seas are no more than gatherings together of the waters in the depths of valleys between the mountains [...] the Ocean does not surround the earth, as is commonly thought. Rather it is surrounded on all sides by mountains [...] From this it follows that the centre of the earth should not be shifted away from the centre of the universe; rather its centre should be at an equal distance from all the celestial orbs.'[33]

There is nothing to show that Perez' ptolemaic critique of the pseudo-aristotelian

---

30 Claudius Ptolemy, *Géographie*, French translation by the Abbé Halma (Paris, 1828), p. 11. It is true that a much less precise version of Ptolemy's definition was to reach the Latin West as early as the twelfth century, through the work of Alfraganus (al-Farghānī) (died *c*. 861) translated into Latin by John of Seville (1135) and by Gerard of Cremona (*c*. 1175). Alfraganus wrote that 'the earth together with all its parts, both terrestrial and marine resembles a sphere.' *Cf* Alfraganus, *Differentie scientie astrorum* (edited by Frank J. Carmody, Berkeley, Calif. 1943), p. 6. Alfraganus' definition does not seem in any way to have shaken the Aristotelian scholastics before the appearance of Ptolemy's *Geography* at the beginning of the fifteenth century.

31 Las Casas, Bartolomé de, *Historia de las Indias* (ed. cit.) I, Cap. V, p. 27, Fernando Columbus' text is almost identical. *Cf* Columbus, F., *Historie* (ed. cit.) Cap. VI, p. 41. It is true that during his third voyage, Columbus had doubts on the subject and returned for a moment to the pseudo-aristotelian doctrine, even admitting that the earth might be pear shaped. *Cf* Columbus' letter of 18 Oct 1498, in *Select Documents illustrating the Four Voyages of Columbus*, pp. 29–31 (English and Spanish texts).

32 Pérez, Jacob (de Valencia), *[Commentaria in psalmos]*, Valencia, 1484, sign. ii f. ii (v.)

33 *Idem*. sign. ii f. iv (r.–v.)

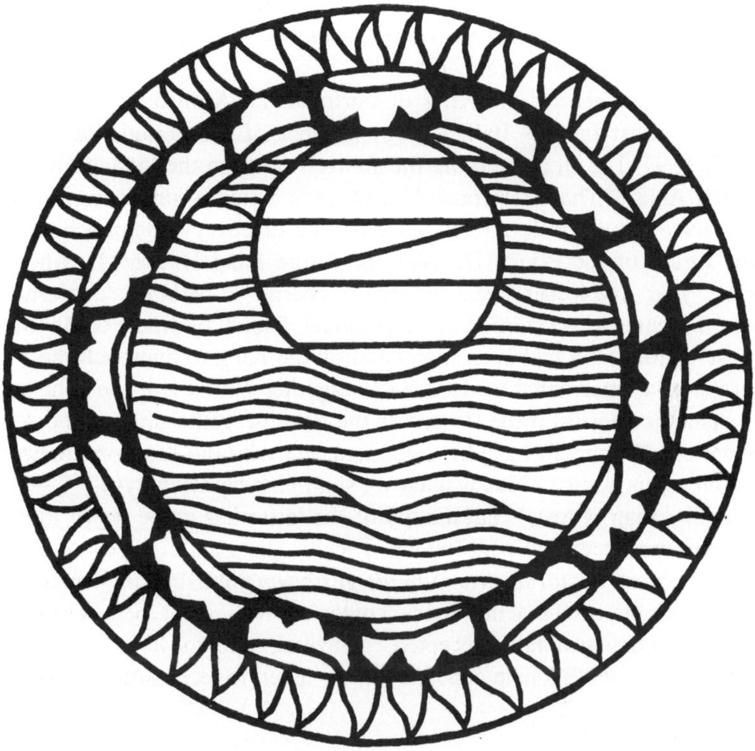

Fig. 2. Engraving showing the sphere of the earth just emerging from the sphere of the water, both being surrounded by the spheres of air and fire in the 1485 and 1490 Venice editions of Johannes de Sacrobosco, *Sphaera*. It is probably at such a representation of the four Aristotelian elements that Jacob Perez de Valencia directed his criticism when he wrote that 'some imagine that the earth lies in the water like a light ball or an apple in a basin full of water.'

doctrine was known in Portugal in the fifteenth century,[34] nor did the Spanish cosmographers of the *Juntas* of 1486–7 and of 1491 seem to have been aware of it in view of their reiterated defense of Paul de Burgos, but the critique did indeed exist and it came from a thinker inside the Iberian peninsula. It is symptomatic of a changing climate of opinion.

34 The Portuguese soldier-sailor-scientist, Duarte Pacheco Pereira does expound it, but at the beginning of the sixteenth century. *Cf* Pereira, Duarte Pacheco, *Esmeraldo de Situ Orbis* (*c.* 1508), translated and edited by G.H.T. Kimble, Hakluyt Society, 2nd series, No. LXXIX, London, 1937, reprint Kraus, 1967, Bk I, Chap. 2, p. 13.

There was however another aspect of the ptolemaic 'revolution', presented in quantitative rather than qualitative terms, which was to reach Portugal in 1474, more than a decade before Perez wrote.

In 1474, or just before, the King of Portugal, D. Afonso V, had solicited the advice of the most famous Italian cosmographer of the period, Paolo dal Pozzo Toscanelli (1397–1482), concerning the shortest route to the lands of the spices, in other words Asia.[35] That D. Afonso V took such a step is surprising, since he was known to have shown little interest in maritime discovery. Behind him no doubt, stood the figure of the Prince D. João, the future King D. João II, who from the very same year 1474 had been placed in charge of Atlantic exploration.[36] In response, Toscanelli, forwarded to a personal friend of his, Fernão Martins, a Portuguese canon, who had been mandated by the King to approach the Florentine cosmographer, a letter dated 25 June 1474 together with a map. In it he stated that to reach Asia the shortest route lay across the Atlantic, rather than round Africa.

On what did Toscanelli base his argument and why did the Portuguese not take any account of his advice? This is what we shall now try to analyse.

Toscanelli's map is lost,[37] but a copy of the letter has come down to us, copied by Columbus on an empty page attached to his personal copy of Aeneas Silvius Piccolomini's work *Historia rerum ubique gestarum* published in Venice in 1477.[38]

The authenticity of the letter, once contested with great persistence by Henri Vignaud,[39] is today generally accepted.[40] Several scholars have sought, from the description of Toscanelli's map contained in the letter, to reconstruct it. The most con-

---

35 Toscanelli in his reply stated: 'Whereas I have spoken [...] elsewhere concerning a shorter way of going by sea to the lands of the spices, than that which you are making by Guinea, the most Serene King [D. Afonso V] now wishes that I should give some explanation thereof, or rather that I should so set it before the eyes of all'. Text in Vignaud, H., *Toscanelli and Columbus* (London, 1902), Appendix A, pp. 276–278.

36 *Cf* Peres, Damião, *História dos Descobrimentos Portugueses* (3rd. ed., Oporto, 1983), pp. 118, 193 and for the official document appointing the prince, Bensaude, J., *L'astronomie nautique au Portugal* (Berne, 1912–1917. Reprint Amsterdam, 1967) pp. 271–274 and Brásio, A., *Monumenta Missionária Africana* (Lisbon, 1954), IV, pp. 12–15.

37 Sebastiano Crinò's alleged identification of Toscanelli's map with the so-called Genoese planisphere of 1447 has been vigorously refuted by a number of scholars. *Cf* Crinò, Sebastiano, *La Scoperta della Carta originale di Paolo dal Pozzo Toscanelli* (Florence, 1941). His identification has been refuted by R. Biasutti, 'Il mappamondo del 1457 [sic] non è la carta navigatoria di Paolo dal Pozzo Toscanelli', in *Rivista Geografiche Italiano*, 49 (1942), pp. 44–54; by H. Winter, 'Die angebliche Toscanelli Karte', in *Koloniale Rundschau,* 33 (1942), pp. 228–38; and by A. Magnaghi, *Tutto è chiaro finalmente* (Turin, 1941).

38 The letter is reproduced with an English translation in Vignaud, H., *Toscanelli and Columbus* (London 1902), pp. 275–292.

39 Vignaud, *op. cit.*, pp. 36–51.

40 *Cf* the criticisms of Vignaud's thesis by Hermann Wagner, in the *Göttingische gelehrter Anzeiger*, 2 (Berlin, 1902) pp. 108–13 1; by Lucien Gallois, 'Toscanelli et Christophe Colomb', in *Annales de Géographie*, 11e Année, No. 56 (1902), pp. 97–110; *Idem.* in *Annales de Géographie*,

vincing reconstruction still remains that of Hermann Wagner in 1894.[41] According to Wagner, the map was drawn on the projection of Marinus of Tyre, i.e. with the meridians parallel to one another, cutting rectilinear parallels at right angles.[42]

All that we know of Marinus of Tyre comes to us in the criticisms of his cartographical method given by Ptolemy in his *Geography*. Of all the cartographical projections described by Ptolemy, only Marinus' was capable of being used by sailors of the period and of this Toscanelli seems to have been fully aware. Thus did Toscanelli write of his map: '... the straight lines, therefore marked lengthwise in the chart [i.e. meridians] show the distances from east to west, but those which are transverse [i.e. parallels] show the spaces from south to north'.[43] He gave the distance in a straight line from Lisbon to China as almost a third of the circumference of the earth.[44]

From the data contained in the letter, Wagner deduced that the straight line from Lisbon to China is the same as the median parallel, of the map (in terms of the Marinus projection) and he further deduced that Toscanelli, following Regiomontanus (1436–1476), had attributed a value of 41 degrees for the latitude of Lisbon[45] (the correct value being 38° 44') (Fig. 3). On the parallel of Lisbon, each degree, according to Toscanelli, corresponds to 50 miles, which implies, at the latitude of 41 degrees, an equatorial degree of 66²/₃ miles or of 16²/₃ leagues (the nautical league being equal to four miles.[46]

For Wagner, the origin of the value of 66²/₃ miles to a degree is to be found in the *De Crepusculis*,[47] a work once attributed to Alhazen (ibn al-Haytham)

XXXVII (1928), pp. 207–8; S. Ruge, 'Die Echtheit des Toscanelli-Briefs', in *Zeitschrift des Ges. für Erdkunde zu Berlin* (1902), pp. 498–511. The only writers who persist in defending Vignaud are Th. Simart, 'Le crépuscule d'une légende, in *Bulletin de la Société Royale de Géographie* (Antwerp, 1914), xxxviii, pp. 5–62 and A. Cioranescu, 'Portugal y las cartas de Toscanelli', in *Estudios Americanos*, XIV (1954), pp. 1–17. Neither is convincing and the latter's claim that the Toscanelli correspondence was fabricated by the Portuguese carries not a shred of evidence.

41 Wagner, Hermann, 'Die Rekonstruktion der Toscanelli Karte vom Jahre 1474 ... ', in *Nachrichten der K. Gesellschaft der Wissenschaften zu Göttingen* (Phil.-Hist. Kl., 1894), pp. 208–312. The first to attempt to reconstruct the map (with converging linear meridians) was O. Peschel in 1867 (*Cf* [O. Peschel], 'Die Schiffbuch des Entdeckers von America bei seiner uberfahrt über das atlantische Meer' in *Das Ausland*, 40 (1867), pp. 1–4. Peschel's same reconstruction was adopted by K. Kretschmer, *Die Entdeckung Amerikas-Atlas* (Berlin, 1892), Tafel, 1.

42 Wagner, H., *art. cit.*, pp. 238–242.

43 Toscanelli's letter in Vignaud, H., *op. cit.* p. 280.

44 Idem., *op. cit.* pp. 287 and 289.

45 Wagner, H., *art. cit.* pp. 250–251; Regiomontanus, *Ephemerides sive Almanach perpetuum* (Venice, 1496). We have not been able to consult the earlier editions of Regiomontanus' work.

46 Wagner, H., *art. cit.*, pp. 257–258.

47 *Cf* the edition of Alhazen's *De Crepusculis* in Nunes, Pedro, *Obras* (Lisbon, 1943, II, p. 131. *Quantitas circuli magni continentis [terram] secundum quod dixerunt sapientes [...] est vigenti quatuor mille milliaria* (24,000 divided by 360 = 66²/₃).

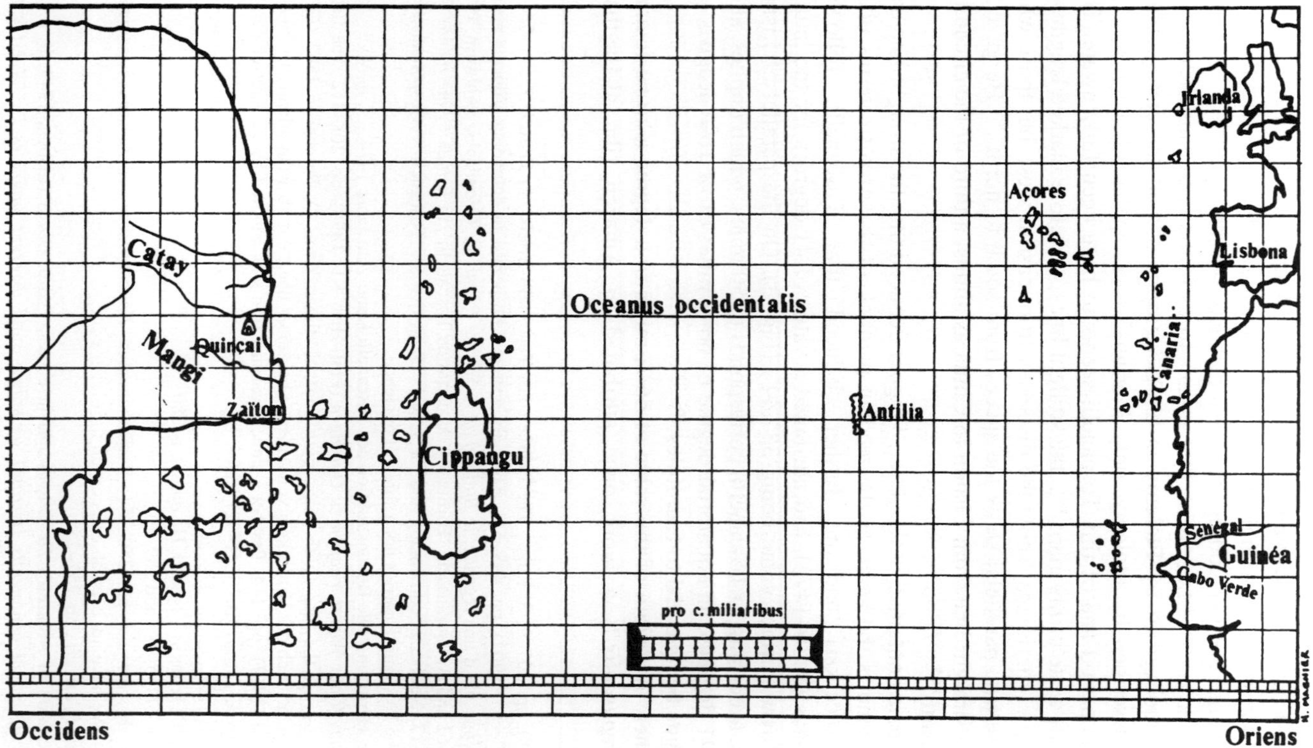

Fig. 3. Reconstruction of Toscanelli's map by Hermann Wagner, 'Die Rekonstruktion der Toscanelli-Karte vom Jahre 1474 ... ' in *Nachrichten der K. Gesellschaft der Wissenschaften zu Göttingen*, Phil.-Hist. Kl. (1894), p. 313. Within the framework of Marinus of Tyre's cylindrical projection, Toscanelli placed the European coastline and the Atlantic islands drawn from portulan charts and the coastline of Asia with Japan (Cypangu) taken from the account of Marco Polo's travels.

(965–1039), though now attributed to Abu ibn Muʿadh of Seville (2nd half of the eleventh century).[48]

Three different values are commonly found among the Arabs for the value of a degree of the meridian: 56²/₃ miles, 66²/₃ miles and 75 miles, the miles being Arab miles, which the Latin translators had carelessly translated as Roman miles, considerably shorter.[49]

The Venetian cartographer Fra Mauro in his World Map of 1459, quotes various opinions on the value of the degree of the meridian: 56²/₃ miles, 66²/₃ miles and 62²/₃ miles.[50]

The latter value is attributed to Ptolemy by Martianus Capella (sixth century A.D.)[51] and was often repeated during the Middle Ages.

If the value of 56²/₃ miles to a degree was adopted by Columbus, who took it from Pierre d'Ailly,[52] who in turn took it from Alfraganus,[53] that of 66²/₃, or 16²/₃ leagues, was a value commonly used by Portuguese sailors in the fifteenth century and in particular by Bartolomeu Dias.[54] Thus the value used by the theoretically minded Toscanelli can be seen to have coincided with that used in practice by Portuguese sailors, though there is no way of knowing whether there may have been influence of the one on the other, either way.

Between Lisbon and Quinsay, a city which Marco Polo had placed on the coast of China (identified as the modern city of Hangzhou, Lat. 30° 18' W. and Long 120° 07' E.), Toscanelli wrote that his map showed 26 'spaces', each measuring

---

48 *Cf* Sabra, A. I., 'The Authorship of the *Liber de Crepusculis*' in *Isis*, LVIII (1967), pp. 77–85.

49 Nallino, C.A., 'Il valore metrico del grado di meridiano secondo i geografi arabi', in Idem, *Raccolta di scritti editi ed inediti* (Rome, 1944), I, pp. 409–59.

50 Fra Mauro, *Il Mappamondo di Fra Mauro* (facsimilé edition by T.G. Leporace and R. Almagià, Rome, 1956), pp. 19 and 62.

51 *Cf* Martianus Capella, 'The Marriage of Philology and Mercury,' in Stahl, W.H., and Johnson, R., *Martianus Capella and the Seven Liberal Arts*, II (New York, 1977), p. 228.

52 *Cf* the marginal notes of Columbus reproduced by George E. Nunn *The Geographical Conceptions of Columbus* (New York, 1924, Reprint 1977), pp. 9–10 and Pierre d'Ailly, *Imago Mundi*, ed. E. Buron (Paris, 1930) II, p. 531 (*Epilogus Mappe Mundi*).

53 Alfraganus (al-Farghānī) (d. after 861), *Differentie scientie astrorum*, ed. Frank C. Carmody, Berkeley (California, 1943), Cap. 8 pp. 13–14. A degree of 56²/₃ miles corresponds to 83.87 kms. *Cf* J. Keuning, 'Anschauungen über die Grösse der Erde im Zeitalter der Entdeckungen und ihre Beziehungen zu den älteren Gradmessungen', in *Comptes rendus du Congrès International de Géographie* (Amsterdam, 1938), II, p. 31.

54 *Cf O Livro de Marinharia de André Pires* (*c.* 1530), ed. Luís de Albuquerque (Lisbon, 1963), pp. 218–19, and A. Teixeira da Mota, 'Bartolomeu Dias e o valor do grau terrestre' in *Congresso Internacional de História dos Descobrimentos* (Lisbon, 1961) II, pp. 299–309. The degree of 16²/₃ leagues is equal to 96.666 kms, whereas the true value of the degree is 111.111 kms (*Cf* Fontoura da Costa, A., 'La lieue marine des Portugais aux XVe et XVIe siècles', in *Comptes rendus du Congrès Internacional de Géographie* (Amsterdam, 1938), II, p. 7.

250 miles. From these figures Wagner deduced that a 'space' is equal to 5 degrees and that the coast of Asia is situated at 130° from Lisbon, or 10 degrees more than 'almost a third of the circumference of the earth', as Toscanelli had said in his letter, the difference being, according to Wagner, explainable by Toscanelli's use of the Latin *fere* ('almost'). As a result, Toscanelli's *oikumene* stretched over 230 degrees, just 5 degrees more than that of the 225 degrees given it by Marinus of Tyre. Such a close correspondence led Wagner to call Toscanelli 'Marinus redivivus'.[55]

As Wagner noted, Toscanelli's letter and map constituted the first attempt to put at the disposition of sailors, a chart on which the circumference of the earth and the distance between Europe and Asia were expressed in units of distance which they were accustomed to use in practice.[56]

Was Toscanelli's advice to the Portuguese in favour of a route westward to Asia, advice that undoubtedly had its effect on Columbus (even if he did not follow it to the letter), ever at any time taken seriously by the Portuguese authorities? This seems to us doubtful. According to the account of the Portuguese historian João de Barros, D. João II's cosmographers, when they examined Columbus' project in 1483–84, seem to have directed their criticisms less against a cartography based on Marinus of Tyre, than on what they regarded as the absurdities of Marco Polo, since all three agreed that Columbus' ideas were based on 'the imaginary things that Marco Polo recounted of the island of Cypango [Japan]'.[57]

The Portuguese thus put aside Toscanelli's proposal and concentrated all their attention on the route round Africa, but their progress in this direction was hampered by repeated setbacks, to the extent that there followed, after Diogo Cão's second voyage along the African coast (1486), a period of discouragement and uncertainty over whether the route round Africa was really practicable.

Up until then (1486), the Portuguese were persuaded that the shortest and the only route to Asia was round Africa, since their goal was neither China nor Cypango, but the shores of the Indian Ocean between Ethiopia and the coast of Malabar in India, a region where it was believed lay the Kingdom of Prester John. The historian João de Barros recalls the considerable outlay in expenses that the King D. João II had incurred in order to discover the King of the Abyssinians, identified as Prester John. In the instructions later given to Vasco da Gama, ' ... almost all consisted in learning of the political power of this Prince'.[58]

In Ptolemy's *Geography*, D. João II's cosmographers could have seen that the distance between the meridian of the Fortunate Islands (the Canaries) and the

---

55 H. Wagner, *art. cit.* pp. 225–30 and pp. 236–42.

56 H. Wagner, *art. cit.* p. 226.

57 João de Barros, *Ásia*, Lisbon, 1552, Déc. I, Liv. III, Cap. IX, ed. (Coimbra, 1932), p. 113.

58 João de Barros, *Terceira Década* (Lisbon, 1562), Liv. IV, Cap. I éd. (Lisbon, 1946), III, p. 165.

meridian passing through the northern tip of the island of Taprobana (Sri Lanka) was hardly more than 125 degrees,[59] five degrees less than the distance evaluated by Toscanelli between Lisbon and China.

Between 1460 and 1472, the discovery of the east-west extension of the Gulf of Guinea had considerably reduced Ptolemy's 125 degrees. At the end of 1485, or at the beginning of 1486, probably before the return of Diogo Cão from his second voyage (1486),[60] João Afonso de Aveiro returned to Lisbon with the news that at the bottom of the Gulf of Guinea he had discovered the Kingdom of Benin, where he had learnt that 250 leagues to the east, lay a potentate named the Ogané. The historian João de Barros, who describes this discovery, tells how the King D. João II and his cosmographers examined Ptolemy's map of Africa and taking account of the way in which the discoverers had placed the *padrões* (stone pillars used to mark the limits of Portuguese discovery) along the coast, as well as the distance of 250 leagues to the east, where the inhabitants of Benin said lay the kingdom of Ogané, they considered that he must be Prester John,[61] since both lived hidden behind silk curtains (an allusion to a ritual of divine kingship) and held the Cross in great veneration. 'And it seemed to them that, if the King's ships continued to follow the coast that they had been discovering, they could not but reach the *Promontorium Prassum*, which lay at the extremity of the continent.[62]

Ptolemy had placed the *Promontorium Prassum* on the East Coast of Africa in 15 degrees Lat. S, and here it should be emphasized that in this period no one in Europe any longer believed the Ptolemaic doctrine of a closed Indian Ocean without communication with the Atlantic. Fra Mauro on his World Map of 1459 specifically stated that the Indian Ocean was not a *stagnon* (a closed sea).[63]

The optimistic calculations of the Portuguese cosmographers were to be belied by the failure of Diogo Cão to round the extremity of the continent on his second voyage (1486), which revealed that the African continent extended much further southward, and far beyond what had been expected, as far as the *Serra Parda* (Ponta dos Farilhões in 22° 10' Lat. S.)[64] As a result, the distance to be sailed to reach India became considerably increased. This could only but encourage those who believed the route across the Atlantic to be shorter.

---

[59] Claudius Ptolemy, *Géographie*, Bk. 1, Chap. XIV, French translation by the Abbé Halma (Paris, 1828), p. 40.

[60] Fontoura da Costa, A., *Às Portas da Índia em 1484*, Lisbon, 1936, p. 10 and p. 29.

[61] Prester John had been placed in Ethiopia from 1306 onward. *Cf* R.A. Skelton, 'An Ethiopian Embassy to Western Europe in 1306', Appendix III in O.G.S. Crawford, *Ethiopian Itineraries, c. 1400–1524*, Hakluyt Society, 2nd series, 109 (London, 1958), pp. 212–15.

[62] João de Barros, *Ásia*, Lisbon, 1552, Déc. 1, Liv. III, Cap. IV (ed. Coimbra, 1932), p. 84.

[63] Fra Mauro, *Il Mappamondo di Fra Mauro*, Facsimilé ed T.G. Leporace and R. Almagià (Rome, 1956) XI, p. 27.

[64] Fontoura da Costa, A., *op. cit.*, p. 20.

The result of the next Portuguese voyage south along the African coast, that of Bartolomeu Dias (1488) proved even more discouraging for D. João II and his cosmographers. Though it revealed that the Cape of Good Hope represented the southern extremity of the continent, (Dias turned back just beyond it), there still remained no proof that the continent did not continue to develop eastwards to end in a *Promontorium Prassum* as yet undiscovered, perhaps even more difficult to round than the Cape itself. This hypothetical undiscovered eastwards pointing cape, a relic of Ptolemy's *terra incognita* bordering the southern edge of the Indian Ocean, can be seen outlined on the map of Henricus Martellus *c.* 1490, and it was only with Vasco da Gama's voyage nine years later in 1497, that its non-existence was finally settled.

In an Oration given in Beja (Portugal) in 1489, the year after Dias' voyage, João Teixeira declared that ' ... each day we strive to reach the Promontories *Raptum* and *Prassum* [...] and from there to the approaches of the Indian Ocean,'[65] a proof that at that date, it was admitted that the promontories had not yet been discovered.

Dias had passed the Cape in a storm which had left terrible memories among his crew, and he himself had named it the *Cabo Tormentoso* (Cape of Storms), a name which according to Barros, the King D. João II changed to 'Cape of Good Hope', by reason of the promise it held for the discovery of India.[66]

Yet as Barros notes, such was ' ... the fury of the seas' round the Cape of Good Hope that in the opinion of the sailors, ' ... there arose a new myth of the dangers as there once had been concerning Cape Bojador.'[67]

Further evidence of the climate of discouragement and uncertainty, perhaps more among the sailors, than with the King himself, regarding the practicability of the sea route to India, appears in Fernando Columbus' ironic comment that D. João II, ' ... had not been able to go beyond the Cape of Good Hope and that it had been thus called because it marked the end and termination of his hopes of discovery and conquest (*per esser quello il capo e il fine della buona speranza della sua con-*

---

65 João Teixeira, *Oratio habita ab insigni viro Joanne Teyxeira* (1489), Coimbra, 1562, sign. A. viii (r). It is true that Duarte Pacheco Pereira wrote *c.* 1508, that after having passed the Cape of Good Hope, Dias saw that ' ... the coast here turned northwards and northeastwards towards Ethiopia [...] giving great hope of the discovery of India' (*Cf* Duarte Pacheco Pereira, *Esmeraldo de Situ Orbis*, Bk III, Chap. VII, translated by G.H.T. Kimble, Hakluyt Society, 2nd series, 79 (London, 1937) p. 154. But Pereira wrote after Vasco da Gama's voyage which had proved that the Cape was the only cape of importance. For Dias the way to India was still not so clearly established.

66 Barros, João de, *Ásia*, Déc. 1, Liv. III, Cap. IV, *ed. cit.* p. 87.

67 Barros, João de, *op. cit.*, Déc. 1, Liv. Ill, Cap. I *ed. cit.* p. 112. Gaspar Correa and Fernão Lopes de Castanheda confirm Barros on this point. Gaspar Correa, *Lendas da Índia* (Lisbon, 1858), I, p. 8 and Fernão Lopes de Castanheda, *História do Descobrimento & Conquista da India pelos Portugueses* (Lisbon, 1555) Liv I, Cap. 1, p. 2. (Copy of the 1st edition consulted on a microfilm from the British Library).

quista e discoprimento)'. Others, according to Fernando Columbus, did agree that the Cape had been given the name because of the hope it brought of 'pleasanter lands and of easier sailing'.[68]

At this juncture, and even before Dias' return in December 1488,[69] D. João II seems to have turned anew to Columbus, for he wrote him an affectionate letter dated 20 March 1488, guaranteeing him a safe conduct back to Portugal, since he had left secretly for Spain without the King's authorization. D. João II added that 'he needed his energy and intelligence' and that 'it would give great pleasure if he were to come and in that which concerned him, dispositions would be taken which would make him content'.[70] It is unlikely that in thus encouraging Columbus, D. João II was preparing to help him realize his dream of sailing westward to Asia. It is probable that he was thinking more explicitly of the discovery of unknown lands in the Atlantic. Columbus, on his third voyage (1498), was to write that he wanted to sail southwards (into the southwestern Atlantic) to see 'what had been the intention of D. João II of Portugal, who had said that to the south there was mainland'.[71]

Columbus accepted the invitation to return to Portugal and was present in Lisbon when Dias made the report of his voyage to the King. In a marginal note in his copy of Pierre d'Ailly's *Imago Mundi*, Columbus described how Dias had reached a cape named by him Cape of Good Hope, 'and that by the astrolabe he had found himself in that place to be 45 degrees beyond the equator'.[72] The Cape of Good Hope being in Lat S. 34° 50', the figure of 45 degrees is a manifest error and has given rise to much discussion.[73] It is however quite possible that the error was in fact Dias' and

---

68 Columbus, Fernando, *Historie, ed. cit.*, I, pp. 92–93.

69 Fontoura da Costa, A., *Às Portas do Índia en 1484* (Lisbon, 1936) p. 37.

70 Letter from D. João II to Christopher Columbus, 20 March 1488 in João Martins da Silva Marques, *Descobrimentos Portugueses*, III (1461–1500), Lisbon, 1971, p. 371.

71 Las Casas, Bartolomé de, *Historia de las Indias, ed. cit.* Liv. I, Cap. CXXX, p. 350.

72 *Cf* the reproduction of the note in, Pierre d'Ailly, *Imago Mundi*, reproduced by Nunn, George E., *The Geographical Conceptions of Columbus*, pp. 7–8. The authenticity of this marginal note has been the object of much polemic. Las Casas claimed that it had not been written by Columbus, but by his brother Bartolomé. (Las Casas, *Historia de las Indias, ed. cit.* Liv. I, Cap. XXIX, p. 110). Las Casas' opinion is judged to be groundless by A. Magnaghi, in Questioni colombiane (Naples, 1939), pp. 18–19), and by S.E. Morison, in *Admiral of the Ocean Sea* (Boston, 1942), I p. 106, note 29). Rinaldo Caddeo takes, without reason in our opinion, the opposite view claiming that not only did Bartolomé Columbus write the note, but even sailed with Dias on his voyage! *Cf* Caddeo, Rinaldo, Appendix H in Colombus, Fernando, *Historie, ed. cit.* II pp. 366–72.

73 Ravenstein accepted the figure of 45 degrees as a latitude observed at sea beyond the Cape (Ravenstein, E.G., 'The voyages of Diogo Cão and Bartolomeu Dias (1482–1488) in *Geographical Journal*, 1900 (off print of 31 pages). Bensaude and Fontoura da Costa held the figure to be a gross error imputable to Columbus. (*Cf.* Bensaude, J., *L'astronomie nautique au Portugal* (Bern, 1912/17. Reprint Amsterdam, 1967) p. 109 and Fontoura da Costa, A., *A Marinharia dos Descobrimentos*, 2nd ed. (Lisbon, 1939), p. 37 n. 48). Jaime Cortesão claimed that the figure was deliberately invented by D. João II to mislead Columbus into believing that the Cape route was really much longer than it was! *Cf*

that at this date the Portuguese had still not thoroughly mastered the art of measuring latitude from the height of the sun.[74]

Columbus, at the news that the African landmass extended 45 degrees into the southern hemisphere, would have seen in this a proof of his conviction that the distance between Europe and Asia across the Atlantic was small, for if the *oikumene* had such a large north/south extension, it must necessarily have the same east/west extension, given that in the cartography of the time it was assumed to be circular in shape.

It may then have been the figure of 45 degrees, discouraging for the Portuguese, but encouraging for Columbus, that made him hurry off again to Spain, where his project was examined once more at the *Junta* of Santa Fé in 1491. He was to hear again brought out against him the arguments based on Paul de Burgos, which the Italian Geraldini, who was present, relates. Then when all seemed lost, suddenly the Spaniards became completely won over to Columbus' ideas. The Cardinal Pedro Gonzalez de Mendoza, to whom Geraldini at the *Junta* had remarked that the Portuguese in their voyages into the southern hemisphere had invalidated the doctrine of Paul de Burgos, now 'accepted', according to Gonzalo Fernandez de Oviedo, 'to give audience to Columbus and he saw that he was knowledgeable and spoke well and that he supported his ideas with sound reasoning. The Cardinal judged, that he was a man of intelligence and very clever, and having reached this opinion, he held him in high esteem and chose to support him [...] As a consequence of this he was listened to by the King and the Queen who began to accord credit to his memoranda and petitions'.[76]

What new factor in the situation had emerged for the Spaniards to become so suddenly persuaded that Columbus' project was feasible and that he should be given men and ships to accomplish it? This point is crucial and obscure, and in the absence of documents, we can but make conjectures.

Nothing in the geographical science of the time could justify to the Spaniards Columbus' choice of $56^2/_3$ miles to a degree, rather than that of $66^2/_3$, adopted by

---

Cortesão, Jaime, *A Política do Sigilo* (Lisbon, 1960), pp. 50–51. That the Cape of Good Hope stood in a latitude of 45 degrees was a notion that spread into European cartography. The Cape is thus shown on the gores of Waldseemüller's globe in his *Cosmographiae Introductio* (St Dié, 1507). The Polish cosmographer Matthaeus Shamotuly read this very figure off the map and placed it in the text of his commentary on Sacrobosco (Shamotuly, Matthaeus, *Ioannis de Sacrobosco astronomi celeberrimi sphericum opusculum* ... (Cracow, 1521) f. 54 (r).) Only known copy in the Huntingdon Library, San Marino, California.

[74] The first clear proof of their mastery of the technique of measuring latitude comes in the figures given by Duarte Pacheco Pereira (*c.* 1508). Pereira, Duarte Pacheco, *op. cit. passim.*

[75] Geraldini, Alexandro, *op. cit. loc. cit.* (see note No. 23).

[76] Fernandez de Oviedo, Gonzalo, *Historia General y Natural de las Indias* (Seville, 1535), Liv. II, Cap. IV (ed. B.A.E., Madrid, 1959), 117, p. 22).

the Portuguese and by Toscanelli.[77] Nothing in contemporary knowledge of Atlantic navigation could justify to the Spaniards his conviction that he would find mainland between 700 and 750 leagues west of the Canary islands.[78]

No direct information has reached us concerning the kind of world map that Columbus used.[79] Roberto Almagià suggested in 1940 in a clever article, that a map similar to Henricus Martellus' World map of *c.* 1490 is likely to have circulated at the end of the fifteenth century and that it probably served as a model for Columbus.[80] Almagià wrote before the discovery and acquisition in 1961 by the Yale University Library, of a hitherto unknown world map by Henricus Martellus (Fig. 4), which, unlike the other known copies, is graduated in latitude and longitude (Fig. 5). R.A. Skelton, the only expert to have examined this map carefully, described it as representing an *oikumene* of 270 degrees in extent, leaving a distance of 135 degrees between Quinsay and Lisbon and a distance of 90 degrees from the base meridian in the Canary islands to Cypango (Japan). These distances, according to Skelton, 'coincide almost exactly with the distances which Toscanelli deduced from Marinus of Tyre and from Ptolemy'.[81] There is no direct proof that Columbus ever saw a copy of the Martellus map, but when he reached Cuba, he

---

77 Columbus in one of his marginal notes reproduced by George Nunn claims to have checked experimentally that the degree of the meridian corresponded to $56^{2}/_{3}$ miles and that Master Joseph [Vizinho] whom the King D. João II had sent to Guinea to make astronomical observations of latitudes in the region, had also confirmed this figure. See Nunn, George E., *Geographical Conceptions of Columbus* (New York, 1924, Reprint, New York, 1977) pp. 9–10. Nunn's demonstration of how Columbus had performed the verification of the figure is vitiated by his misreading of the latitude of the Los Idolos islands referred to in one of Columbus' notes: '1° 5' N', instead of '5° ... minutes N'. His mistake was pointed out by A. Magnaghi, 'I presunti errori che vengono attributi a Colombo nella determinazione delle latitudini', in *Bollettino Soc. Geog. Ital.* (1928), pp. 462–65. We do not necessarily follow Magnaghi in the other conclusions of his article. R.A. Laguarda Trias, in *El enigma de las latitudes de Colón* (Valladolid, 1974), p. 37, is also of the opinion that Columbus could not have confirmed experimentally the figure of $56^{2}/_{3}$ milles/degree.

78 Columbus, Fernando, *Historie, ed. cit.* I, Cap. XXI, p. 159, (700 leagues) and Las Casas, Bartolomé de, *Historia de las Indias*, Liv. I, Cap. XXXIX, ed. B.A.E. Tome XCV, p. 139 (750 leagues). A repetition of this 'mysterious' distance can be found in a Portuguese rutter dated to *c.* 1495: 'Este livro he de rotear', reproduced in *O Manuscrito de Valentim Fernandes* (Lisbon, 1940), p. 230. (*Sabe q̃ jazẽ as Antilhas com o Ferro leste e hoeste E ha na rota vij*^{c} *leguas largas*).

79 Morison and Nunn's attempts to analyse the cartographical bases of Columbus' conviction of the short distance he had to sail are not entirely convincing, especially Morison's, and lead to differing conclusions (*Cf* Morison, S.E., *op. cit. ed. cit.* pp. 86–91 and pp. 103–104 and Nunn, George E., *op. cit. ed. cit.* pp. 25–30). Much depends on the estimation of the number of degrees of longitude by which interpreters of Marco Polo's travels are said to have extended Asia eastwards beyond the limits assigned by Ptolemy.

80 Almagià, Roberto, 'I mappamondi di Enrico Martello e alcuni concetti geografiche di Cristoforo Colombo', in *La Bibliofilia*, XLII (1940), pp. 228–311.

81 Skelton, R.A., note in Destombes, Marcel, *Mappemondes A. D. 1200–1500* (Amsterdam, 1964), pp. 229–33.

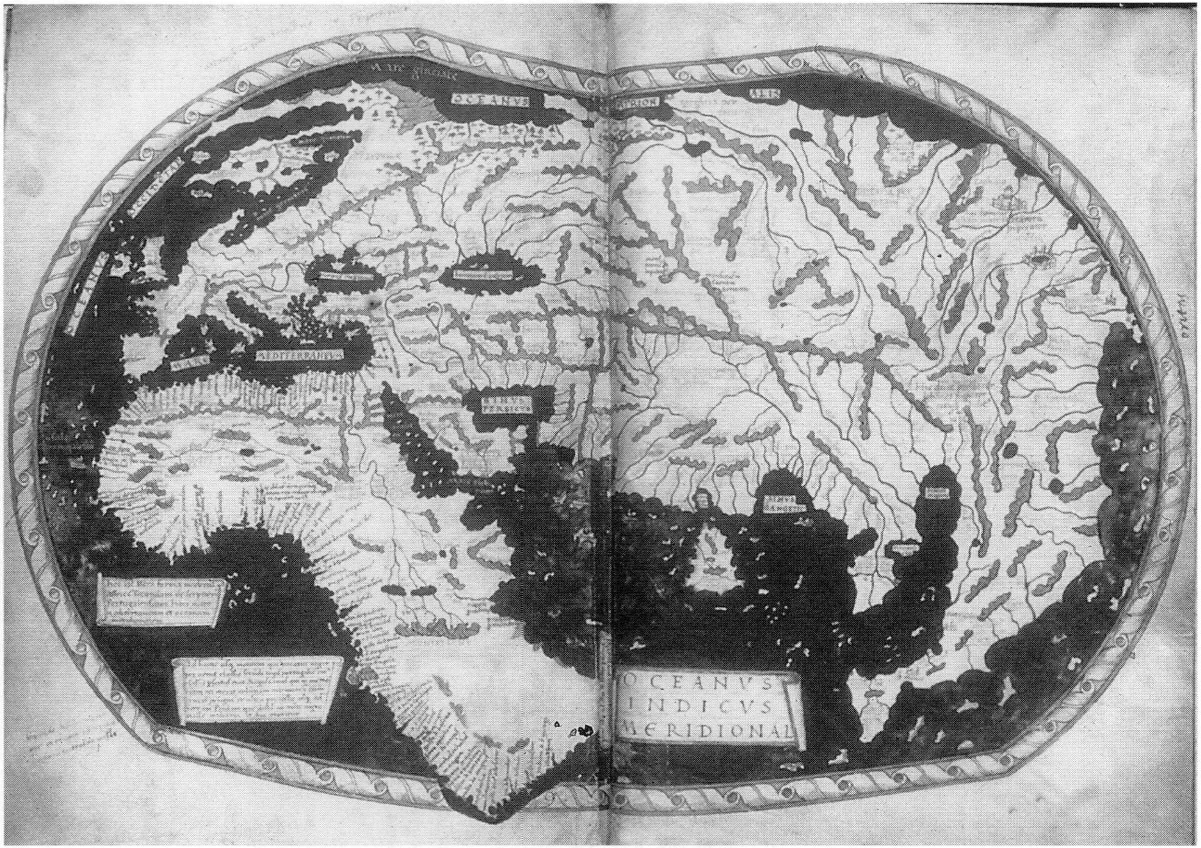

Fig. 4. World Map (*c.* 1490) of Henricus Martellus. *By courtesy of Yale University Library New Haven.*

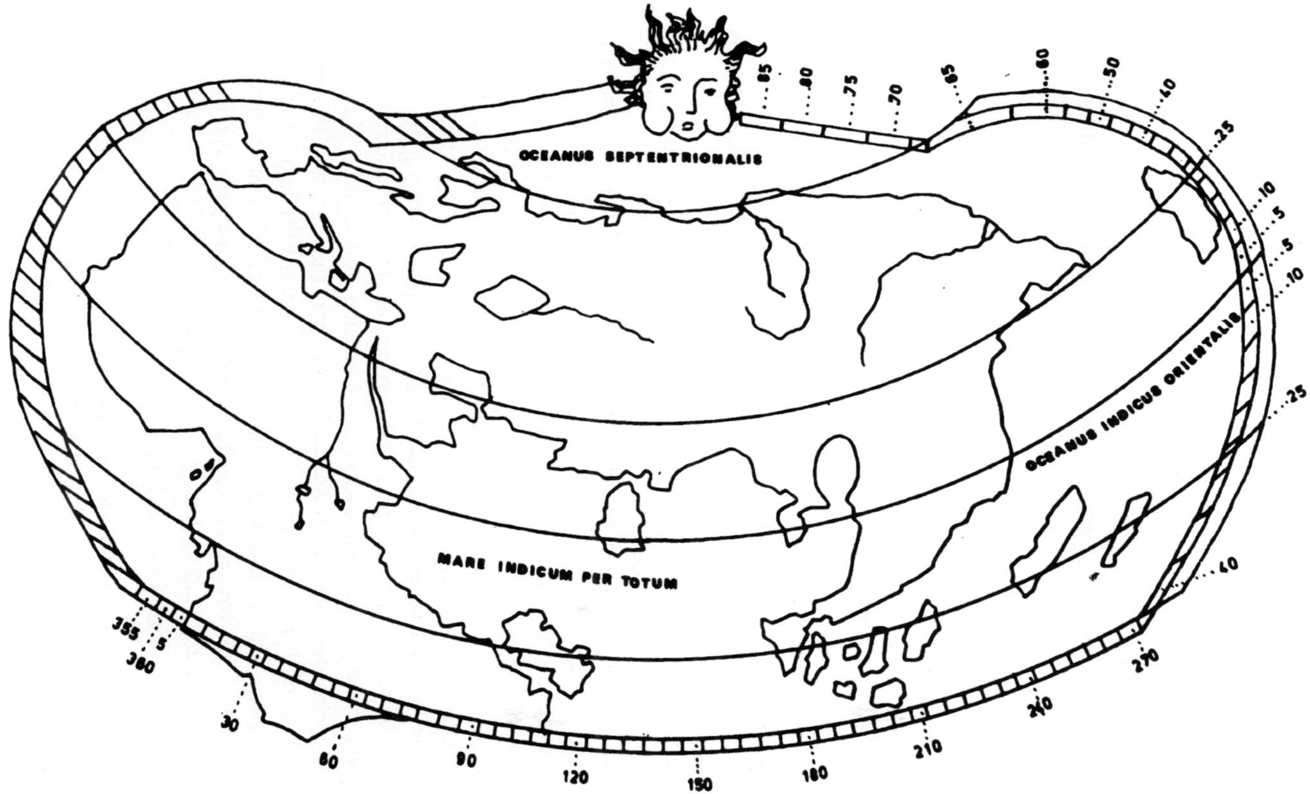

Fig. 5. A simplified reconstruction of the Yale Marrellus Map to show scales of latitude and longitude.

became convinced that the Caribbean island was in fact Cypango,[82] and he estimated that he had sailed 1142 leagues from the Canaries to Cuba.[83] On the Marinus projection of Toscanelli's map, on which the median parallel contains 50 miles to a degree, a distance of 1142 leagues becomes equivalent to 91¹/₃ degrees, just one degree and a third more than the distance between the Canaries and Cypango on the Yale Martellus Map.

This would all fit together very well, were it not that the Yale Martellus map is drawn on Ptolemy's second projection, quite unsuitable for sailors and that Toscanelli's map, according, to Wagner, was based on a module of 66²/₃ miles to a degree, and not 56²/₃, which was Columbus' choice. Yet it remains possible that Columbus only repeated his preference for the figure of 56²/₃, miles/degree to emphasize his conviction of a small globe and of a short distance between Europe and Asia. It is exceedingly doubtful if he ever applied the module in practical cartography, probably contenting himself with using, in traditional Mediterranean fashion, the kind of chart that Toscanelli had prepared. He may not even have realized that its module was different from his.

Whatever kind of chart Columbus had with him, when he faced the Salamanca and Santa Fé *Juntas*, it is doubtful if the Spanish cosmographers would have found in it anything sufficient to shake their scholastic prejudices.

With Bartolomeu Dias' voyage there emerged however new empirical evidence concerning the size of the *oikumene* with experimental proof that it covered a very large surface of the globe and that as a consequence the distance between Europe and Asia was conceivably much smaller than the scholastic philosophers had been claiming it to be. It is this fact that may have swung the minds of the Spanish authorities and led them to give their support to Columbus, support all the more readily given in view of the increasing discouragement of the Portuguese that they would ever achieve their hope of reaching India by circumnavigating the African continent.

Neither the Portuguese nor the Spanish cosmographers were ignorant or stupid men.[84] They were aware of the principal currents of scientific thought in Europe, even if some were slower than others in embracing new trends. They could not know, even when they were disposed to follow their theories, that Ptolemy and *a fortiori* Marinus of Tyre had extended the *oikumene* some 50 degrees too far to the east and that even without the interposition of the American continent, unforeseen and unforeseeable, the voyage across the Atlantic from Europe to Asia would have proved too long for any ship of the time.

---

82 *The Journal of Christopher Columbus*, translated by Cecil Jane, revised and annotated by L.A. Vigneras, with an appendix by R.A. Skelton (London, 1960), pp. 42–43.

83 *Ibid.* p. 51.

84 Such as Washington Irving described them. *Cf* Irving, Washington, *A History of the voyages of Christopher Columbus* (London, 1828), pp. 119–126.

In the end it was the Portuguese cosmographers who made the most careful and prudent evaluation of the geographical situation and it was the Portuguese who, after nine years of persistent preparation first reached, round Africa, the India they had so long sought, while the Spaniards came upon what they were to call New Spain.[85]

Acknowledgement

My deep gratitude is here expressed to Professor Léon Bourdon of the Sorbonne for his inexhaustable patience in helping me in the translation of the Latin, Portuguese and Spanish texts, often of treacherous difficulty.

---

85 The term 'America' was not used by the Spaniards until the eighteenth century.

X

LA CARTOGRAPHIE DE L'ATLANTIQUE A LA VEILLE DU VOYAGE DE CHRISTOPHE COLOMB

Au dernier tiers du XVe siècle, dans les milieux maritimes atlantiques, deux traditions cartographiques fondées sur des principes très différents et difficilement conciliables s'affrontaient.

L'une este celle de la carte portulan méditerranéenne, instrument de navigation des marins de la mer intérieure, dont on avait déjà prolongé, depuis le XIVe siècle, le domaine pour cartographier, au-delà des colonnes d'Hercule, les rives européennes de l'Atlantique et les îles adjacentes. La carte portulan est une carte sans projection, fondée uniquement sur des données de distance à l'estime et de direction fournie par l'aiguille aimantée, toutes étant rapportées directement à la surface de la terre.

L'autre tradition cartographique est celle de la *Geographie* de Ptolémée, inconnue au Moyen Age, et qui ne fut mise en circulation en Europe qu'au début du XVe siècle, lorsque le mécène florentin Palla Strozzi fit apporter de Byzance un manuscrit grec, qu'il fit traduire en latin en 1406 (1).

La carte ptoléméenne est fondée sur des coordonnées de latitude et de longitude, données fournies par l'observation du ciel. Pour reporter ces données à la surface de la terre, il faut disposer d'un module fixant le rapport entre distance terrestre et différence de distance angulaire séparant l'horizon d'un objet céleste déterminé: l'étoile polaire ou le soleil.

En Europe, au XVe et au XVIe siècle, il n'existait encore aucune méthode de vérification expérimentale pour établir ce rapport en unités de distance terrestre de l'époque, et l'on continuait à se fonder sur des modules hérités d'astronomes arabes à travers des traductions du Haut Moyen Age, sans se soucier du fait que, en raison de l'ignorance des traducteurs, on avait traduit l'unité de distance terrestre arabe (le mille arabe) par l'unité de distance terrestre romaine (le mille romain), beaucoup plus petite (2).

---

(1) G. R. Crone, *Maps and their Makers*, Londres, 5e éd., 1978, p. 35.

(2) C. A. Nallino, «Il valore metrico del grado di meridiano secondi i geografi arabi», in *Raccolta di Scritti editi ed inediti*, Vol. 5, Rome, 1944, pp. 295-305 et 413-416.

Au XVe siècle, deux modules hérités des arabes circulaient dans les milieux maritimes: 56 2/3 milles/degré (3) et 66 2/3 milles/ degré (4). Le premier sera adopté par Christophe Colomb (5), le second par les Portugais et notamment par Bartolomeu Dias (6).

Le premier à tenter d'élaborer une synthèse des deux types de cartographie évoqués fut l'astronome florentin Paolo dal Pozzo Toscanelli: il adoptera le module de 66 2/3 milles/degré (7).

Interrogé par l'intermédiaire du chanoine portugais Fernão Martins, Toscanelli répond à une demande de renseignements du roi de Portugal D. Afonso V sur la meilleure route à suivre pour atteindre le pays des épices, autrement dit l'Orient. Toscanelli adresse à son ami Fernão Martins (8) une lettre et une carte. La carte est perdue (9), mais la lettre avec une des-

(3) L'origine de ce module est chez Alfragan (al-Farghani) (fl. c.861), *Differentie in quibusdam collectis scientie astrorum*, trad. lat. Jean de Séville, 1135, ed. mod. F. J. Carmody, Berkeley, Calif., 1943, pp. 13-14. On retrouve le même module dans les milieux nautiques vénitiens au XVe siècle: cf. Andrea Bianco, *Atlas* (1436), éd. facsimilé d'Oscar Peschel, Venise, 1869 (dernière planche), et Fra Mauro, *Mappamondo* (1459), éd. facsimilé de T. G. Leporace, Rome, 1956, p. 19 (la circonférence de la terre = 20.400 milles, c'est-à--dire 56 2/3 milles/degré). Fra Mauro ajoute cependant plus loin (p. 62) qu'il y a diverses opinions à propos de la circonférence de la terre: 22.500 ou 24.000 milles; le premier chiffre correspond au module attribué à Ptolémée au Moyen Age, c'est-à-dire 62,5 milles/degré (cf. Martianus Capella, *The Marriage of Philology and Mercury*, in W. H. Stahl & R. Johnson, *Martianus Capella and the seven Liberal Arts*, Columbia University Press, N.Y., 1977, Vol. II, p. 228; le second chiffre correspond au module d'Alhazen (ibn al-Haytham), *De Crepusculis*, ed. lat. in Pedro Nunes, *Obras*, Lisbonne, 1943, Vol. II, p. 131. Aujourd'hui l'ouvrage d'Alhazen est attribué à Abu ibn Mu'adh de Séville (2e moitié du XIe siècle), cf. A.I. Sabra, «The authorship of the *Liber de Crepusculis*», in *Isis*, Vol. LVIII (1967), pp. 77-85.

(4) La source de ce module est l'astronome arabe Alhazen, ou plus précisément Abu ibn Mu'adh de Séville (voir note 3). Cf. H. Wagner, «Die Rekonstruktion des Toscanelli-Karte vom J. 1474», in *Nachrichten der K. Gesellschaft der Wissenschaften zu Göttingen*, Phil.-Hist. Kl. (1894), p. 260.

(5) Colomb l'avait repris de Pierre d'Ailly, *Imago Mundi*, Louvain (?) 1483 (?), qui à son tour, l'avait repris d'Alfragan. Cf. Apostilles de Colomb in *Raccolta Colombiana*, Rome, 1892-96, Pt. I, Vol. 2, p. 374, n.º 4, pp. 376-77, n.º 23, p. 378, n.º 30 et n.º 31.

(6) A. Teixeira da Mota, «Bartolomeu Dias e o valor do grau terrestre», in *Congresso Internacional de História dos Descobrimentos*, (1961), Vol. II, pp. 299-309.

(7) Cf. Hermann Wagner, «Die Rekonstruktion...», pp. 208-312.

(8) Qu'ils se connaissaient personnellement est attesté par le fait qu'ils ont tous deux assisté à la mort de Nicolas de Cuse en 1464. Cf. G. Uzielli, «Toscanelli, Colombo e Vespucci», in *Atti del IV Congresso Geográfico Italiano*, Milan, 1902, pp. 575-576.

(9) La thèse de Sebastiano Crinò, selon laquelle la carte de Toscanelli serait la même que la mappemonde dite gênoise de 1457, conservée à la Bibliothèque Nationale de Florence (Cf. S. Crinò, *La Scoperta della Carta originale di Paolo dal Pozzo Toscanelli*, Florence, 1941), a été réfutée par R. Biasutti, «Il mappamondo del 1457 non è la carta navigatoria di Paolo dal Pozzo Toscanelli», in *Rivista Geografica Italiana*, Vol. 49 (1942), pp. 44-54; par R. Almagià, «Cristoforo Colombo davanti la scienza» (1948, in *Scritti Geografici*, Rome, 1961, p. 589; par A. Magnaghi, *Tutto è chiaro finalmente*, Turin, 1941; et par Heinrich Winter, «Die angebliche Toscanelli-Karte», in *Koloniale Rundschau*, Vol. 33 (1942), pp. 228-238.

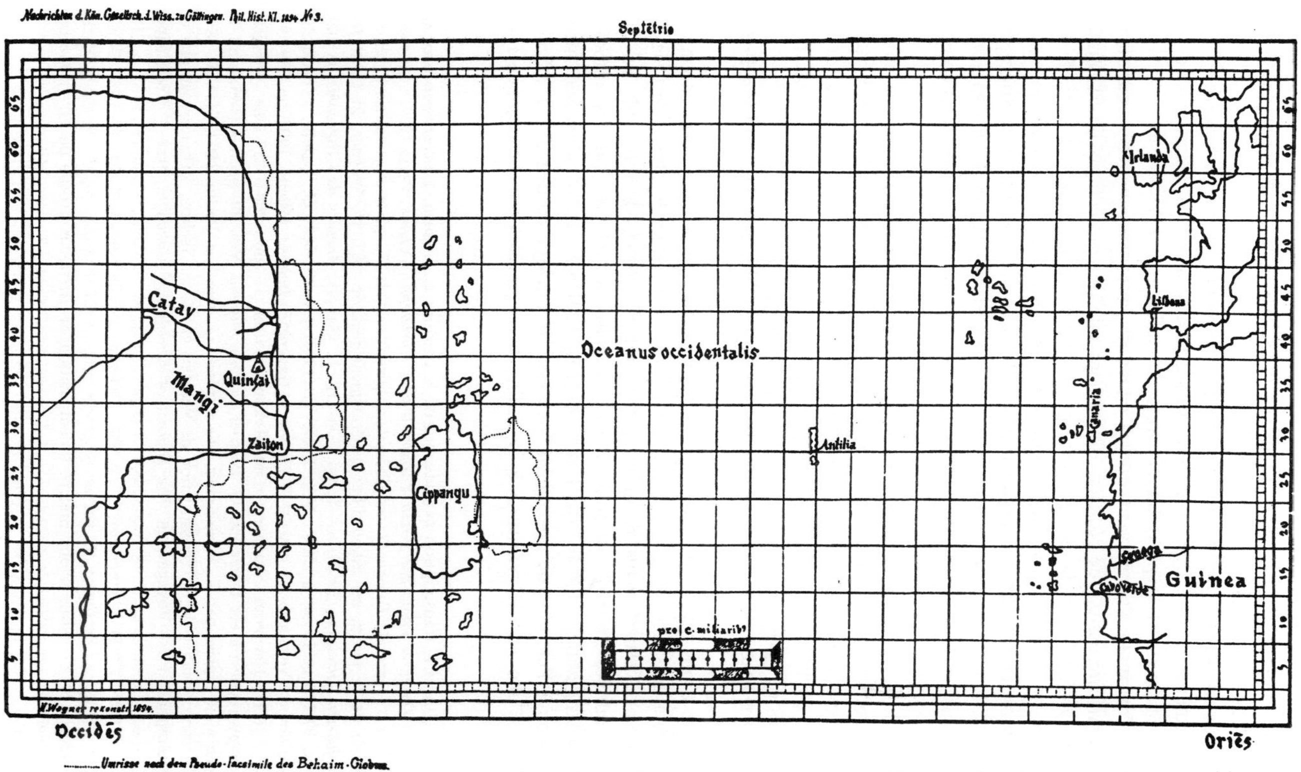

Reconstruction de la carte de Toscanelli par H. Wagner,(cf. note 4).

cription de la carte a été recopiée par Christophe Colomb sur l'une des feuilles vierges d'un cahier attaché à l'un de ses livres, l'*Historia rerum ubique gestarum*, du Pape Aeneas Silvius Piccolomini (10).

A partir de la description de la carte, Hermann Wagner a tenté, en 1894, de réaliser ce que l'on considère aujourd'hui comme la reconstruction la plus plausible et la plus convaincante de cette carte (11).

Toscanelli s'apercevant que, à cause de la convergence des méridiens, il serait impossible de réaliser, moyennant une des projections de Ptolémée, une synthèse avec la carte portulan, avait choisi la projection cylindrique de Marin de Tyr, où les méridiens sont tous parallèles les uns aux autres et coupent les parallèles à angles droits (12). Les principes de la cartographie de Marin de Tyr ne nous sont connus qu'à travers le texte de la *Géographie* de Ptolémée, où celui-ci les critique pour n'avoir précisément pas tenu compte de la convergence des méridiens.

Dans le cadre de la projection marinienne, Toscanelli a dessiné sur sa carte, entre les rivages des continents asiatique et européen, séparés par l'Atlantique sur une distance de 130 degrés, les îles de l'Atlantique figurant sur les cartes portulans (notamment celle d'Antilia) et des éléments de la géographie de l'Orient pris dans le récit de voyage de Marco Polo.

Au cours de son retour par bateau de la Chine aux îles de l'Insulinde, le voyageur vénitien avait démontré l'existence de la façade maritime de la Chine, alors que Ptolémée avait tout simplement étendu la terre inconnue au-delà de l'Asie indéfiniment vers l'est. La ville située le plus à l'est, selon Ptolémée, était Cattigara, qu'il a placée par 177 degrés de longitude E.. Si l'on adopte l'identification de Berthelot, selon lequel Cattigara correspond à Zhang-zhou, 117° 40' de Long. E. (13), Ptolémée aurait ainsi indûment prolongé l'étendue de l'*oecumène* vers l'est de quelque 60 degrés. L'adjonction d'une façade maritime à la Chine avait nécessairement augmenté ce

---

(10) Cf. transcription et traduction de la lettre in N. Sumien, *La correspondance du savant florentin Paolo dal Pozzo Toscanelli avec Christophe Colomb*, Paris 1927. La thèse d'Henri Vignaud, *Toscanelli and Colombus*, New York, 1917, qui met en cause l'authenticité de la lettre, est aujourd'hui généralement récusée. Cf. H. Wagner, in *Göttingen gelehrten Anzeigen*, (1902), n.° 2; pp. 108-131; L. Gallois, *Annales de Géographie*, (1902), pp. 97-110 et Idem, in *Ibid.*, Vol. XXXVII (1928) pp. 207-208; G. Uzielli, «Toscanelli, Colombo e Vespucci», in *Atti del IV Congresso Geografico Italiano*, Milano, 1902, pp. 539-591.

(11) Hermann Wagner, «Die Rekonstruktion...», pp. 208-312. Une première reconstruction a été tentée par O. Peschel, in *Das Ausland*, Vol. 40, pp. 1-4 (1867), n.° 1; elle a été reprise par K. Kretschmer, *Die Entdeckung Americas*, Berlin, 1892, Pl. VII. Sur la vraisemblance de la reconstruction de Wagner, cf. R. A. Skelton, «The cartography of Columbus' First Voyage», in *The Journal of Christopher Columbus*, trans. Cecil Jane, révisée et annotée par L. A. Vigneras avec appendice de R. A. Skelton, Londres, 1961, p. 218,

(12) Cf. H. Wagner, *art. cit.*, p. 238. Wagner appelle Toscanelli «Marinus redivivus».

(13) A. Berthelot, *L'Asie ancienne centrale et sud-orientale d'après Ptolémée*, Paris, 1930, p. 414. Nous avons préféré l'identification de Berthelot à celle de G. E. Gerini. *Researches on Ptolemy's Geography of Eastern Asia*, Asiatic Society Monographs, n.° 1, Londres, 1909, pp. 302-304. Gerini identifie Cattigara à Hangchou (120° 07').

prolongement erroné. Toscanelli avait-il préféré suivre Marin de Tyr (*oecumène* de 225 degrés) plutôt que Ptolémée (*oecumène* de 180 degrés) précisément pour tenir compte de l'apport polien? Marco Polo avait situé l'île de Chipangu (le Japon) à 1500 *li* de la Chine. Il a confondu le *li* chinois, dont il y a 250 dans un degré de latitude, avec le mille italien (14). Si l'on admet que le Japon fait partie de l'*oecumène*, il en résulte un nouvel accroissement de 22° 30' vers l'est, en adoptant le module de 66 2/3 milles/degrée.

Toscanelli avait pris comme parallèle médian de sa carte le parallèle de Lisbonne, qu'il situait, d'après les tables d'Alfonso X et de Regiomontanus à 41 degrés de latitude (15), et il avait établi, à partir de ce parallèle, dans l'hypothèse d'un voyage direct, la distance en milles séparant Lisbonne de la ville de Quinsay, actuellement Zhang-zhou.

De l'ensemble des éléments dont Toscanelli avait composé sa synthèse, il résultait néanmoins, comme l'a bien noté Wagner, une carte destinée, par sa conception sinon par ses données exactes, à l'usage des marins. De plus, et à la différence des cartes portulans, l'introduction d'un module spécifiant le rapport distance terrestre/distance angulaire permettait d'établir, en unités de distance terrestre, la distance séparant l'Europe de l'Asie à travers l'Atlantique. De sorte qu'on a vu s'esquisser un projet concret et non pas imaginaire de circumnavigation du globe, chose jusqu'alors totalement inédite dans l'histoire de la découverte de la terre.

Pourquoi alors, en cette année 1474, lorsque la lettre et la carte de Toscanelli sont parvenues entre les mains du roi D. Afonso V, n'a-t-on donné aucune suite aux conseils du Florentin? (16) Sans doute parce que les Portugais cherchaient à atteindre, non pas l'Extrême-Orient révélé par Marco Polo, mais l'Ethiopie et l'Inde cis-gangétique où, à l'époque, on localisait le royaume du Prêtre Jean, but principal de leur entreprise de découverte. La route la plus courte devait être, ils en étaient convaincus, celle qui contournait l'Afrique (17). L'une des raisons invoquées par les cosmographes

---

(14) Sir Henry Yule, *The Book of Ser Marco Polo*, Londres, 1875, Vol. II, p. 253 p. 193 n, p. 194 n, et p. 209.

(15) Alfonso X, *Alfontij regis Castellae Illustrissimi caelestium motuum tabulae...* Venise, 1483; Regiomontanus, *Ephemerides sive Almanach*, Venise, 1498. Nous n'avons pas trouvé à la Bibliothèque Nationale de Paris d'éditions antérieurs de Regiomontanus qui, selon Wagner, contiennent également la *tabula regionum*. Cf. H. Wagner, *art. cit.*.

(16) On peut s'étonner que D. Afonso V ait pensé à interroger Toscanelli sur un pareil sujet, car il est notoire qu'il s'intéressait peut aux découvertes. Derrière lui se profilait sans doute le prince D. João, le futur roi D. João II, qui, dès cette même année 1474, avait été chargé des affaires maritimes. Cf. Damião Peres, *História dos Descobrimentos, Portugueses*, 3e éd. Porto, 1983, pp. 118 et 193; J. Bensaude, *L'astronomie nautique au Portugal*, Berne, 1912-1917, Reprint Amsterdam, 1967, pp. 271-274, et A. Brásio. *Monumenta Missionária Africana*, Lisbonne, 1954, Vol. IV, pp. 12-15.

(17) L'historien João de Barros rappelle les frais considérables qu'avait engagés le roi D. João II pour découvrir le roi des Abyssins, identifié au Prêtre Jean. Dans les instructions données à Vasco da Gama, «presque tout consistait à connaître l'état et les

du roi D. João II, en 1483-84, pour rejeter le projet asiatique de Colomb (18), était le peu de crédit qu'ils accordaient aux «choses imaginaires que Marco Polo racontait de l'île de Cypango», choses qui faisaient partie du projet de Colombo (19).

Si aucun document ne nous est parvenu, qui confirme que les Portugais aient jamais partagé la conviction de Toscanelli ou de Colomb qu'il était possible d'atteindre l'Asie en traversant l'Atlantique (20), il existe trois indices permettant de croire qu'ils ne négligeaient pas tout à fait la possibilité de trouver des terres, ou de la terre ferme, outre Atlantique.

L'un de ces indices est le sauf-conduit, rédigé le 20/3/1488 en termes particulièrement chaleureux et adressé à Colomb, alors en Espagne, par le roi de Portugal D. João II, l'invitant à revenir au Portugal (21). Une apostille de Colomb, dans son exemplaire de l'*Imago Mundi* de Pierre d'Ailly (22), faisant état de sa présence lorsque, devant le roi, Bartolomeu Dias fit le récit de son voyage autour du Cap de Bonne Espérance, prouve que Colomb a donné suite à l'invitation qui lui avait été faite. Les Portugais ne semblent pas s'être totalement désintéressés, à cette date, du Gênois et de ses idées.

Le second indice est l'autorisation accordée par D. João II à Fernão Dulmo, le 24 juillet 1486, de tenter de découvrir une grande île ou *terra firma per costa* dans l'Atlantique (23).

Un troisième indice est l'indication donnée par Christophe Colomb, lors de son troisième voyage, en 1498, «... qu'il voulait voir quelle était l'intention du roi D. João II de Portugal, qui disait que, vers le midi, il y avait de la terre ferme» (24).

---

affaires de ce prince, cf. João de Barros, *Terceira Década*, Lisbonne, 1563, Liv. IV, Cap. I, éd. Lisbonne, 1946, Vol. III, p. 165.

(18) Damião Peres, *História dos Descobrimentos Portugueses*, 3e éd., Porto, 1983, p. 256.

(19) João de Barros, *Ásia*, Déc. I, Lisbonne, 1552, Liv. III, Cap. IX, éd. Coimbra, 1932, p. 113.

(20) La thèse bien connue d'Henri Vignaud, selon laquelle Colomb ne cherchait pas à atteindre l'Asie, mais seulement des terres à l'ouest, est contredite par la découverte récente d'un document important: la *Carta comendaticia* du 17 avril 1492, écrite par les rois catholiques, et où l'on peut lire la phrase: «Mittimus in presenciarum nobilem virum Christoforum Colon, cum tribus carauelis armatis, per maria oceana ad partes Indie». Reproduit par Antonio Rumeu de Armas, *Nueva luz sobre las capitulaciones de Santa Fé de 1492, concertadas entre los reyes catolicos y Cristobal Colón*, Madrid, 1985, p. 94.

(21) Reproduit in João Martins da Silva Marques, *Descobrimentos Portugueses*, Vol. III, (1461-1500), Lisbonne, 1971, p. 341.

(22) Apostille in Pierre d'Ailly, *Imago Mundi*, éd. et trad. française de E. Buron, Paris, 1930, Tome I, pp. 206-208.

(23) Reproduit in *Alguns Documentos do Archivo Nacional da Torre do Tombo*, Lisbonne, 1892, p. 58.

(24) Bartolomé de las Casas, *Historia de las Indias*, éd. B.A.E., Tome XCV, Madrid, 1957, Cap. CXXX, p. 350.

Cependant aucune carte portugaise du XVe siècle ne nous est parvenue, qui fasse état, ni de l'hypothèse, ni de la réalité, d'une terre ferme outre-Atlantique.

C'est évidemment dans ce contexte, où s'entremêlent théorie mathématique de la cartographie, connaissances poliennes de l'Orient, expérience pratique des Portugais de la navigation dans l'Atlantique et rumeurs évoquant l'existence de grandes terres au-delà de l'Océan, que s'est constituée une tradition cartographique, que nous sommes réduits à conjecturer.

Que Colomb possédait des cartes et des mappemondes, nous le savons, bien qu'aucune ne nous soit parvenue (25). Selon André Bernaldez, il avait montrée aux rois catholiques «une mappemonde» *de manera que les puso en deseo de saber de aquellas tierras* (celles qu'il voulait découvrir) (26). Durant son voyage, il avait montré à Pinzon une carte avec *ciertas islas por aquella mar* (27).

Sur quoi Colomb s'était-il appuyé pour représenter l'Atlantique, ses îles et ses rivages? Il est généralement admis que, si Colomb avait eu connaissance de la carte de Toscanelli et reçu d'elle sa première inspiration, ce n'est pas la route préconisée par l'astronome florentin qu'il suivit en définitive (28).

Grâce à une analyse du récit du quatrième voyage de Colomb, l'historien de la cartographie Roberto Almagià, dans un article paru en 1940, proposait comme exemples d'un type de carte qui circulait en Europe à la veille du premier voyage de Colomb, les mappemondes d'Henricus Martellus, dont la plus complète est celle, datée de c. 1489, qui est conservée à la British Library (29). De la vie de Martellus, nous ne savons presque rien, sauf qu'il était d'origine allemande et qu'il a travaillé à Florence (30).

Dans sa lettre écrite de la Jamaîque, le 7 juillet 1503, Colomb affirmait qu'au fond du golfe du Mexique, près de Veragua [côte du Panama] et à une distance de neuf jours de voyage par terre vers l'ouest, se trouve la province de Ciguare. «On dit, écrit Colomb, que la mer entoure Ciguare et de là

(25) Henri Harrisse, «Cartographia Americana Vetustissima», in *The Discovery of North America*, Paris, 1892, pp. 365-648.

(26) André Bernaldez, *Memorias del Reinado de los Reyes Católicos*, (comp. c. 1513), éd. Madrid, 1962, p. 270.

(27) Christophe Colomb, «Journal» (25 septembre), in Consuelo Varela, *Cristobal Colón, Textos y Documentos completos*, Madrid, 1984, p. 24.

(28) Colomb n'a pas eu connaissance du globe de Martin Behaim, comme l'ont prétendu certains, car le globe a été complété après le départ pour son premier voyage. Cf. Ilaria Luzzana Caraci, «Il planisfero di Enrico Martello delle Yale University Library e i fratelli Colombo», in *Rivista Geografica Italiana*, Annata LXXXV, fasc. 2 juin 1978, p. 132. R. A. Skelton, «The Cartography of Columbus First Voyage», in *The Journal of Christopher Columbus*, éd. cit., p. 219.

(29) Roberto Almagià, «I mappamondi di Enrico Martello e alcuni concetti geografici di Cristoforo Colombo», in *La Bibliofila*, Vol. XLII (1940), pp. 288-311.

(30) Idem, *art. cit.*, p. 290.

[par mer], à dix jours de voyage, se trouve le fleuve Ganges. Il paraît que ces terres [de Ciguare] sont par rapport à Veragna comme Tortosa par rapport à Fuenterravía, ou Pise par rapport à Venise» (31).

Par cette analogie, Colomb évoquait le concept d'isthme, isthme, en l'occurence, séparant l'Atlantique de l'Océan Indien et, selon Almagià, il se rapportait à une carte qu'il avait sous les yeux, une carte qui représentait une péninsule de l'Asie méridionale, située plus à l'est que les trois péninsules connues de Ptolémée (32). Il s'agissait d'une quatrième péninsule asiatique, nécessairement très étroite et tout à fait imaginaire. Elle doit son origine à la modification introduite dans la carte de Ptolémée par les nouvelles données rapportées par Marco Polo lors de son voyage de retour de la Chine par mer, c'est-à-dire l'existence d'une façade maritime de la Chine et la reconnaissance du fait que la mer de Chine communique avec l'Océan Indien. Pour tenir compte de cette communication des mers, le (ou les) cartographe (s) avait (avaient) coupé la terre australe qui, pour Ptolémée, entourait l'Océan Indien. C'est cette coupure qui a donné naissance à la quatrième péninsule (33).

Tout cela apparaît clairement sur la carte d'Henricus Martellus, mais on ignore si c'est lui ou un autre qui, le premier, procéda à cette modification.

Ni la mappemonde de Martellus de la British Library, notons-le bien, ni les autres exemplaires dont Almagià fait l'inventaire, ne comportent des échelles de latitudes ou de longitudes.

Dans les années 60, une vingtaine d'années après la publication de l'article d'Almagià, est entrée à la Bibliothèque de l'Université de Yale une autre mappemonde de Martellus, datée de c. 1490 et jusqu'alors inconnue (34).

La projection de cette mappemonde est un développement de la seconde projection, homéotère, de Ptolémée, mais à la différence des autres mappemondes de Martellus, celle de Yale comporte (et cela représente une nouveauté exceptionnelle) une échelle de latitudes et de longitudes; y figure également l'île de Cypangu (le Japon de Marco Polo) (35). Ainsi, la mappemonde de Yale devient la première carte non Ptoléméenne du XVe siècle qui nous soit parvenue, et qui indique de façon explicite en degrés la largeur attribuée par son auteur à l'océan séparant l'Europe de l'Asie orientale.

Sur la mappemonde de Yale, l'*oecumène* s'étend sur 270° et les distances

(31) «Relacion del Cuarto viaje» (7/7/1503), in Consuelo Varela, *op. cit.*, p. 319.

(32) Roberto Almagià, *art. cit.*, p. 289.

(33) Cf. George E. Nunn, *The Columbus and Magellan Concepts of South American Geography*, Glenside (California), 1932, et Roberto Almagià, *art. cit.*, p. 205.

(34) Cf. la description et la datation de R. A. Skelton, in Marcel Destombes, *Mappemondes A.D. 1200-1500*, Amsterdam, 1964, pp. 229-234; cf. aussi Ilaria Luzzana Caraci, «Il planisfero di Enrico Martello della Yale University Library e i fratelli Colombo», in *Rivista Geografica Italiana*, Annata LXXXV, fasc. 2, (juin 1978), pp. 132-143.

(35) Cette île est absente des autres mappemondes de Martellus, à l'exception de celle du codice Laurenziano. Cf. R. Almagià, *art. cit.*, p. 298.

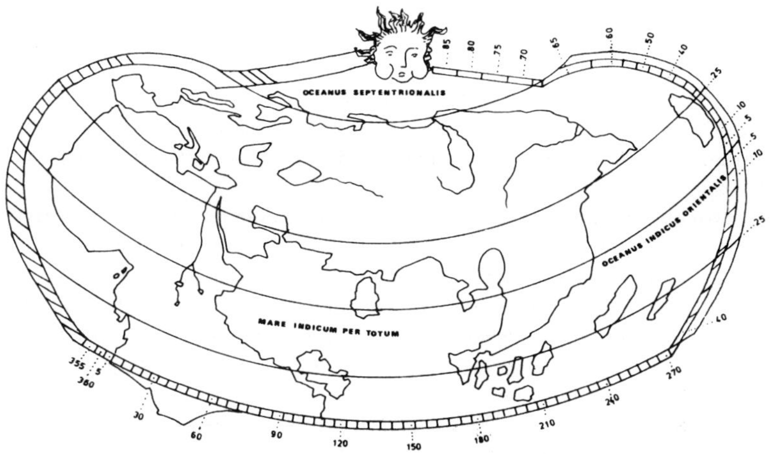

Henricus Martellus, *Mappemonde*, (c.1490)
Conservée à l'Université de Yale, U.S.A.

Reconstruction simplifiée pour montrer les échees de latitude et de longitude.

de Lisbonne à Quinsay (135°), de Lisbonne à Cipangu (105°) et du méridien d'origine (îles Canaries) à Cypangu (90°) coïncident, selon Skelton, «presque exactement avec celles que donne Toscanelli dans la description de sa carte» (36). Une différence sensible réside dans le fait que la carte de Yale donne une distance de 30° séparant Cypangu de la Chine, alors que, nous l'avons vu, on déduirait plutôt du récit de Marco Polo l'équivalent de 22° 30'.

La mappemonde de Martellus, de Yale, confirme l'intuition d'Almagià lorsque, parlant de l'exemplaire de Londres, il disait qu' «une mappemonde du type de celle de Martellus représente la carte qui correspond le plus fidèlement aux concepts de Colomb» (37).

Comparons donc les longitudes sur la carte de Martellus, de Yale, avec la distance qu'avait parcourue Colomb, selon ses propres dires.

(36) R. A. Skelton, in *op. cit.*, p. 233. Rappelons que Toscanelli avait suggéré une distance de 130° séparant Lisbonne de la côte de la Chine. Cf. H. Wagner, «Die Rekonstruktion...», p. 236. Notons également que Toscanelli, dans la description de sa carte, ne spécifie pas la distance entre les Canaries et Cypangu.

(37) R. Almagià, *art. cit.*, p. 307. Que la carte de Yale represente un exemplaire d'un modèle dont il en existait d'autres semble confirmé par un document des *Pleitos* (1515) de Colomb, qui affirme que Martin Alonso Pinzon avait, lors d'une visite à Rome, vu dans la bibliothèque du Pape Innocent VIII, une *escritura* (sic) qui indiquait la terre de Sypanso (= Cypangu) à 95° à l'ouest de l'Espagne. Cf. «Los Pleitos de Colón», in *Colección de Documentos ineditos relativos al Descobrimiento, Conquista y Organización de las antiguas posesiones españoles de ultramar*, 2e série, Tome 8 (II), Madrid, 1894, p. 126.

Colomb, dans son *Journal*, était persuadé que Cuba et Cypangu étaient une seule et même chose (38); il situait Cuba à 1142 lieues des Canaries et à 42° de Lat. N. (39) (Cuba est en réalité à 21° de Lat. N.)

Comment expliquer une erreur aussi grossière que cette latitude de 42°? (40) Et comment expliquer que la distance estimée de chemin parcouru (1142 lieues) l'ait conforté dans sa conviction que Cuba était Cypangu?

Colomb affirmait sa préférence pour le module de 56 2/3 milles/degré, mais il est peu probable qu'il l'ait adopté dans un autre but que de soutenir sa conviction que la circonférence de la terre n'était pas très grande. Il nous semble vraisemblable qu'il ait pensé son voyage, non pas dans le cadre d'une carte du type de Martellus, mais plutôt dans celui d'une carte de type toscanellien de projection marinienne, où les degrés de tous les parallèles sont égaux aux degrés sur le parallèle de Lisbonne.

Si maintenant, pour expliquer la distance de 1142 lieues, nous appliquons le module de Toscanelli de 66 2/3 milles/degré et si nous opérons, comme l'avait fait l'astronome florentin, la réduction nécessaire (66.67 × cosinus 41° = 50 milles/degré) pour le parallèle de Lisbonne, ville placée par Toscanelli à 41° de Lat. N. (presque la même latitude que celle attribuée par Colomb à Cuba: 42°) correspondant au parallèle sur lequel Toscanelli avait imaginé qu'aurait lieu un voyage vers l'Orient, notre distance de 1142 lieues donne un arc de longitude de 91 1/3 degrés, ce qui est presque exactement, à un degré un tiers près, le nombre de degrés séparant les Canaries de Cypangu sur la mappemonde de Martellus, de Yale, Dans l'hypothèse d'une carte de projection marinienne avec le parallèle médian placé à la latitude de Lisbonne, la distance de 1142 lieues donnera, bien ententu, le même nombre de lieues sur le parallèle des Canaries que sur celui de Lisbonne.

Et en effet, selon Las Casas, la distance parcourue par Colomb jusqu'à Cuba *concordaba cuasi al justo* avec l'endroit et la région où Toscanelli avait placé la grande île de Cypangu (41).

Si forte était, dans l'esprit de Colomb, la conviction que Cuba était réellement Cypangu, qu'il alla jusqu'à lui attribuer la latitude proposée pour l'île nippone par Toscanelli, latitude qu'il prétendra même avoir observée avec un quadrant! (42) Dans les premiers moments de sa découverte, la

(38) Colomb, «Journal» (24/10/1492), in Consuelo Varela, *op. cit.*, p. 44.

(39) Idem, «Journal» (2/11/1492), p. 50. Colomb répétait à deux autres reprises la même latitude de 42°. Idem, p. 48 (30/10/1492) et p. 61 (21/11/1492).

(40) R. A. Laguarda Trias (*El enigma de las latitudes de Colón*, Casa-Museo de Colón, Valladolid, 1974, pp. 18-28), propose une série d'explications aux erreurs de latitudes de Colomb, et notamment celle qui serait due à une méconnaissance de la variation magnétique, mais ses arguments ne nous paraissent pas satisfaisants en ce qui concerne une si énorme erreur, dans le cas de Cuba.

(41) Bartolomé de las Casas, *Historia de las Indias*, éd. B.A.E., Tome XCV, p. 154.

(42) Colomb, «Journal» (2/11/1492), in Consuelo Varela, op. cit., p. 50.

carte, pour Colomb, semblait plus réelle que la réalité, celle-ci devant se) conformer à l'image qu'il s'en faisait (43).

Mais comment expliquer que la distance parcourue de 1142 lieues ait si commodément concordé avec la situation sur la carte? C'est là une coïncidence extraordinaire, qui est à l'origine de l'étonnant paradoxe que l'on ait pris une île dans la mer des Antilles pour le Japon, et que cette découverte ait semblé confirmer toute une fausse cartographie fondée sur une multiplication d'erreurs (erreur de Ptolémée concernant les limites de la Chine; erreur dans les déductions faites à partir du récit de Marco Polo à propos de l'étendue de la façade maritime de la Chine; et erreurs héritées des mauvaises traductions d'auteurs arabes, au sujet du nombre de milles dans le degré du méridien).

Seuls les Portugais ne furent pas dupes de cette fausse représentation du globe. Leur scepticisme à l'égard du projet toscanellien leur a valu, après de longs efforts, d'atteindre une Inde réelle et attendue. Le lot des Espagnols fut une entité géographique inattendue et imprévisible, que l'on mit de longues années à connaître.

(43) Dans les lettres que Colomb devait écrire plus tard à Luis de Santangel (15/2/1493 et à Rafael Sanchez (25/4/1493), toutes deux écrites après qu'un délai de réflexion l'eût conduit à se conformer à une vision plus réaliste des choses, il situait Cuba à 26° de Lat. N. — une erreur beaucoup moindre de 5 degrés. Cf. la lettre à Luis de Santangel in Consuelo Varela, *op. cit.*, p. 144, et la lettre à Rafael Sanchez, facsimilé, Johnson Publishing C. Murphreesboro, N. Carolina, 1970, p. 11.

Addendum
La cartographie de l'Atlantique à la veille du voyage de Christophe Colomb

Depuis la rédaction de cet article, l'apparition d'un manuscrit contenant des relations inédites des quatre voyages de Colomb éclaire d'un jour nouveau le problème de ses connaissances en cartographie, sans toutefois le résoudre de façon satisfaisante. Le *Libro Copiador de Cristóbal Colón*, fut retrouvé par un libraire tarragonais après avoir 'dormi trois siècles dans une bibliothèque de Majorque'. Le manuscrit, une copie sur papier italien du XVIe siècle, conservé désormais à l'Archivo General de Indias, a été édité par Antonio Rumeu de Armas, de la Real Academia de la Historia, et imprimé en 1989 sous l'égide du Ministerio de la Cultura.[1] Parmi les documents inédits du manuscrit un seul nous apporte un éclairage nouveau sur les connaissances cartographiques du navigateur.

Il s'agit en effet d'une relation du second voyage de Colomb, redigée en janvier 1494 et adressée aux Rois Catholiques. Y figure la description d'une carte représentant les îles qu'il venait de découvrir. Il est intéressant de comparer la description de cette carte (aujourd'hui perdue) à la description par Toscanelli de la carte destinée au roi portugais D. Afonso V et envoyée au chanoine portugais Fernão Martins en 1474.

Voici la description de sa carte par Colomb:

> 'Toutes ces îles, dont on vient de parler, je les envoie représentées avec celles de l'année dernière, le tout sur une carte que j'ai faite [...].
>
> Vos Altesses y verront les terres d'Espagne et d'Afrique et, face à elles, toutes les îles rencontrées et découvertes durant ce voyage et le précédent; les lignes dans le sens de la longueur donnent la distance de l'Orient à l'Occident, les autres transversales, la distance du Septentrion au Sud. Les espaces entre les lignes correspondent à un degré, soit cinquante-six milles deux tiers, équivalant à quatorze lieues un sixième de nos lieues de la mer; et ainsi vous pouvez compter de l'Occident à l'Orient aussi bien que du Septentrion au Sud ce nombre de lieues, en utilisant le

---

[1] Antonio Rumeu de Armas, *Libro Copiador de Cristóbal Colón*, Tome I, Estudio histórico-crítico y edición; Tome II, Transcripción del *Libro Copiador*, Colección *Tabula Americae*, Testimonial Compañia Editorial, s.a., (Madrid, 1989).

> calcul de Ptolémée qui, comparant les degrés de longitude à ceux de l'équateur, dit que quatre degrés de l'équateur correspondent à cinq sur le parallèle de Rhodes, qui se trouve à trente-six degrés; chaque degré, sur la dite carte, correspond à quatorze lieues et un sixième, aussi bien du Septentrion au Sud que de l'Orient à l'Occident.'[2]

Voici maintenant comment Toscanelli décrit sa carte:

> 'Paul medecin, à Fernão Martins, chanoine de Lisbonne, salut. [...] J'envoie donc à Sa Majesté une carte faite de ma main sur laquelle sont dessinés vos rivages et les îles à partir desquelles vous commencerez à prendre la direction de l'Ouest [...] Ainsi donc les lignes droites tracées dans le sens de la longueur de la carte indiquent la distance de l'Orient à l'Occident; tandis que les transversales indiquent les espaces qu'il y a du Midi au Septentrion'.[3]

Il est évident que Colomb décrit sa carte en termes qui rappelle ceux de Toscanelli (parallèles et méridiens se coupant à angle droit); mais, s'éloignant de la méthode que le Florentin avait emprunté à Marin de Tyr, Colomb ne se sert pas d'un parallèle médian. Sa carte est une carte plate carrée, le parallèle de référence étant l'équateur et les degrés du méridien étant égaux à ceux de l'équateur. Colomb introduit alors, au milieu de sa description, une parenthèse apparemment sans rapport avec ce qui précède, et qui évoque le parallèle médian de Marin de Tyr passant par l'île de Rhodes, ce qui permettrait aux Rois Catholiques, dit-il, 'd'utiliser le calcul de Ptolémée qui a établi un rapport entre les degrés sur cette ligne et ceux de l'équateur, et qui dit que quatre degrés sur cette ligne correspondent à cinq sur le parallèle de Rhodes, qui se trouve à 36 degrés'.

Si les Rois Catholiques avaient voulu 'utiliser' la méthode de Marin de Tyr, ils n'auraient pu opérer le calcul proposé qu'après la lecture de la carte, et indépendamment d'elle, puisque la carte elle-même était une 'carte plate carrée'.

On peut donc douter que Colomb lui-même ait jamais maitrisé la méthode de Marin de Tyr, ou su utiliser, pour naviguer, la superposition, imaginée par Toscanelli, de la carte nautique méditerranéenne sur la projection marinienne. Rien d'étonnant à cela puisque, pour tenir compte de la convergence des méridiens et de la coubure de la terre, il fallut aux navigateurs attendre l'invention par Gerard Mercator, au milieu du XVIe siècle, de la carte réduite avec latitudes croissantes. Par contre la cartographie marino-toscanellienne permettait à Colomb (elle lui était même indispensable) de parvenir à une évaluation (qui s'est néanmoins révélée erronée) de la distance, qu'il entendait parcourir pour atteindre l'Orient, depuis la façade maritime de l'Europe.

Dans cette perspective, si nous voulons connaître les idées de Colomb en matière

---

2 *Ibid.* Tome II, Doc. II, pp. 451–452.

3 Lettre de Paolo dal Pozzo Toscanelli à Fernão Martins (25/6/1474) in N. Sumien, *La correspondance du savant florentin Paolo dal Pozzo Toscanelli*, (Paris, 1927), pp. 14-16.

de cartographie avant son premier voyage, que nous apprend la description qu'il donne, dans la relation de son second voyage, de la carte qu'il envoie aux Rois Catholiques?

Il nous semble peu probable qu'il n'ait pris connaissance de la cartographie marino-toscanellienne qu'après son premier voyage et juste avant son second.[4] Il serait plus raisonable d'admettre qu'il avait acquis cette connaissance avant son premier voyage, et qu'elle l'ait aidé à le préparer. Mais cette conclusion, nous insistons sur ce point, ne saurait être que conjecturale; sans autre témoignage, nous ne pouvons rien affirmer avec certitude. La question des connaissances cartographiques de Colomb lors de la préparation de son premier voyage reste toujours un problème ouvert.

---

4 Bien que l'historien Bartolomé de las Casas ait affirmé que la carte de Toscanelli a beaucoup influencé Colomb et que 'tout son voyage était fondé sur elle', (Bartolomé de las Casas, *Historia de la Indias*, ed B.A.E., tome XCV, Cap. XII, p. 47), Ilaria Luzzana Caraci a cru pouvoir démontrer en 1985 que la lettre n'avait joué aucun rôle dans la préparation de son entreprise, puisque que le texte de la lettre de Toscanelli n'a été copié par Colomb qu'après son premier voyage, au plus tôt en 1498. (See Idem, 'La cultura de Colombo', in *Atti del IV Convegno Internazionale di Studi Colombini*, Genova, 21–23 Oct. 1985, Gênes, 1987, pp. 209–228). On pourra répondre que rien n'empêchait Colomb de connaître le contenu de la lettre avant qu'il n'ait eu l'idée de la copier. Mais dans ce cas pourquoi n'avait-il éprouvé le besoin de la copier qu'après son premier voyage? Ce problème reste à présent sans solution.

XI

Colomb découvreur: perceptions contemporaines de son projet et de sa réalisation

Les historiens modernes ont prêté à Colómb des projets divers: les uns soutiennent la thèse classique de la découverte de l'Asie; d'autres n'y voient que la recherche d'îles dans l'Atlantique; ou encore l'espoir exaltant de découvrir un Nouveau Monde.

Dans cette variété d'interprétations, on rencontre trop souvent la volonté de créer du sensationnel autour d'un sujet rebattu, plus que d'exposer les résultats de recherches solides et approfondies.

Plus intéressante peut être l'étude de la diversité des analyses, par des auteurs du XVIème siècle, de la réalisation du projet du navigateur génois car, à la différence des historiens de nôtre époque, leur culture n'avait pas prévu le fait américain.

Par sa volonté de faire de l'histoire un instrument susceptible de créer une conscience nationale, l'historiographie du XIXème siècle a contribué plus que toute autre à déformer le déroulement de la découverte du nouveau continent.

Un bel exemple est celui de l'autodidacte américain Henri Vignaud. Mû par l'ambition d'exalter l'histoire de son pays, il tenta d'établir un lien causal et prestigieux entre une terre, que l'on sait aujourd'hui immense, et un grand navigateur (qui d'ailleurs n'en avait jamais atteint qu'une infime partie). Il s'est donc vu obligé d'évacuer du projet de la découverte l'hypothèse asiatique du premier voyage de Colomb, et fut contraint, par la même occasion, d'éliminer le rôle de Toscanelli, de sa lettre et de sa carte comme éléments ayant pu inspirer le Génois.

« La découverte de l'Amérique », écrivait Vignaud en 1921, « ne s'est pas faite en cherchant à gagner les Indes asiatiques par la voie de l'ouest [...] mais elle fut la conséquence logique, inévitable, prévue d'avance, d'une expérience organisée expressément pour chercher la terre même qui reçut ensuite le nom d'Amérique »[1]. Vignaud insistait également sur « .... la légende qui fait de Toscanelli l'initiateur de la découverte de l'Amérique »[2].

Les travaux de Vignaud ne convainquaient que très peu de monde, et la découverte, dans les années 20, d'un document inédit détruisit définitivement d'un seul coup toute sa thèse. Le document en question est la *Carta comendaticia* du 17 avril 1492, où les Rois Catholiques stipulent clairement qu'ils investissent Colomb de la mission d'atteindre l'Inde en traversant l'Océan. *Mittimus in presenciarum nobilem virum Christoforum Colon cum tribus carauelis armatis, per maria oceana ad partes indie...*[3].

Si le projet de Colomb ne comporte aujourd'hui plus aucune ambiguité, nous pensons que l'analyse des appréciations contemporaines de son accomplissement, mérite toujours un examen.

Durant les huit dernières années du XVème siècle, les interprétations de la découverte de Colomb se révèlent plus ou moins les mêmes que celles prêtées par les auteurs modernes à son projet, c'est-à-dire la découverte de quelques îles[4], de l'Inde[5], d'un Nouveau Monde[6], des Antipodes[7].

---

1 Henri Vignaud, *Le vrai Christophe Colomb et la légende*, Paris, 1921, p. 2.

2 Idem, *loc. cit.*

3 Antonio Rumeu de Armas, *Nueva luz sobre las capitulaciones de Santa Fé de 1492 concertadas entre los Reyes Catolicos y Cristobal Colón*, Madrid, 1985, p. 94.

4 Pierre Martyr d'Anghiera, Lettre au Comte de Tendilla et à l'Archevêque de Grenade, Barcelone, 13/9/1493, in P. Gaffarel, « Lettres de Pierre Martyr Anghiera », in *Revue de Géographie*, Vol. 15 (1884), p. 461; Bernardino de Carvajal, *Oratio superpraestanda solenni obedientia Sanctissima D.N. Alexandro Papae VI*, 19/6/1493, Bibl. Nat. Paris, Rés. H 1508; Giuliano Dati, *La lettera dell'isole che ha trovato nuovamente il Re di Spagna*, Florence, 1493, éd. G. Uzielli, Bologne, 1873, p. 26; Gaubert Fabricius Vagad, *Cronica de Aragon*, Saragosse, 1499, sign. B.iiij (r.).

5 Pierre Martyr d'Anghiera, Lettre à l'Archevêque de Braga, Barcelone, 1/10/1493, *éd. cit.*, p. 464; Anon. (1493), Biblioteca del Monasterio de El Escorial, Q. II, fol. 9 uta, cité par Antonio Bofarull, *Historia critica de Cataluña*, Barcelone, 1876-78, Vol. VI, p. 393.

6 Pierre Martyr d'Anghiera, Lettre au Cardinal Ascanio, au Palais, 1/11/1493, in *éd. cit.*, p. 465 (« ... *Colonus ille novi orbis repertor* »); Gaubert Fabricius Vagad, *op. cit., loc. cit.*

7 Pierre Martyr d'Anghiera, Lettres à Jean Borromée, Barcelone, 14/5/1493 et au Cardinal Ascanio Sforza, Barcelone, 13/9/1493, in éd. cit., p. 460 et p. 463.

L'humaniste italien Pierre Martyr d'Anghiera évoquera, dans ses lettres publiées à Alcalá en 1530, ces diverses hypothèses[8]. D'autres témoins se limiteront, par contre, à parler soit de l'Inde (Anon., 1493, Biblioteca del Monasterio de El Escorial)[9], soit tout simplement d'îles (Bernardino de Carvajal, 1493)[10], soit des deux à la fois.

L'interprétation, sans doute au début la plus courante, de la découverte par Colomb d'îles inconnues est qu'elles n'étaient que d'autres îles Canaries « en direction de l'Inde ».

C'est ainsi que Bernardino de Carvajal, dans un discours d'obédience devant le Pape Alexandre VI, en 1493, parlait des Canaries (*fortunatas insulas*) et « d'autres [îles] inconnues vers les Indiens » (*alias incognitas versus Indos*)[11]. Giuliano Dati, dans son édition de la première lettre de Colomb sur sa découverte, évoquait dans le colophon « les nouvelles îles des Canaries indiennes » (*nuove isole de Cannaria indiana*)[12].

Sans faire expressément allusion aux Canaries, Jacopo Filippo Foresti consacrait à la découverte un paragraphe, dans son *Supplementum Chronicarum* (1503). Il y évoquait « ... les quatre îles dans l'Inde hors de l'*oecumène* » (*quattuor ... insulis in India extra orbem*)[13].

Certains essaient de situer les nouvelles îles à des distances précises de l'Europe.

Francisco Nuñez de la Yerva, dans son édition de Pomponius Mela, publiée en 1498, les avait placées à quarante-cinq degrés à l'ouest de l'Espagne[14]. En revanche, Raffaele Maffei da Volterra, dans son encyclopédie, *Commentariorum urbanorum libri XXXVIII*, publiée à Rome en 1506, les localisait à vingt degrés de l'Europe[15].

---

8 Sur Martyr d'Anghiera, cf. P. MARIÉJOL, *Un lettré italien à la cour d'Espagne, Pierre Martyr*, Paris, 1887.

9 Anon. (1493), Biblioteca del Monasterio de El Escorial, in *éd. cit.*, *loc. cit.*

10 BERNARDINO DE CARVAJAL, *op. cit.*

11 BERNARDINO DE CARVAJAL, *op. cit.*

12 GIULIANO DATI, *éd. cit.*, p. 26.

13 JACOPO FILIPPO FORESTI, *Supplementum Chronicarum*. Venise, 1503, ff. 441 (v.).

14 FRANCISCO NUÑEZ DE LA YERVA, *Cosmographia Pomponii cum figuris*, Salamanque, 1498, f. 1 (v.).

15 RAFFAELE MAFFEI DA VOLTERRA, *Commentariorum urbanorum libri*, XXXVIII, Rome, 1506, f. 139 (v.) (*plures invenerunt insulas parum inter se distantes ultra fortunatas xx fere partibus sitas...*).

Ces chiffres, obtenus à partir d'estimations de distances en lieues converties en degrés de longitude, sont tout à fait fantaisistes et ne reposent sur aucune donnée sérieuse.

Dès les premiers voyages de Colomb, il s'est trouvé des Espagnols qui refusaient de croire que les îles, ou la terre, que le génois avait découvertes se trouvaient en « Inde ».

Vers 1497 déjà, Francisco de Cisneros assurait les Rois Catholiques que « ... les îles dont on a parlé récemment [...] ne sont pas en Inde, mais dans la mer océane atlantique éthiopienne »[16], et Francisco Nuñez de la Yerva, dans son édition de Pomponius Mela déjà citée, écrivait que la terre découverte à l'ouest par les Rois Catholiques, « ... est appelée abusivement « Inde »[17].

Le fondateur de l'Université de Séville, Rodrigo de Santaella, déclarait dans la préface à son édition des voyages de Marco Polo, publiée à Séville en 1503, que ceux « qui pensent que Antilia [sic] et les îles récemment découvertes sur l'ordre des Rois Catholiques sont aux Indes, sont trompés par ce nom des Indes qu'on leur a donné. [...] ces îles se trouvent dans la grande mer occidentale »[18].

Reste donc l'hypothèse de la découverte par Colomb d'un « nouveau monde », exploit d'ordinaire attribué à Améric Vespuce, bien qu'il n'ait lui-même jamais employé le terme dans ses lettres jugées authentiques.

A propos des découvertes, non seulement celles des Espagnols, mais également celles des Portugais, on rencontre des expressions comme *novus mundus*, *novus orbis* et, *alter orbis*, employées au début dans des contextes qui ne concernent pas spécifiquement le continent américain.

L'Italien Alvise Cadamosto, parlant de la terre des Noirs qu'il avait visitée sur la côte de l'Afrique vers 1456, emploie le terme *altro mondo*[19], peut-être parce qu'il y avait aperçu la Croix du Sud visible, on le sait déjà, depuis la Haute Egypte[20].

---

16 Francisco de Cisneros, « Memorial de Zamora sobre las Indias » (c. 1497), éd. Zamora, 1983, cité par Demetrio Ramos, *Algunas cuestiones de fondo en el Descubrimiento de America*, Granada, 1990, p. 97.

17 Francisco Nuñez de la Yerva, *op. cit., loc. cit.*.

18 Rodrigo de Santaella, *Cosmographia breve introductoria en el libro de Marco Polo*, Séville, 1503, sign. A.ij. (v.).

19 Alvise Cadamosto, « Proemio, Navigation prima », in G.B. Ramusio, *Delle Navigationi et Viaggi*, Vol. I, Venise, 1550, f. 105 (r.).

20 Sur l'identification des étoiles observées par Cadamosto, cf. Alexandra Mauro, *O « Carro do Austro » de Alvise Cà da Mosto: observaçoes astronómicas e fortuna editorial*, Centro de Estudos de História e Cartografia Antiga, Lisbonne, 1988, Vol. 210, p. 15.

Le portugais João Teixeira félicitait son Roi, dans un discours prononcé en 1489, d'avoir par ses voyages, organisés le long de la côte africaine, « ...ouvert presque un autre *orbis terrarum*. Ce fut une chose », ajoute-t-il, « comme on n'en avait jamais vu, car nos marins passèrent le cercle equinoxial et virent le soleil droit au-dessus de leurs têtes »[21].

Dans son *Liber Chronicarum* de 1493, l'Allemand Hartmann Schedel écrivait, à propos des Portugais qui accompagnaient le voyage de Diogo Cão de 1493, qu'ils pénétrèrent dans « ...l'autre monde (*alterum orbem*) où, lorsqu'ils étaient tournés vers l'est, leur ombre tombait vers le Sud et à leur droite »[22].

L'expression « nouveau monde », employée dans ces témoignages, évoque un ciel nouveau, nouveau par l'aspect des astres qui s'y trouvent, plutôt que la rencontre d'une nouvelle masse terrestre.

Il se peut qu'il soit tout simplement question, du moins en ce qui concerne les deux premiers témoignages, d'un écho des vers de Lucain dans son *Pharsale*, à propos des Arabes venus à Rome pour aider Pompée:

Ignotum vobis Arabes venistes in orbem
Umbras mirati nemorum non ire sinistras[23].

C'est à dire: « Vous autres Arabes êtes venus dans un monde qui vous est inconnu [le monde romain] et vous vous émerveillez que les ombres des arbres ne tombent pas vers la gauche ».

Il faut croire que les Arabes abordaient Rome en regardant vers l'ouest.

Les expressions « Nouveau Monde » et « Autre Monde » sont il est vrai également employées par certains dans un contexte qui évoque plutot une région géographique qu'une zone délimitée par ces critères astronomiques.

Valentim Fernandes, Allemand établi au Portugal, appellera en 1502 toute la région au-delà du Cap de Bonne Espérance un « nouveau mon-

---

21 João Teixeira, *Oratio habita insigni viro Ioanne Teixeira...* Coimbra, 1562. Texte latin et traduction portugaise (sans pagination).

22 Hartmann Schedel, *Liber Chronicarum*, Nuremberg, 1493, f. CCXC (v.).

23 Lucain, *Pharsalia*, iii, 247-248, in Johannes de Sacrobosco, « Tractatus de Spera », in Lynn Thorndike, *The Sphere of Sacrobosco, and its Commentators*, Chicago, 1949, p. 106 et 136.

de »[24]; un an plus tard, il dira du Brésil qu'il s'agit d'un « autre monde » [...] « terre des Antipodes »[25].

Pour en revenir à Christophe Colomb, voici ce qu'écrivait, dans ses *Enneades*, l'humaniste Marco Antonio Coccio Sabellico des découvertes du Génois: « ...un autre monde, en quelque sorte, a été ouvert aux peuples, sous les auspices des Rois [Catholiques] par la flotte espagnole. Colomb, prénommé Christophe, homme rompu à la navigation, osa le premier de tous tenter l'entreprise »[26].

Écoutons encore l'ambassadeur espagnol Fernando Tellez déclarer devant le Pape, en 1508, que les Espagnols « ... avaient reconnu et exploré un autre monde (*alterum orbis*). Nos ancêtres doutaient de l'existence, dans l'autre hémisphère, de gens qui seraient nos Antipodes »[27].

« Nouveau Monde » et « Autre Monde » sont-ils l'équivalent de « la terre des Antipodes »? Si l'on s'en tient à l'Acte Notarial de Valentim Fernandes et au discours de Fernando Tellez, que l'on vient de citer, il semble bien que tel soit le cas.

Traditionnellement, on situe les Antipodes au Sud de l'équateur, et c'est dans cette région que le concept d'un « nouveau monde » émerge au début chez les humanistes. Lorsque la continuité du rivage atlantique de la nouvelle terre sera établie, la notion d'un « nouveau monde » subira une rotation de 180 degrés pour se situer verticalement à l'ouest, au-delà de l'Atlantique. Mais cette nouvelle orientation n'entrera dans la culture européenne que vers la fin du XVIème siècle.

On peut donc conclure de l'analyse de ces divers témoignages qu'il régnait, durant les premières années qui suivirent les voyages de Colomb, une diversité d'appréciations de la nature de leur aboutissement. La plu-

---

24 VALENTIM FERNANDES, « A Epistola sobre a tralladaçam », in *O Livro de Marco Polo*, Lisbonne, 1502, f. A ij (v.). Rééd. par Francisco Maria Esteves Pereira, Libonne, 1922.

25 VALENTIM FERNANDES, Acto Notarial, 20 Mai 1503, éd. António Alberto de Andrade, in *Arquivos do Centro Cultural Português,* Fondation Gulbenkian, Paris, Vol. V, 1972, p. 538.

26 MARCO ANTONIO COCCIO SABELLICO, *Secunda pars Enneadum,* Venise, 1504, Enneadis Decimae, Lib. Octavus, ff. CLXX (v.)-CLXXI (v.).

27 FERNANDO TELLEZ, *Razonamiento de los embaxadores de España en la obediencia que dieron al Papa*, Rome, 1508. Texte reproduit in H. HARRISSE, *Biblioteca Americana Vetustissima, Additiones*, n. 30.

part de ceux qui y font allusion reconnaissent que la découverte de Colomb est importante, mais aucun n'est en mesure de la décrire autrement qu'en termes d'une géographie traditionnelle. L'émergence de l'Amérique ne pourra être qu'un processus s'étalant sur une longue durée et nullement une révélation totale et instantanée.

LA SCIENCE UNIVERSITAIRE EN EUROPE ET LES DÉCOUVERTES PORTUGAISES:

ARISTOTÉLISME DOCTRINAIRE ET EXPÉRIENCE DES NAVIGATEURS

La science de la terre et de l'Océan, telle qu'elle est enseignée dans les universités européennes à la fin du Moyen Age, consiste pour une large part en une interprétation par les commentateurs scolastiques des doctrines d'Aristote.

Au XIVe siècle, l'aristotélisme médiéval évolue vers un réaménagement des principes fondamentaux énoncés; la conséquence en fut l'émergence, chez des universitaires parisiens notamment, de théories tout à fait originales.

L'expérience des navigateurs portugais au cours de leurs grands voyages dans l'hémisphère austral, à la fin du XVe et au début du XVIe siècle, invalida brutalement une de ces théories, amorçant la ruine de la science aristotélicienne universitaire, ruine qui devait aboutir à la révolution copernicienne et à l'avènement de la science moderne.

Nous allons donc analyser ici l'accueil réservé, dans les milieux académiques européens, aux nouvelles données apportées par les Découvertes et les étapes de la lente restructuration de la science universitaire.

Pour comprendre la portée de cette mutation culturelle, il faut revenir en arrière et esquisser rapidement ce qu'est, selon Aristote, la situation de la terre dans le cosmos, puis la modification imposée à cette représentation par les savants du Moyen Age.

Dans sa volonté persévérante de rendre compte, dans une théorie unique, de tous les phénomènes de la nature, Aristote avait organisé les quatre éléments sublunaires, la terre, l'eau, l'air et le feu, en un système absolument concentrique de sphères autour d'un centre unique, le centre de l'univers[1].

Les éléments les plus pesants, les plus denses et ayant le moins de volume, sont entourés concentriquement par les éléments les plus légers, les moins denses et ayant le plus de volume[2]. Selon sa nature propre, chaque élément cherche à se placer dans le lieu qui lui revient selon l'ordre décroissant de sa densité, les plus lourds le plus près possible du centre de l'univers, les plus légers le plus près de la surface extérieure du cosmos.

Pour que la terre repose paisiblement à sa place naturelle[3], il faut non seulement que son centre coïncide avec le centre de l'univers, mais également qu'elle soit partout en contact avec la surface concave de l'élément qui la suit en densité: l'eau.

Dans sa volonté d'unifier en une théorie totalisante les quatre éléments de sa physique, ainsi que la terre et l'Océan de sa «géographie», Aristote fut conduit par la rigueur de sa logique à recouvrir la terre tout entière d'une couche uniforme d'eau et à inonder l'habitat de l'homme sous les flots de l'Océan[4].

Il fallut attendre le Moyen Age et jusqu'au XIIIe siècle pour qu'un astrologue, Michel Scot[5], s'aperçoive de l'absurdité de la représentation du Stagyrite, jusqu'au XIVe siècle pour qu'un physicien de l'Université de Paris, Jean Buridan (c. 1300-1358) trouve une solution inspirée de la physique d'Archimède.

Empruntant à ce grec célèbre le concept du centre de gravité d'un corps, Buridan postulait que la terre n'a pas une densité uniforme et que, de ce fait, son centre de gravité ne coïncide pas avec son centre de volume. Le premier coïncide, disait-il, avec le centre du volume de l'eau, ainsi qu'avec le centre de l'univers. Grâce à son *imaginatio*, comme Buridan l'appelait, il a pu faire émerger de la sphère de l'eau une faible partie de la surface de la terre[6].

Albert de Saxe (c. 1316-1390), un contemporain de Buridan, en récapitulant sa théorie, y ajouta une dimension proprement cartographique. Sur la partie de la terre émergée, il situait la carte médiévale de l'*oecumène*, la carte dite «T & O», représentant les trois parties de la terre: Europe, Afrique et Asie[7].

Si la représentation de ces deux physiciens offrait une solution à la difficulté léguée par Aristote, elle éliminait toute possibilité de l'existence de terres émergées dans l'hémisphère austral ou en dehors du continent euro-asiatique, et constituait de ce fait un obstacle épistémologique [8] aux projets de navigation transatlantique.

Si Jean Buridan et Albert de Saxe se gardaient d'établir un rapport quantitatif entre les volumes des sphères des éléments de l'eau et de la terre, une tradition médiévale remontant aux commentateurs grecs d'Aristote (mais jamais énoncée par le Stagyrite lui-même)[9] postulait un rapport de 1 a 10 entre les volumes des éléments; aussi de nombreux savants du Moyen Age, dont Michel Scot[10], admettaient-ils que l'eau (ou l'Océan) avait dix fois le volume de la terre.

L'espagnol Pedro Ciruelo, professeur à l'Université de Paris à la fin du XVe siècle, expose en 1498, dans son commentaire sur la *Sphere* de Sacrobosco, cette doctrine du décuplement des volumes des éléments, doctrine qui aggravait encore l'obstacle épistémologique créé par Buridan et Albert de Saxe.

Ciruelo va jusqu'à invoquer les deux premiers voyages de Christophe Colomb comme preuve du bien-fondé de la doctrine. Rappelons que Colomb n'a atteint le continent américain que lors de son troisième voyage, en 1498, voyage que Ciruelo ne pouvait pas connaître quand il écrivait:

> «En effet, la mer est dix fois plus grande que la terre [...]. Par conséquent ce trajet [entre l'Espagne et l'Orient] ne pourrait pas être accompli en [...] peu de temps, quelque favorables qu'aient pu être les vents. Bien plus, nos compatriotes [les Espagnols] (à ce qu'on dit), lorsqu'ils cherchaient des îles en naviguant directement vers l'ouest, ne purent en deux ans [il s'agit des voyages de Colomb de 1492 et de 1494] atteindre aucun rivage. Donc [cet argument] montre que l'eau est placée au-dessus de la terre et qu'elle l'entoure.»[11].

Le premier démenti apporté à cette science scolastique universitaire viendra, non pas de l'expérience des navigateurs, mais de l'autorité d'un mathématicien grec, que redécouvrira la Renaissance italienne.

C'est la *Géographie* de l'alexandrin Claudius Ptolémée (c. 90-168), ignorée des savants médiévaux et introduite en Occident au début du XVe siècle par le mécène florentin Palla Strozzi[12] (traduction latine en 1406), qui fournira une alternative à la doctrine des éléments des scolastiques et qui, l'expérience des Découvertes aidant, finira par transformer en une science mathématique appelée «Géographie», ce qu'auparavant on nommait platement «image du monde».

Indifférent à la doctrine d'Aristote des sphères des éléments de la terre et de l'eau, Ptolémée définit le globe terrestre en ces termes:

> «... La surface continue de la terre et des mers forme une sphère dont le centre est le centre même de la sphère des corps célestes.»[13].

L'expression «surface continue de la terre et des mers» constitue la condition première d'une cartographie mathématique fondée sur une grille de parallèles et de méridiens, recouvrant toute la surface du globe séparés par des espaces réguliers. De cela, Ptolémée fut l'inventeur.

Dans les universités d'Europe, scolastiques aristotéliciens et mathématiciens humanistes vont, tout au long du XVe siècle, s'affronter à propos de ces deux doctrines rivales de la représentation de la terre, et ce seront les navigateurs, forts de leur expérience *de visu*, qui arbitreront leur querelle.

Dejá, lors de l'examen, en 1483-84, par les cosmographes de D. João II de son projet de circumnavigation du globe, Colomb avait soutenu, selon l'historien João de Barros, que la découverte des Açores et d'une grande étendue de la côte africaine, permettait de croire que

> «... la Nature ne devait pas se révéler désordonnée dans la composition de l'orbe universel, au point d'avoir voulu y accorder à l'élément de l'eau une plus grande part qu'à la terre sèche pour la vie et la reproduction des êtres vivants.»[14].

Les convictions des scolastiques, au Portugal, ont peut-être été ébranlées par cette émergence, au loin dans l'Océan, de terres comme les Açores, ainsi que par la découverte par Diogo Cão en 1483-84 d'une vaste étendue de la côte africaine, jusqu'à la latitude de 13º26′ S[15], mais nous n'avons aucun texte qui l'atteste.

En Allemagne, par contre, en 1493, Jérôme Münzer écrira de Nuremberg à D. João II pour le conjurer d'abandonner la doctrine de Buridan et d'Albert de Saxe (sans toutefois les nommer), de ne pas écouter des

> «... auteurs sans expérience, qui disent qu'un quart seulement de la terre est au-dessus du niveau de la mer, et que la terre, pour ses trois autres quarts, est noyée sous la mer.»[16].

Le Roi de Portugal avait-il encore besoin, à cette date, de tels conseils?

Les attaques des humanistes contre les aristotéliciens, surtout en Italie et en Allemagne, avaient pour cible ce bastion de la scolastique que restait l'Université de Paris, d'où d'ailleurs, au début du XVIe siècle, nombre d'Espagnols devaient ramener, en regagnant la péninsule, un aristotélisme sclérosé et désuet. Tel sera le cas d'un Gaspar Lax (1487-1560) qui, dans ses *Quaestiones physicales*, publiées à Saragosse en 1527, expose encore la doctrine du décuplement des volumes des éléments[17].

L'humaniste italien Antonio de Ferrariis (1444-1517), auteur du *Liber de situ elementorum*, composé en 1501 et publié seulement en 1558 à Bâle, s'en prend avec vigueur à la «philosophie parisienne», cette «philosophie des accidents» *(accidentali philosophia)*, puis accable de ses sarcasmes la théorie de Buridan, dont il avait pris connaissance à travers l'oeuvre d'un certain Johannes Magistris[18], lequel, dit Ferrariis, s'applique à imiter Duns Scot *(scotizare)* et, en «scotisant», suit Albert le Grand (c. 1193-1280)[19].

A propos d'Albert, Ferrariis déclare que ses opinions concernant la terre et l'eau sont «[tellement] obscures qu'elles me forcent à penser que lui-même ne voulait pas les comprendre». «Mais», ajoute-t-il avec une certaine indulgence pour les scolastiques, «que la faute soit imputée aux circonstances, car la *Cosmographie* de Ptolomée [il s'agit de sa *Géographie]*, n'était pas encore parvenue aux Latins et [les oeuvres de] Strabon et de Pline étaient lues par peu de gens»[20].

Ferrariis se montre pleinement acquis à la théorie du globe de Ptolémée.

> «... la terre et l'eau font un seul globe ...»[21] et plus loin: «... ces deux éléments [l'eau et la terre] ont, comme Dieu l'a ordonné, leurs étendues imbriquées et jointes ... (*intermixtas et consitas*)»[22].

Ferrariis s'était-il inspiré, pour sa définition de cette imbrication des éléments de la terre et de l'eau, de celle de Pline, qui disait qu'ils «... s'unissaient en un mutuel enlacement» *(mutuo implexu iugerentur)*?[23]. Cela nous semble fort probable: les hommes de la Renaissance reprenaient presque toujours des schémas de l'Antiquité, sur lesquels ils bâtissaient leurs concepts.

Ferrariis s'étend par ailleurs assez longuement sur les voyages de découvertes, ceux des Portugais en particulier, mais en défendant sa préférence pour la théorie ptoléméenne du globe terrestre contre la doctrine des éléments des scolastiques, il n'invoque jamais en termes explicites l'expérience des navigateurs.

Ce sera le fait d'un autre Italien, Raffaele Maffei da Volterra (1451-1522), auteur d'une encyclopédie plutôt médiocre, *Commentariorum urbanorum libri XXXVIII*, publiée à Rome en 1506 et suivie, au cours du XVIe siècle, de nombreuses autres éditions à Paris, Bâle, Lyon et Francfort.

Bien renseigné sur les voyages des Portugais dans l'hémisphère austral (ceux de Cabral (1500) et de Gonçalo Coelho (1501) par le Cardinal portugais Jorge da Costa, alors résidant à Rome[24], Maffei est le premier à relever l'incompatibilité entre l'expérience des navigateurs et la doctrine scolastique du décuplement du volume des éléments, dont l'eau et la terre.

> «Il n'est pas légitime, écrit Maffei, «de croire à autre chose que ce qui est observé [par des marins portugais]. En effet [devant l'expérience de ces marins] le lieu de l'eau [selon la représentation aristotélicienne des éléments] n'existerait pas [il n'y aurait pas de place pour tant d'eau] si selon l'opinion du philosophe [le volume] de l'eau dépassait de dix fois celui de la terre.»[25].

Le passage de Maffei cité fut entièrement recopié, mot à mot, par l'ancien étudiant portugais de l'Université de Paris Pedro Margalho (1474-1556) et inséré dans son *Physices Compendium*, qu'il publia en 1520 à Salamanque, ville où il enseigna de 1517 à 1529, avant de regagner définitivement le Portugal[26].

Le plagiat, fréquent chez les auteurs de l'époque, devient excusable chez Margalho [27] lorsqu'on s'aperçoit que, immédiatement après le passage en question, à la ligne suivante, il ajoute une phrase d'une bien plus grande originalité, et qui est de son propre cru:

> «C'est pourquoi il est manifeste», écrit-il, «que les éléments de l'eau et de la terre sont imbriqués l'un dans l'autre, et il ne paraît pas probable, comme le disent [les navigateurs du roi de Portugal] qu'il existe un degré de l'eau, auquel en opposition ne corresponde pas un degré de la terre, et il semble probable que ces deux éléments sont à égalité étant donné que, à l'une quelconque des parties de l'un, correspond par opposition, une partie de l'autre.»[28].

Si le concept de l'imbrication des éléments de la terre et de l'eau avait déjà été formulé par Ferrariis, les expressions «degrés de la terre» et «degrés de l'eau», définies par Margalho comme étant identiques, apportent une bien plus grande précision et illustrent la confrontation en cours entre la pensée scolastique et la représentation ptoléméenne, confrontation qui devait aboutir, grâce à l'arbitrage déterminant des navigateurs portugais, au triomphe de cette dernière.

Margalho omet cependant de spécifier explicitement, en termes de distance terrestre, l'identité des «degrés de la terre» et des «degrés de l'eau». C'est ce que fera

Pedro Nunes (1502-1578) dans son *Tratado da Sphera* (1537):

> «... à un degré du ciel correspond un même nombre de lieues ou de milles sur terre et en mer: ce ne pourrait pas être le cas, si tous deux ne formaient pas une seule sphère.»[29].

Au tournant du XVIe siècle, Christophe Clavius (1537-1612), le grand mathématicien du Collège Romain (il avait séjourné à Coimbra), consacre à cette question, dans son commentaire sur la *Sphère* de Sacrobosco (1581), (17 éditions), un long chapitre, analyse exhaustive mettant définitivement fin, pour l'Europe tout entière, aux hésitations d'aristotéliciens attardés[30].

Jusqu'ici, nous pensons avoir montré comment, dans l'Europe du sud, en Italie, au Portugal et en Espagne, les principes mathématiques de la *Géographie* de Ptolémée, dont la justesse s'était vue confirmée par l'expérience des navigateurs, rendirent entièrement caduque, au cours des deux premières décennies du XVIe siècle, la science scolastique de ces pays, en ce qui concerne les proportions de la terre et de l'eau.

Voyons maintenant quel fut l'impact en France des nouvelles données apportées par les Découvertes.

A l'Université de Paris, les seuls, à notre connaissance, au début du XVIe siècle, à s'intéresser à la question des sphères de la terre et de l'eau, sont l'Ecossais Jean Mair (1467/68-1550), le Flamand Jean Dullaert (1470-1513), tous deux professeurs au Collège de Montaigu, et le médecin français Jean Fernel (1497-1558) professeur au Collège de Sainte Barbe[31].

Dans son *Commentaire* sur les *Sentences* de Pierre Lombard, Jean Mair expose la théorie de Jean Buridan, selon laquelle le centre de magnitude de la terre se trouve décalé par rapport au centre de l'univers, auquel correspond le centre de gravité de la terre. Puis, tout de suite après, Mair expose les preuves astronomiques de la sphéricité de la terre que donne Ptolémée dans l'*Almageste* et, pour les illustrer, il note que les habitants d'Edimbourg observent la Grande Ourse et le Dragon à une plus grande hauteur que ne les voient les Parisiens. Sans se rendre compte que, dans l'hypothèse où la terre ne se trouverait pas au centre de l'univers, il serait impossible de tirer quelque conclusion que ce soit d'observations astronomiques, Mair ne craint pas d'invoquer «les navigations des Portugais et des Espagnols, qui ont rapporté que, lors de leurs voyages «à la recherche des Indiens», des étoiles qui sont toujours visibles pour nous, cessaient de l'être. Il ne semble pas s'apercevoir que ces mêmes voyages avaient déjà invalidé la théorie de Buridan, qu'il venait d'esquisser[32].

Jean Dullaert, qui fut l'élève de Mair, se révèle plus conscient que son maître de la signification de l'expérience des navigateurs portugais, et un peu plus au courant de la nature de leurs voyages.

Dans ses *Questions* sur la *Météorologie* d'Aristote, Dullaert évoque le voyage portugais de 1501 commandé par Gonçalo Coelho, le long de la côte est du Brésil,[33] voyage qu'il connaît à travers le récit attribué à Améric Vespuce et publié dans la *Cosmographiae introductio* de Matthias Ringmann[34].

Selon Dullaert, le voyage portugais avait non seulement apporté la preuve que, contrairement à ce que prétendait Averroès, les étoiles de l'hémisphère austral sont plus nombreuses et plus brillantes que celles de l'hémisphère septentrional, mais également, et contrairement à ce qu'affirmait Paul de Venise (c. 1370-1429) – seul un quart de la terre ne serait pas couvert par les eaux[35] –, que la découverte de nombreuses îles vers le sud et vers l'ouest et d'une quatrième partie du monde, l'Amérique, avait démontré qu'une grande partie de la terre est émergée et «sans doute autant ou davantage vers le pôle austral que vers le pôle septentrional.»[36].

Dullaert sera effectivement le premier, à l'Université de Paris, à tirer une conclusion nette de l'expérience des Portugais pour invalider la doctrine de Buridan.

Comparé à Mair et Dullaert, le médecin Jean Fernel apparaît immédiatement comme un esprit bien supérieur aux deux scolastiques, et très au fait des Découvertes portugaises, comme le prouve la dédicace très élogieuse à D. João III qui précède sa *Cosmotheoria* (1528).

Fernel, qui connaît bien les principes des cosmographies d'Aristote et de Ptolémée, soutient sans hésitation la doctrine de l'alexandrin concernant la terre et l'eau; il refuse la théorie de Buridan qu'il «estime découler d'une fausse hypothèse». «La face habitée de la terre», affirme-t-il, reprenant la formule de Ptolémée dans sa *Géographie*, «constitue avec l'eau qui la relaie une seule et même surface de forme convexe, dont le centre coïncide avec le centre de l'univers»; et il invoque comme preuve «... les écrits des hommes de science et le fidèle témoignage des navigateurs»[37].

Adoptant la représentation ptoléméenne des océans situés dans des bassins isolés, Fernel décrit la terre comme ayant «... l'aspect d'un globe de bois parsemé d'un grand nombre de creux où l'eau puisse être recueillie»[38].

S'il ne cite pas nommément Fernel, l'astronome polonais Nicolas Copernic reprendra en 1543, dans son célèbre ouvrage sur l'héliocentrisme, cette définition:

> «L'Océan», écrit-il, «entourant donc la terre [...] remplit les creux les plus profonds. C'est pourquoi il devait y avoir moins d'eau que de terre pour éviter

que l'eau n'engloutît toute la terre. Et toutes deux [l'eau et la terre]», ose affirmer Copernic face aux aristotéliciens, «tendent vers le même centre en raison de leur gravité [...] Et nous ne devons pas écouter certains péripatéticiens, qui ont enseigné que l'eau [...] est dix fois plus grande que toute la terre»[39].

Ces énoncés péremptoires, qui battent en brèche les principes fondamentaux de la philosophie scolastique, permettront à Copernic de formuler sa doctrine de l'héliocentrisme, où la terre devient une planète parmi d'autres. Il est évident qu'il ne lui était pas possible d'envisager une planète tournant autour du soleil, si elle était composée de sphères de terre et d'eau, ayant les centres décalés l'un par rapport à l'autre. Sans le choc des Découvertes maritimes et l'impact de l'expérience des navigateurs, jamais Copernic n'aurait réussi à faire sauter l'obstacle épistémologique que fut la théorie médiévale des deux sphères de l'eau et de la terre. Grâce à cette victoire sur les aristotéliciens, Copernic a pu déclarer que «... la terre est entraînée dans son mouvement circulaire avec l'élément aqueux qui lui est conjoint»[40].

Il nous semble donc évident, d'après l'analyse que nous venons de faire des auteurs cités, que sans la révolution géographique des Découvertes, la révolution astronomique du XVIe siècle n'aurait pu se réaliser de la manière que nous connaissons.

NOTES

1 Aristote, *De Caelo*, II, 4, 287a; *Physique*, IV, 5, 212b; *Météorologiques*, II, 2, 354b.

2 P. Duhem, *Le Système du monde*, vol. IX, 1958, p. 89.

3 Tout déplacement hors de son lieu naturel ne pourrait être que la conséquence d'une violence et aucune violence, selon Aristote, ne pourrait se perpétuer.

4 P. Duhem, *op. cit.*, p. 88.

5 Michel Scot, «Super auctore spere cum questionibus», in Lynn Thorndike, *The Sphere of Sacrobosco and its commentators*, Chicago, 1949, pp. 248-342.

6 Jean Buridan, *Questiones super libris quattuor de Caelo et Mundo*, Lib. II, Qu. 7, éd. E. A. Moody, Cambridge, Mass., 1942, pp. 159-160.

7 Albert de Saxe, *Quaestiones et decisiones physicales insignium virorum: tres libros de Caelo et Mundo*, Paris, 1518, f. cxix (v.). Cf. la traduction française du texte dans notre étude: *La naissance d'un concept nouveau à l'époque des Grandes Découvertes maritimes: le globe terraqué*, Centro de Estudos de História e Cartografia antiga, Lisbonne, vol. 171, p. 337.

8 Sur la notion d'obstacle épistémologique, cf. Gaston Bachelard, *La Formation de l'esprit scientifique*, Paris, 1977, Chap. 1.

9 Cf. P. Duhem, *Le Système du monde*, vol. IX, Paris, 1958, pp. 92-96.

10 Michel Scot, *op. cit.* in *éd. cit.*, p. 263.

11 Pedro Ciruelo, *Uberrimum sphaerae mundi comentum...*, Paris, 1498, Sign. e. iiij (r.).

12 Vespasiano da Bisticci, *Vite di Uomini illustri*, éd. Bologne, 1893, vol. 3, pp. 9-10 et pp. 216-217.

13 Claudius Ptolémée, *Géographie*, éd. et trad. de l'Abbé Halma, Paris, 1828, Livre I, Chap. II, p. 11.

14 João de Barros, *Asia*, Lisbonne, 1552. Dec. I, Liv. III, Cap. XI, éd. Coimbra, 1932, p. 112.

15 A. Fontoura da Costa, *Às portas da Índia em 1484*, Lisbonne, 1935, p. 17.

16 Jérôme Münzer, Lettre au roi D. João II de Portugal, Nuremberg, 14/7/1493, in Henri Vignaud, *Histoire critique de la Grande Entreprise de Christophe Colomb*, Paris, 1911, vol. II, pp. 620-25.

17 Gaspar Lax, *Quaestiones Physicales*, Saragosse, 1527, sign. L. vi. (r.).

18 Johannes Magistris, *Quaestiones perutiles super tota philosophia naturali*, Parme, 1481; Venise, 1487; Heidelberg, 1488; Bâle 1490. Dominicain, Magistris fut professeur de théologie sacrée à Rouen et Inquisiteur Général de France. Cf. Quétif et Echard, *Scriptores Ordinis Praedicorum* (1719), vol. I, p. 782.

19 Le texte d'Albert, repris par Magistris et cité par Ferrariis, est le «De Meteoris». Cf. trad. franç. in Youssouf Kamal, *Monumenta Cartographica Africae et Aegypti*, reprint Francfort, 1987, vol. IV, pp. 396-397.

20 Antonio de Ferrariis, *Liber de situ elementorum*, Bâle, 1558, pp. 56--58. Sur Ferrariis, cf. Roberto Almagià, «Le opinione e le conoscenze di Antonio de Ferrariis», in *Rivista Geog. Ital. (1905), fasc.* VI-VIII, pp. 450-463, et Lynn Thorndike, *Science and Thought in the Fifteenth Century*, New York, 1929, pp. 211-216. Strabon, autre redécouverte de la Renaissance italienne (édition princeps 1469-1473), avait, à l'instar de Ptolémée, posé en principe que «la terre avec la mer, est de forme sphérique, ne constituant qu'une seule et même surface». Cf. Strabon, *Géographie*, II, 55, éd. et trad. de Germaine Aujac, Paris, 1969, T. I, 2e partie, p. 83.

21 Antonio de Ferrariis, *op. cit.*, p. 38: «... *ex terra & aqua unum facit globum...*»

22 Id. p. 72: «... *Haec duo elementa, ut Deo institutum est, intermixtas habent regiones, & consitas...*»

23 Pline, *Hist. Nat.*, Livre II, Cap. LXVI.

24 Cf. Vsevolod Slessarev, «Raphael Maffei's contribution to the Portuguese Discoveries», in *Actas do Congresso Internacional de História dos Descobrimentos*, Lisbonne, 1961, vol. III, pp. 551-575, et G. Caraci, «Americo Vespucci al recente Congresso di storia delle scoperte di Lisbona, 4-12 Settembre 1960, in *Memorie Geografiche*, vol. VII, Rome, 1961, pp. 140-158.

25 Raffaele Maffei da Volterra, *Commentariorum urbanorum libri XXXVIII*, Rome, éd. 1530, f. 4 (v.) *(neque sane aliud credere par est quam hoc apparens, quod aquae locus non esset, si decuplo terram ex ipsius philosophi sententia superaret)*. Je suis très reconnaissant au Professeur Léon Bourdon pour sa traduction de ce texte particulièrement obscur.

26 Sur Margalho, cf. Francisco Leitão Ferreira, *Notícias cronológicas da Universidade de Coimbra*, Coimbra, 1729, vol. I, pp. 475-482; Luís de Matos, *Les Portugais à L'Université de Paris entre 1500 et 1550*, Coimbra, 1950, p.11; Armando Cortesão, *Cartografia e Cartógrafos portugueses dos séculos XV e XVI*, Coimbra, 1935, vol. I, pp. 75-77; J. S. da Silva Dias, *A política cultural da época de D. João III*, Coimbra, 1969, vol. I, pp. 287-297; Idem, *Os Descobrimentos e a problemática cultural do século XVI*, Coimbra, 1973, pp. 30-32; Luís Ribeiro Soares, «O insólito doutoramento de Pedro Margalho em Valhadolid 1517», in *Academia Portuguesa de História, Anais*, II Série, vol. 25, 1979, pp. 87-157.

27 Nous ne nous sommes aperçu du plagiat de Margalho qu'après la publication de notre livre *De la terre plate au globe terrestre*, et cela grâce à la connaissance de l'étude de G. Caraci, citée ci-dessus, dont Madame Ilaria Caraci nous a obligeamment fourni une photocopie.

28 Pedro Margalho, *Physices Compendium*, Salamanque, 1520, f. vii (r.) C'est également au Professeur Bourdon que je dois la traduction de ce texte difficile.

29 Pedro Nunes, *Tratado da Sphera*, Lisbonne, 1537, éd. fac-similé de J. Bensaude, Munich, 1915, f.a.iiii (r.).

30 Christophe Clavius, *In Sphaeram Ioannis de Sacrobosco commentarius*, Rome, 3.e éd. 1581, p. 118 («Tous les navigateurs affirment qu'ils ont pu constater maintes fois par expérience qu'il y a autant de stades ou de milles en un degré à la surface de la terre qu'en un degré à la surface de la mer»).

31 Sur Mair et Dullaert, cf. Charles H. Lohr, *Latin Aristotle Commentaries. II Renaissance Authors*, Florence, 1988, pp. 237-239 et pp. 128-129. Sur Fernel, cf. J. Quicherat, *Histoire de Sainte Barbe*, Paris, 1860-1863, vol. I, pp. 176-186, et Charles Sherrington, *The Endeavour of Jean Fernel*, Cambridge, 1946 (surtout sur ses travaux sur la médecine).

[32] Jean Mair, *In Secundum librum Sententiarum*, Paris, 1510, éd. Paris, 1519, f. lxxxiiii (v.). Dans son *Commentaire* sur le *De Caelo* d'Aristote (1526), Mair ne fera aucune allusion aux Découvertes des peuples ibériques. Cf. Jean Mair, *Octo libri Physicorum*, Paris, 1526, ff. 66 (r.) - 68 (v.) (L'ouvrage contient son *Commentaire* sur le *De Caelo)*. Sur Mair, voir également Edward Grant, «In Defense of the Earth's Centrality and Immobility», in *Transactions of the American Philosophical Society*, vol. 74, pt 4, (1984), p. 16.

[33] Sur le voyage de Coelho, cf. Max Justo Guedes, «As primeiras expedições portuguesas e o reconhecimento da costa brasileira», *Revista Portuguesa de História*, T. XIV (1970), pp. 252-265.

[34] [Matthias Ringmann], *Cosmographiae introductio*, Saint-Dié, 1507, cap. IX. L'ouvrage est souvent attribué à Martin Hylcomilus Waldseemüller. Selon Franz Laubenberger («The naming of America», in *Sixteenth Century Journal*, vol. XIII, N° 4 (1982), p. 99), le vérita ble auteur en est Ringmann.

[35] Paul de Venise, *De Compositione mundi*, Venise, 1498, éd. Pari s. d. (sans pagination).

[36] Jean Dullaert, *Habes humanissime lector librorum metheororur Aristotelis facilem expositionem & questiones super eosdem*..., Pari 1514, sign. C.i (r.).

[37] Jean Fernel, *Cosmotheoria*, Paris, 1528, f. 1 (V.). Fernel démontr également que la doctrine du décuplement des proportions des él ments est insoutenable. Idem, f. 4 (v.)–5 (r.).

[38] Idem, *loc. cit.*

[39] Nicolas Copernic, *De revolutionibus orbium coelestium*, Nurember 1543, Lib. I, Cap. III.

[40] Idem, Lib. I, Cap. VIII.

XIII

PORTUGUESE AND SPANISH ATTEMPTS TO MEASURE LONGITUDE IN THE SIXTEENTH CENTURY

Attempts to measure longitude were made in the sixteenth century for three different purposes: astrology, cartography, and navigation. Here we shall only be concerned with cartography and navigation. The act of measurement was undertaken in three very different circumstances: a) on repeated occasions at a permanent station, b) at a temporary landing place where a ship had stopped, c) at sea. Four different methods were used in the sixteenth century with differing degrees of success (or failure): the hourglass, eclipses, the lunar distance method, and magnetic variation of the compass. The latter method rapidly proved to be illusory.[1]

The two astronomical methods were practised from the very earliest voyages of discovery at temporary landing spots, but were found to produce huge errors. As regards their use at sea, the Portuguese Simão de Oliveira and the Spaniard Garcia de Cespedes, in their navigation manuals both published in 1606, were emphatic that astronomical methods were useless because of the lack of sufficiently accurate tables and because of their impracticability for the navigator.[2]

Spaniards and Portuguese appear to have regularly navigated across the Atlantic in the sixteenth century relying simply on estimation of distance run. Martin Fernandez de Enciso wrote in 1519:

> Sailors calculate distance in an East-West direction in nights and days and with an hourglass and the calculation is reasonably correct for those who know their ship well and how much it sails in an hour; [...] Because their estimation is approximate [...] they over-estimate rather than under-estimate the number of leagues [of their voyage] so as to be warned of their approach to land, rather than running upon it suddenly (*porque hagan con la tierra antes que llegan a ella*).[3]

Estimated distance in leagues was often converted into degrees of longitude (rather than the reverse), and compared with data in Ptolemy's *Geography* in order to situate newly discovered coastlines within the framework of the latter's *oikoumene*. Thus Verrazano in 1524, having sailed 1,200 leagues to a previously undiscovered land in the West, converted this distance into degrees of longitude by elaborate calculation in order to allow for the convergence of the meridians at the latitude at which he sailed. This he did to see how close he had got to China, the goal of his voyage.[4]

Both the astronomical methods mentioned (eclipses and lunar distance) were vitiated, as has been noted, by the lack of sufficiently accurate tables to provide the base line data necessary. The two tables most commonly used in the first half of the sixteenth century were Regiomontanus' *Ephemerides,* first published in 1474, and Abraham Zacuto's *Almanach Perpetuum* published at Leiria (Portugal) in 1496. The errors in both these tables have been calculated by Hermann Wagner.[5]

Columbus' two observations of eclipses during his voyages of 1494 and 1504, while anchored at islands in the Caribbean, would appear to have been made not for navigational purposes to determine the position of his ship but, like Verrazano, with the object of situating his newly discovered land in the context of the world map.[6] He appears to have used both Regiomontanus' and Zacuto's tables; his observations produced errors of longitude of 22°3′ and 38° respectively.[7] The tables, though, were not the only source of error.

Vespucci's alleged determination of longitude on the coast of Venezuela in 1499, in which he claimed to have used the lunar distance method, has been shown to be a fiction in which he probably extrapolated from Columbus' longitudes in the Caribbean.[8] It has been suggested that Vespucci may have learnt of the lunar distance method from Regiomontanus (1436-76), who outlines a very similar idea for calculating the distance of comets.[9]

Methods for determining longitude were given particular importance during the preparations for Magellan's voyage (1519-22). Magellan had proposed to take with him a Portuguese 'astrologer', Rui Faleiro, who had drawn up a set of instructions (*Regimento*) containing 30 chapters on the measurement of longitude.[10] Faleiro became insane and never sailed with the expedition. He was replaced by a Spaniard, Andrès de San Martin, whom the Portuguese historian Barros describes as a man 'learned in astronomy'.[11] Magellan and San Martin took with them Faleiro's *Regimento,* which was later to come into the hands of Barros.[12] Long thought to be lost, it was recently identified by the late Admiral Teixeira da Mota as an anonymous undated Spanish text published in 1893.[13] Teixeira da Mota, from internal evidence, dated it to between 1516 and 1519.[14]

Faleiro's *Regimento* contains three methods for calculating longitude: a) from the latitude of the Moon (*por la latitude de la luna saber la longitude*), b) by the lunar distance method, c) by magnetic variation.

It is commonly held that Johann Werner was the first to describe the lunar distance method in 1514.[15] That Faleiro drew on Werner is doubtful, since the two descriptions are very different. Where, then, did Faleiro get the method from? Hermann Wagner has shown that the source is not, as was claimed by Joaquim Bensaude, the *Libros del Saber* of Alfonso X (second half of the thirteenth century).[16] That the principle of the method was indeed known in the Middle Ages is suggested by the fact that the *Theorica Planetarum* (late twelfth century), attributed to Gerard of Cremona (*c.*1114-87),[17] describes the lunar distance method, though somewhat elliptically as follows:

> When the Moon is on the meridian, if you compare her position with that given in the lunar tables for some other locality, you may determine the difference in longitude between the place where you are and that for which the lunar tables were constructed by noting the differences in the position of the moon as actually observed and as recorded in the tables. It will not be necessary for you to wait for an eclipse.[18]

During Magellan's voyage, the pilots, when asked by Magellan to evaluate the usefulness of Faleiro's *Regimento*, all, with the exception of Andrès de San Martin, declared that it was useless for navigational purposes. San Martin, however, stated that in his opinion the lunar distance method was practicable and he endeavoured to practise it on shore during landings.[19] According to Barros, quoting San Martin's papers, the latter, using Zacuto's and Regiomontanus' tables, made five observations along the eastern coast of South America. Only the figures for the first and fourth observations have reached us. One was out by 231°3′, an error which San Martin fully recognized; the other was only 37′ out. Laguarda Trias has, however, shown that the latter concerned a solar eclipse which was not visible in Europe and therefore no base line data for it would have been available in the tables supposedly used.[20] How such a small error was achieved remains a mystery.

The need to measure longitude for cartographical purposes came to be of vital importance during the Badajoz conference in 1524, at which Portuguese and Spanish delegates sought agreement on whether the islands of the Moluccas in the Pacific lay within the Portuguese or Spanish sphere of control as defined by the treaty of Tordesillas. By the treaty of Tordesillas of 1494 an imaginary meridian line situated 370 leagues west of the Cape Verd islands, and passing through a part of Eastern Brazil, defined the separation of the Portuguese and Spanish rights in the Atlantic.[21] The central question (unresolved) at the Badajoz Conference turned upon the means to be used to determine an imaginary line in the Pacific (the anti-meridian), 180° from the Tordesillas line.

While both sides admitted the superiority of astronomical methods over sea charts and globes, the Portuguese sought to defer a decision till such methods could be successfully applied. The Spaniards wanted a quick decision and resigned themselves to relying on charts and globes. Both sides distorted their maps to their advantage. The Conference meandered on for two months and broke up without any agreement.[22]

Magellan had already stated the Spanish claim modestly before his departure in a memoir (*Membranza*) to the Spanish crown, dated September 1519. In it he laid down the estimated distance in longitude between the Tordesillas line and the Moluccas, placing three of the islands 2½° on the Portuguese side of the anti-meridian, and two of them 4° on the Spanish side.[23] Another more ambitious Spanish claim, also stated before Magellan's voyage, made the anti-meridian pass through the mouth of the Ganges, thus pushing the Moluccas, as well as Java, China and Japan, far into the Spanish sphere.[24] This claim was repeated at the Badajoz Conference.[25]

The Portuguese claim was recounted half a century later by the Portuguese historian Jerónimo Osório in his work *De Rebus Emmanuelis,* published in 1571. Osório states, in referring to the controversy over the Moluccas, that experts, trained by the Portuguese mathematician Pedro Nunes (1502-78), had, with the aid of astronomical methods (which are not described), determined the angular distance between Lisbon and the mouth of the Indus to be 90°. Between the latter and the Moluccas lay a distance of 42° (not measured astronomically). According to Osório, these figures, added to the 36° separating Lisbon from the Tordesillas line, put the Moluccas 12° inside the Portuguese sphere.[26] The mouth of the Indus, identified in contemporary texts as the Portuguese stronghold of Diu in India (71°03′ Long. E),[27] would be actually 80° East of Lisbon, but as Ptolemy had put it at about 107° East of Lisbon, the figure of 90° was an improvement for the Portuguese, as they wanted to bring the Moluccas as far westward as possible, into their sphere.

The question yet remains: was the observation at 'the mouth of the Indus' made before or after the Badajoz Conference? It is possible that it was made before. An undated document believed to have been drawn up for the Badajoz Conference of 1524[28] refers to two undated observations of longitude from eclipses, one by Diogo Lopes de

Sequeira 'between India and Arabia' (*sic*), the other by Bernaldo Pirez '20 or 25 leagues east of Malacca'. No figures are given.[29] These longitude observations may have been the earliest made outside Europe. No reference to them, however, appears in the proceedings of the Badajoz Conference, such as have been published to date.

An unpublished, undated memoir addressed to the Portuguese King D. João III by the cartographer Lopo Homem does, however, throw some light on the progress of longitude measurement in the East in the years between 1529 and 1557 (the year of D. João III's death), albeit with ambiguous results. A transcription of the document, which is in the Paris Bibliothèque Nationale, was generously offered to the author by Professor Léon Bourdon of the Sorbonne, to whom grateful thanks are here expressed.[30] In it, Lopo Homem declares that Pedro Nunes had drawn up a master chart of the route to India, on which longitudes had been determined by observation of lunar and solar eclipses (the meridians of Lisbon and Chaul on the west coast of India are specifically mentioned) and that Portuguese pilots had been ordered by the crown to use charts copied from it. Use of the new chart, writes Homem, had caused shipwrecks and accidents; Portuguese pilots were having their charts secretly made in Spain and elsewhere according to traditional methods and Homem therefore entreats the King to rescind the obligation to follow Nunes' chart. From the pilots' point of view, Lopo Homem's protests were fully justified, since the degree of accuracy of the astronomical methods then available for calculating longitude was still quite insufficient for cartography.

A notable improvement in the calculation of longitude for cartographical purposes was achieved progressively over a period of years, with a series of observations of eclipses at the fixed station of Mexico City. In 1541 the Viceroy wrote that Mexico City was situated at 8 hrs 2 mins 34 secs W from the base meridian of Toledo (Spain).[31] In 1544 Joanoto Duran observed two lunar eclipses in Mexico and, improving on his predecessor, put Mexico City in 103° from Toledo.[32] From observations of the lunar eclipses of 29 September 1577 and of 15 September 1578, the figure was reduced to 100° W of Toledo,[33] an error of 4°52′. The correct figure is 95°8′ W of Toledo. By 1584 the error had been brought down to putting Mexico City only 23 kms too far west.[34] Was such a close figure a fluke or the result of a real mastery of the process of measurement?

For Peru, Hieronimo de Chaves in 1576 put Puerto Viejo (today Porto Viejo in Ecuador) in 86°30′ W of Greenwich.[35] Garcia de Cespedes, from an observation made 4 September 1588, put it in 82°53′ W,[36] an error of only 2°25′.

For India results were not nearly so good. The famous Jesuit mathematician Christopher Clavius in 1593 put the port of Calicut on the west coast of India in 96° 50′ E of Greenwich.[37] David Origanus in 1609 gave a better figure: 90°32′ E[38] and Kepler in 1627 got closer, though yet with considerable error, with 87°45′ E.[39] Kepler was still 12° too far east; the true longitude of Calicut being 75°45′ E.

In China at the end of the sixteenth century, the Jesuit Matteo Ricci, from two eclipses observed in or just before 1583, put the city of Macao 'more or less' in 124° E of the Canary Islands,[40] i.e. 108°50′ W of Greenwich, the true value being 113°36′ E of Greenwich, an error of 4°46′. From the eclipse of 8 October 1612, another Jesuit Fr Vreman improved on this, using Origanus' tables, and put Macao in 116°32′ E of Greenwich, an error of 2°56′.[41] For the longitude of Peking, the Jesuit missionary Sabatino de Ursis was not so fortunate. From the lunar eclipse of 15 May 1612, he put the city of Peking in 125°52′ E of Greenwich, an error of 9°27′ too far east,[42] the true figure being 116°25′ E. Fr Vreman again improved on this figure and from the eclipse of 8 November 1612 put Peking in longitude 12°47′ E, an error of only 4°22′ too far east.

Lists of cities with their longitudes are common in the sixteenth century and indeed are already found in the European Middle Ages,[43] and in Medieval Portugal.[44]

The first to publish a list of longitudes to include newly discovered places outside Europe was, as far as can be ascertained, Johann Schöner in his *Luculentissima quaedam terrae totius descriptio* (Nuremberg, 1515). Other lists are given by Hieronimo de Chaves in the 1576 Seville edition of his *Chronographia* (only places in the New World are mentioned), by André de Avelar in his *Reportorio dos temps* (Lisbon, 1585), by Christopher Clavius in his commentary on Sacrobosco's *Sphaera*,[45] by David Origanus in his *Novae motuum coelestium ephemerides* (Frankfurt, 1609), and by Johann Kepler in his *Rudolphine Tables* (Ulm, 1627).

Kepler was claimed by Oscar Peschel in 1865 to have been the first to have produced tolerably good values for European cities,[46] and in fact some of them show errors of less than 10′. Kepler's longitudes for places outside Europe have, however, much larger errors of up to 10°, though they are still better than Origanus' values.

It is not clear (Matteo Ricci's observations excepted) to what extent, if at all, in the sixteenth century these lists of longitudes were drawn on for cartographical purposes in any systematic fashion. In the Middle Ages it is well established that they had no influence whatever on map-making.[47] The first world map to be based on Kepler's Rudolphine tables of longitude appears to have been that published by Philipp Eckebrecht in Nuremberg in 1630.[48]

Acknowledgements

Reprinted from *Vistas in Astronomy*, Vol. 28, 235-41, with kind permission from Elsevier Science Ltd, The Boulevard, Langford Lane, Kidlington OX5 1GB, U.K. I am grateful for the suggestion of Lt-Cdrs D.W. Waters and H.D. Howse that this article should be republished to reach a wider audience.

References

1 In 1514 João de Lisboa actually claimed the existence of a law relating magnetic variation to longitude. cf Luís de Albuquerque, *O 'Tratado da Agulha de Marear' de João de Lisboa, Reconstituição do seu texto, seguida de uma versão francesa com anotações*, Centro de Estudos de Cartografia Antiga, Secção de Coimbra, Vol. CL, Junta de Investigações Científicas do Ultramar (Coimbra, 1982).

2 Simão de Oliveira, *Arte de Navegar (*Lisbon, 1606), 144, and Andrés Garcia de Cespedes, *Regimento de Navegación* (Madrid, 1606), fo. 108 (v.).

3 Martin Fernandez de Enciso, *Suma de Geographia* (Seville, 1519), sign.b.viii (r.).

4 Giovanni da Verrazano, *Voyage* (1524), Italian text in Lawrence C. Wroth, *The Voyages of Giovanni da Verrazano* (Yale Univ. Press, New Haven and London, 1971), 131-32.

5 Hermann Wagner, 'Die legende der Längenbestimmung Amerigo Vespucci's nach Mondabständen, 23 Aug. 1499' in *Nachrichten d.K. Gesell. d. Wiss.*, Göttingen, Math-Physik Kl. (1917), Heft 2, 279-80.

6 I. Luzzana Caraci, 'Colombo e i longitudini' in *Bol. Soc. Geog. Ital.* Tome 9 (1980), 519.

7 Hermann Wagner, 'Die legende der ...', 275.

8 *Ibid.*, 284-98, and J.W. Stein SJ, 'Esame critico intorno alla scoperta di Vespucci circa la determinazione delle longitudini in mare mediante le distanze lunari' in *Memorie della Società Astronomica Italiana*, Vol. 21, No. 4 (Rome, 1950), 345-53.

9 Johann Müller (called Regiomontanus), *De Cometae magnitudinis longitudinaeque ac de loco ejus vero problemata XVI* (Nuremberg, 1531), cf. Ernst Zinner, *Leben und Wirken des Joh. Müller von Königsberg (Regiomontanus)*, 2nd revised edition (Osnabruck, 1968), 197 and 207.

10 Fernão Lopes de Castanheda, *História do Descobrimento e Conquista da Índia pelos Portugueses*, Liv. VI, Cap VI (Lisbon, 1554), ed. Oporto, 1977, Vol. II, 161.

11 João de Barros, *Terceira Década da Ásia* (Lisbon, 1563), Liv. V, Cap. VIII, ed. Lisbon, 1946, Vol. III, 286.

12 *Ibid.*, 297.

13 *Anon.*, 'Tratado de Navegación y de la longitude ó altura de Leste á Oeste' in *Revista de Mariña*, Tome XXXII (April, 1893), 379-94.

14 A. Teixeira da Mota, 'O "Regimento da altura de Leste-oeste" de Rui Faleiro' in *Boletim Geral do Ultramar*, Ano 28, No. 331, (Jan., 1953), 163-71, and *idem.*, 'A Contribuição dos Irmãos Rui e Francisco Faleiro no campo da náutica em Espanha' in *A Viagem de Fernão de Magalhães e a questão das Molucas,* Actas do II Colóquio Luso-Espanhol de História Ultramarina (Lisbon, 1975), 317-41.

15 Johann Werner, *In hoc opere haec continentur Nova translatio primi libri geographiae Claudius Ptolemei ...* (Nuremberg, 1514), Note 8, sign. d.v.(r.).

16 Hermann Wagner, 'Die legende der ...', 286-7.

17 Richard Lemay, 'Gerard of Cremona' in *Dictionary of Scientific Biography*, Vol. XI, 189.

18 *Theorica planetarum* (Venice, 1478), Renner ed. 4 pages before the explicit. English translation in J.K. Wright, 'Notes on the knowledge of latitudes and longitudes in the Middle Ages' in *Isis*, Vol. V (1923), 83-4.

19 Fernão Lopes de Castanheda, *História do Descobrimento ...*, Liv. VI, Cap. VII, ed. Oporto, 1977, Vol. II, 163.

20 Barros, *Terceira Década da Ásia*, ed. Lisbon, 1946, 297-9, and Antonio de Herrera y Tordesillas, *Historia General de los Castellanos en las islas y tierra del Mar Oceano* Madrid, 1601-1615, Dec II, Lib. IV, Cap. X (Vol. IV of the Madrid ed. 1936, 338-9), and R.A. Laguarda Trias, 'Las longitudes de la membranza de Magallanes y del primer viaje de circunnavegación' in *A Viagem de Fernão de Magalhães e a questão das Molucas* (Lisbon, 1975), 158-65.

21 F.G. Davenport, *European treaties bearing of the history of the United States and its dependencies*, Vol. I (Carnegie Institute, Washington, 1917), reprint Peter Smith, 1967, 88.

22 The Spanish documents relating to the Badjoz Conference have been published by D. Martin Fernandez de Navarrete, *Colección de los Viajes y Descubrimientos*, Biblioteca de los Autores Españoles, Vols 75-6 (Madrid, 1955). The Portuguese documents have been published in *As Gavetas da Torre do Tombo*, Vols IV, VI, VIII & IX, Centro de Estudos Historicos Ultramarinos, Lisbon, 1964-71.

23 Reproduced from the original in the Seville archives by R.A. Laguarda Trias, 'Las longitudes de la membranza ...', 176-8.

24 Martin Fernandez de Enciso, *Suma de Geographia* (Seville, 1519), sign. viii. (r.).

25 Navarrete, *Colección de los Viajes...*, Vol. 76, 615 (*Parecer* of 15 April 1524).

26 Jerónimo Osório, *De Rebus Emmanuelis Lusitaniae invictissimi virtute et auspicio gestis libri duodecim* (Lisbon, 1571), 423-5.

27 Document dated 31 May 1524, in *As Gavetas da Torre do Tombo*, Vol. VI (Lisbon, 1967), 571.

28 Luís de Matos, 'Antonio Maldonado Hontiveros e a questão das Molucas' in *A Viagem de Fernão de Magalhães*, 560.

29 'Apontamentos que o Duque de Bragança mandou a El-Rei João III ...' in *Alguns Documentos da Torre do Tombo* (Lisbon, 1892), 494.

30 Lopo Homem, *Apontamentos pera el Rey de Portugal que fez Lopo Homem ... en hun padrão de navegar feito sobre e por razam dos eclipses do sol e lua*, Bibliothèque Nationale, Paris, Ms V^c Colbert, 298, fos 6-8.

31 Viceroi Antonio de Mendoza, letter to Gonzalo Fernandez de Oviedo, 6 October 1541, in Gonzalo Fernandez de Oviedo, *Historia General y Natural de las Índias*, Lib. XXX, Cap. 52, Biblioteca de los Autores Españoles, Vol. 120 (Madrid, 1959), 253.

32 Juan Lopez de Velasco, *Geografia y Descripción Universal de las Índias desde el año 1571 al de 1574* (written in 1574) (Madrid, 1894), 10 and 188.

33 Andrés Garcia de Cespedes, *Regimento de Navegación* (Madrid, 1606), fo. 144 (r.).

34 Clinton R. Edwards, 'Mapping by questionnaire: early Spanish attempts to determine New World geographical positions', *Imago Mundi*, Vol. XXIII (1969), 22.

35 Hieronimo de Chaves, *Chronographia o Reportorio de los tiempos* (Seville, 1576), 88. (The 1576 edition is the only one that it has been possible to consult. The work first appeared in 1548.)

36 Andrés Garcia de Cespedes, *Regimento de Navegación*, 154.

37 Christophorus Clavius, *In Sphaeram Joannis de Sacrobosco commentarius* (Lyon, 1593), 310-33.

38 David Origanus, *Novae motuum coelestium ephemerides* (Frankfurt, 1609), 105.

39 Johann Kepler, *Tabulae Rudolphinae* (Ulm, 1627), 33-6.

40 Matteo Ricci, letter to Giambattista Roman, 15 September 1584, in P[e] Pietro Tacchi Ventura, *Opere storiche del P[e] Matteo Ricci* (Macerata, 1913), Vol. II, 38.

41 P[e] Giovanni Vreman, letter to P[e] Griemberger, 20 November 1616, in Georges Fournier, SJ, *L'Hydrographie ...* (Paris, 1643), 577.

42 Sabatino de Ursis, letter from Peking, 1 September 1612, in Pascuale d'Elia, SJ, 'Galileo in Cina' in *Analecta Gregoriana*, Vol. XXXVII (Rome, 1947), 111 (Portuguese text and Italian translation).

43 J.K. Wright, 'Notes on the knowledge of latitudes and longitudes in the Middle Ages' in *Isis*, Vol. V (1932), 87-8.

44 Luís de Albuquerque, *Os Almanaques Portugueses de Madrid*, Centro de Estudos de Cartografia Antiga, Junta de Investigações do Ultramar, Coimbra, Vol. II (1961), 64. (The text is of the first half of the fourteenth century and the list is the same as that given by Wright, together with the longitudes of some Portuguese towns.)

45 Christophorus Clavius, *In Sphaeram Joannis ...* .

46 Oscar Peschel, *Geschichte der Erdkunde* (Munich, 1865), 363.

47 J.K. Wright, 'Notes on the knowledge of latitudes ...', 97-8.

48 Cf Johann Gabriel Doppelmayr, *Historische nachricht von den Nürnbergischen Mathematicis und Künstlern* (Nuremberg, 1730), 171-2.

William Graham Randles is a member of the Centre National de la Recherche Scientifique (Maison des Pays Ibériques) of the University of Bordeaux III and author of articles on sixteenth-century navigation and cartography.

XIV

Pedro Nunes' Discovery of the Loxodromic Curve (1537)

How Portuguese Sailors in the Early Sixteenth Century, Navigating with Globes, had Failed to Solve the Difficulties Encountered with the Plane Chart

In the early sixteenth century, Portuguese navigators on voyages outside the tropics ran into an unexpected problem in navigating with the plane chart since the latter took no account of the convergence of the meridians. Voyages between Brazil and the Cape of Good Hope were accomplished much faster than the chart led them to forsee.[1] To overcome this problem, were Portuguese pilots of the first half of the sixteenth century using globes to set their courses with on great circle courses? The general opinion is that they were not. Yet the mathematician Pedro Nunes (1502–78) describes in a short treatise entitled *Tratado que ho Doutor Pedro Nunez fez sobre certas duvidas na navigação*[2] (published in 1537, but probably written in 1534)[3] how the navigator Martim Afonso de Sousa, on his return from the east coast of Brazil in 1530–2, had asked him to solve several problems of navigation encountered on the voyage.

Concerning one of them, Martim Afonso de Sousa asked Nunes why '... with the Sun over the equator [i.e. at the equinox', [...] '... if we sail to the East or to the West, we always follow a parallel on the same latitude without ever being able to reach the equator towards which we have set our ship on an East magnetic course.[4]

Although the journal of Martim Afonso de Sousa's voyage, written by his brother Pero Lopes de Sousa, has come down to us, there is nothing specific in it about the problem submitted to Nunes and the above text remains our only source.[5]

Nunes was not a navigator and his formulation of Sousa's question, modelled in the theoretical terms of a mathematician, probably does not reflect the latter's actual words as a sailor would have expressed them. His approach is more that of a 'thought experiment'.

In our reading of the words attributed by Nunes to Martim Afonso de Sousa there can be no doubt that the latter is understood by Nunes as seeking to lay a great circle course home from Brazil on a globe, since only on a globe could a course set to the East by the compass lead in theory to the equator. Our interpretation is supported by the Portuguese scholar Rodolfo Guimarães who wrote that '...c'est qu'en partant d'un certain lieu dans une direction rigoureusement perpendiculaire au méridien que [Martim Afonso de Sousa] voyait *sur la sphère* [our italics] qu'il devrait bientôt croiser l'équateur...'[6]

The evidence often cited that Portuguese navigators did not use globes by which to set their courses appears in a decree issued by the King D. Manuel in 1504 fobidding the making of them.[7] His reasons, however, which he does not give, were probably to prevent dissemination of cartographical knowledge to rival nations, rather than to prevent the use of globes for navigating. Approved navigators were probably authorized to carry them.

The sea chart customarily used by pilots at the time was the standard Mediterranean portulan chart constructed from estimated (by sand glass) distances and magnetic rhumb line bearings; to this had been added, from about the beginning of the sixteenth century, a graduated meridian. A graduated meridian had the effect of introducing into the chart celestially determined parallels of latitude (not delineated, apart from the equator and the tropics of Cancer and Capricorn) and, at the same time, *implicitly*, the existence of meridians parallel to one another, not converging at the poles. The chart had thus the implied appearance of being a quadratic plane chart in which degrees of longitude along every parallel contained the same number of units of terrestrial distance as that contained in a degree of the meridian.[8] Since neither the framework of the true quadratic plane chart, nor the basic principles of the portulan charts, took account of the convergence of the meridians, there resulted great exaggeration of east–west distances in high latitudes.[9]

The difficulties inherent in this problem were recognized by the pilot João de Lisboa in his *Livro de Marinharia*, a collection of texts compiled about 1550, though containing texts dated to the first two decades of the sixteenth century. As the text about to be cited appears translated into Castillian by Martin Fernandes de Enciso in his *Suma de Geographia*, Seville, 1519,[10] it can be roughly dated to just before then. João de Lisboa wrote: '...One can, and one should, make charts in the figure of a quadrant [by which he meant a 'quarter' of a sphere drawn on a plane chart showing the convergence of the meridians] so that they conform to the spherical body [of the Earth] which is round. However as the things (*cousas*) [i.e. gores] of every sphere converge (*vão jumtas*), charts are drawn on a plane surface as regards the longitudinal dimension (*por lomgitude*) because sailors are not astronomers (*estrolliquos*) and if one of them is, it is only by chance, and because on a plane surface, they understand the subject better...'[11]

In view of the above, it is therefore possible that, to overcome the problem of converging meridians, navigators may have tried setting their courses using globes. The result, as Martim Afonso de Sousa discovered, was not straightforward, since one cannot lay a great circle course on a globe and then follow it using a magnetic compass without making constant corrections of the course steered at regular intervals. The compass in fact takes no account of the convergence of the meridians (on the globe), maintaining a constant magnetic (true) course and producing, when plotted on the sea chart, a straight line track which necessarily cuts the meridians at constantly equal angles; that is, at the same angle. This is due to the fact that, on the chart, the meridians, even if not actually drawn on it, lie implicitly parallel to one another.[12]

Pedro Nunes continues, in his *Tratado ... sobre certas dúvidas*, to explain why, when setting one's course on a great circle and with a compass course of true East, one will never reach the equator.[13] 'But although the above mentioned great circle course carries us to "East on the equator" (*posto que o circulo grande nos encaminhe ao orient equinocial*) [i.e. the point at which the equator intersects with an East course on a great circle course as determined *at the moment of departure*] and [although] the great circle course be determined [at the moment of departure] by an East magnetic heading and although whoever sets his course along this great circle, will head towards the "East on the equator" (*orient equinocial*), we should not imagine that whoever follows it, will [actually end up] going towards the East [to cross the equator], because as he proceeds, he will find that the East course does not coincide with the direction in which his ship [keeps] pointing, and after going forward enough for this difference to be perceptible, he will find that he is set on another course [from the one he started on]. And thus the helmsman, without understanding why he does so, immediately 'corrects' his course [as] if he wishes to hold to the same latitude. And this happens in such a way that, if we were to steer on an East course [determined at the point of departure] and if we were to tie the rudder so that it could not be moved and if the sea were calm, and if nothing else were to hinder our course, and if, above all, the wind were in our favour, and if it were blowing in the direction in which we had set our course, yet, if we were to proceed thus, for a given distance, and if we were to look at the compass, we would find that we were not sailing East [on a great circle course to cross the equator]'.[14]

The explanation of the above is that if a ship were to steer due east at its point of departure with its rudder tied amidships and with no lateral forces acting on it, then it would follow a great circle track. The course would not remain due east and the ship would cross the Equator at a point determined at the point of departure. On the other hand, if the helmsman were to 'correct' the course to due east after following an arc of the great circle track long enough so that the course became appreciably different from due east, and then repeated this process indefinitely, the ship would follow a track, made up of a sequence of great circle arcs, becoming progressively parallel to the Equator and getting closer and closer to it, though never actually crossing it. A ship following a series of great circle arcs is Nunes' mathematical concept to explain the difference between a great circle and what he was to understand as a loxodromic track.

For a ship starting north of the equator and heading away from it, a steady compass course will, if so indeed obeyed by the helmsman, lead the ship on a spiral course progressively closer and closer to the north pole, though without ever touching it.[15] Such is the loxodromic curve, the track of a ship on a constant magnetic heading cutting all the meridians at the same angle.[16]

In his *Tratado ... em defensam da carta de marear* (1537), Nunes vaguely implies that he had glimpsed the idea of the loxodromic curve in Ptolemy's *Geography*.[17] 'Ptolemy', wrote Nunes, 'was aware that the track made [by a ship] along [a set] course is not on a great circle, which is straight and continuous.'[18] No one else

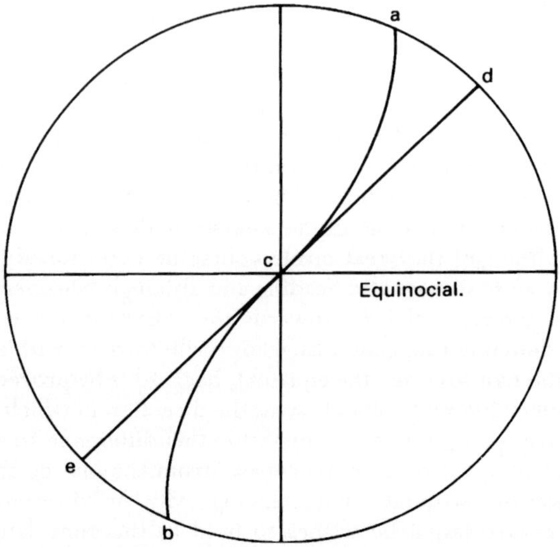

Fig. 1. Pedro Nunes' diagram to illustrate the loxodromic curve which he comments on as follows: 'The North-East towards which they think they are going is the line d.c.e. in this figure, but the track which they truly follow is the curved line a.c.b. which is not a [great] circle.' Pedro Nunes, 'Tratado [...] sobre certas duvidas da navegação', in *op. cit. ed. cit.* vol. 1, p. 168.

has ever suggested that Ptolemy ever got near the concept of the loxodromic curve and there does not seem to be any evidence in support of such an idea. The loxodromic curve was born out of a practical navigator's experience and was in no way a theoretical postulate. Nunes may have sought to tie his discovery to Ptolemy to lend it greater authority.

In his Portuguese texts, Nunes never uses the term 'loxodrome', only referring to the loxodromic curve by using the expressions *fazendo grandes rodeos* (making great swings) or *hua certa maneira de linhas curvas* (a certain kind of curved lines),[19] or *hua linha curva e yrregular*[20] (a curved and irregular track). The word 'loxodrome' is a Latin translation, inspired from the Greek, of a Dutch word *kromstrijk* (curve line) used by Simon Stevin in his work *Wisconstige Ghedachtenisse* (1608) to describe Nunes' discovery. 'Loxodrome' in Greek appears for the first time in Willebrord Snel van Royen's Latin translation of Stevin's Dutch work, entitled *Hypomnemata mathematica*, Lyons, 1605–8.[21]

Simon Stevin developed his analysis of the loxodromic curve, not from a reading of Nunes' Portuguese text, but from an expanded and clearer Latin version in his *Opera* published in Basel in 1566[22] (later republished in Coimbra in 1573).

In Book II, Chapter 21 of his Latin work, Nunes gives a description of how the loxodromic curve is experienced in a ship. 'At the beginning of the previous book,[23] we showed that the line [track] which a ship makes by its course (except for when the course corresponds with a meridian or with the equator) is not circular [i.e. a great circle] but is made up of a number of tiny segments of great circles. We have however pointed out, on good grounds, that the particular curved line [track] is of a different form [from a great circle] and that it is similar to a spiral and that it is made up of two movements. That the onward driving movement (*latio*) of a ship, when it follows a course (other than one corresponding to a meridian [i.e. N.S.] or to the equator [i.e. E.W.]) be derived from two onward driving movements (*lationibus*) and from two thrusts (*motoribus*), can readily be grasped. One onward driving movement is that by which the ship, lying with its axis in the plane of a great circle directed to a given point on the horizon, is carried forward either by the wind or by the force of its oars. The other onward driving movement occurs to the side (*in latus*), or obliquely, by which the helmsman, with his hand on the tiller and guided by the compass needle, turns the ship aside from the direction in which it was heading when it started on its course. This happens because, as the ship proceeds, it crosses new meridians and is headed successively towards new [points on the] horizon, and in this way the ship keeps heading towards [successive] [points on the] horizon that have a constant relation to one another. Thus, given this situation, the ship's track, which we call a rhumb, will neither be a [great] circle, nor made up of parts [i.e. arcs] of great circles. However, to us it seems otherwise. For we observe that the ship is carried slightly forward, before being turned aside,[24] and thus we consider the track to be made up of tiny segments of great circles. For then why is the ship constantly turned aside, when in spite of being carried by the wind a short distance on a great circle course, the helmsman barely notices that the ship is heading in a different direction?'[25]

Instead of allowing the loxodromic curve to distort the navigator's course, the helmsman must, said Nunes, follow a great circle course and make the necessary corrections at regular intervals. This is his explanation of how this must be done in his *Tratado ... sobre certas duvidas*. 'And the advantage of sailing by a great circle [course] is that one can reduce the distance travelled, but he who follows it should know that he must change his course every hour according to the change in the angles of the positions of the places formed by the new meridians with the great circle [course] followed by the ship (*segundo a mudança que fazem nos angulos da posição dos lugares os novos meridianos com o circulo per que ymos*) [by this Nunes means the change in the angle of intersection of the meridians with the great circle course on the point at which the ship lies when the change is measured, compared to the angle of intersection on the point where the ship was previously when the angle was last measured]. The difficulty of measuring this is considerable. It consists of finding by how much these angles increase or decrease as the voyage proceeds, compared to the size of the angle of the course at the point of departure. He who sails in such a way will sail a straight [great circle] course'.[26]

In his other Portuguese treatise *Tratado ... em defensam da carta de marear* (1537), Nunes proposed a mathematical method for introducing the necessary corrections to keep the ship on a great circle course. Imagining a ship starting from the equator on a course headed N.E. Nunes considers a spherical triangle of which one side is the arc of the meridian joining the ship's point of departure to the pole, another side is the great circle course the ship intends to follow, and the third side is the arc of the meridian joining the pole to the point reached by the ship after having raised one degree of latitude along its great circle course. After producing the great circle side of his triangle to form an exterior angle with the second meridian, Nunes then used Geber's theorem that the sines of the angles of a spherical triangle are in inverse proportion to the sines of the arcs opposite them,[27] and assuming that the sine of the obtuse angle inside the triangle is the same as that of its supplement outside, he compares the outer angle with the angle of the course set at the point of departure inside the triangle. The difference between the two gives the necessary correction to be added to maintain the ship on its great circle course.

There is no direct mention of Geber's theorem in the Portuguese text, but his name appears in the Latin one published in 1566, where the explanation is clearer.

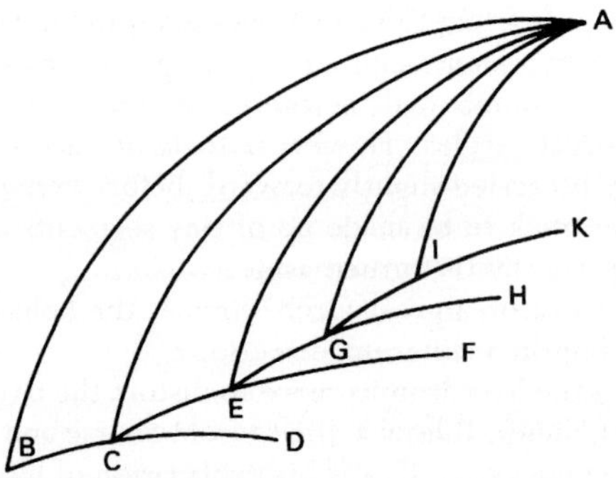

Fig. 2. 'A is one of the poles of the world [i.e. of the globe]. AB is the quarter of a meridian. The segments BC, CE, EG, GI make up the rhumb of the angle of departure ABC. BC is then prolonged to D; CE to F; EG to H; GI to K. From the pole A the meridians AC, AE, AG, and AE are joined to points C, E, G. I. The segments AB, AC, AG, and AI have sines that have the same regular ratio [to one another] as the ratio of the sine of the exterior angle ACD to the sine of the interior angle ABC opposite in the spherical triangle BAC. For in a spherical triangle the ratio of the sine of the side AB to the sine of the side AC is equal to the ratio of the sine of the angle ACB to the sine of the angle ABC according to the 13th proposition of Geber's 1st Book. Since the two angles ACB and ACD have the same sine, so the ratio of the sine of the side AB to the sine of the side AC is equal to the ratio of the sine of the angle ACD to the sine of the angle ABC.' Pedro Nunes, 'Tabulam quandam numerorum edere, cuius adminiculo in in dato globo rumbos quoslibet describamus', Cap. 23, *Opera*, Basel, 1566, pp. 167–168.

Nunes' method of introducing corrections to counter the effect of the loxodromic curve, so as to maintain the ship on a great circle course, was never adopted by Portuguese sailors. The margin of error in the change of latitude of one degree, such as he suggested for the introduction of successive changes of course, would have rendered the whole operation impracticable, given the accuracy of the instruments then available, to say nothing of the winds.

Nunes was a *savant de cabinet* and not a navigator. His treatise in Portuguese, his correction method for great circle navigation and his mathematics were violently attacked by a compatriot Diogo de Sá, no more a navigator than he, in a work published in Paris in 1549, *De Navigatione*. Although it is said that Diogo de Sá had voyaged to India, he shows no proof that the experience had altered his conservative Aristotelian mind or cured his aversion to mathematics. However, some of his remarks about the practicality of Nunes' correction methods are worth quoting. Before doing so, it should be emphasized that Diogo de Sá had only read Nunes' treatises in Portuguese and had never seen the one in Latin.

This is how Diogo de Sá writes of Nunes' proposed method of navigating on a great circle course. The text is in the form of a dialogue between a mathematician and a philosopher. The philosopher expresses Diogo de Sá's point of view, the mathematician, Nunes'.

'*Philosopher*: And the usefulness which you [= the mathematician = Nunes] say a great circle provides, will be for he who wishes to travel at sea without turning to one side or to the other [of the track of a great circle]. And I say that not only is this impossible, but that the futile vanity of the man [doing this] would deserve to be laughed at. Who, I ask you, is going to show me these angles so that I can sail according to whether they are equal or unequal [to the track of a great circle]? If you tell me that the compass needle itself can, within the ship, do no more than show its East, or any other course for me to steer by, and, what is more, if you tell me that the ship will show me these angles [to keep a great circle course], you will be utterly mistaken, for the ship cannot do this, and can only, driven by the wind, proceed on a course given by the compass needle on any one of its courses. Who then is going to show me these [great circle track] angles? Should I ask the sea to show them to me? The sea will tell me that the sea is everywhere, and everywhere the same, and cannot itself show me any of them. Therefore it is only on the chart that I can seek them. Go then to the chart [i.e. the quadratic plane chart] and you will find all of the E.W. courses everywhere perpendicular to the N.S. courses and, along the whole course run, the same will hold for all courses and their counter-courses. [...] Since this is so, how, or by whom, can you prove that anyone going from East to West may know that he should change his course from hour to hour according to the observed change of the [angle of the] meridians in relation to the great circle on which we are sailing? Given such a state of affairs, sailing in this manner can produce nothing more than going about making angles in the sea.'[28]

Had Diogo de Sá been a practical pilot, his criticism framed from the point of view of a pilot would have appeared more pertinent and carried greater weight than the sarcastic obstinacies of his scholastic mind.[29]

Another of Pedro Nunes' works on navigational problems, written in Portuguese, was discovered in 1949 in the National Library of Florence. It is an undated and untitled manuscript and was published in 1952 by Joaquim de Carvalho, who gave it the title of *Defensão do Tratado da Rumação do Globo para a Arte de Navegar*. ('In Defence of the Treatise on the placing of Rhumb Lines on the Globe for the Art of Navigation').[30] The work seems to have been incorporated into Nunes' Latin treatises published in 1566. Without mentioning Diogo de Sá's name, Nunes refers, at the end of the manuscript, to how he had suffered from the criticisms levelled at his works. 'They speak ill of my treatises, making use of them and often using my very words and wanting to cover everything, they ruin all. I have decided, for this reason, to polish up some of the things I have written and set about studying philosophy and abandon mathematics, in the study of which, I have irretrievably lost my health.'[31]

That Nunes' work was known outside Portugal is attested by the existence in the Bibliothèque Nationale, Paris, of a manuscript French translation (never published) of his two Portuguese treatises.[32] The translation, on examination, proves to be so literal, that it is clear that the translator did not understand what he was translating.

The Fleming Michel Coignet, in the French version of 1581 of his Latin work of 1578 translated as *Instruction Nouvelle*...[33] shows that he had read Nunes' Latin texts and understood the concept of the loxodromic curve, which he calls *voies tortues* or *lignes spirales*. Coignet recognized the difficulty in applying Nunes' prescriptions for navigation on a great circle course, of which he wrote: '...toutes ses imaginations [Nunes'] ne sont, pour la plus grande part, que choses peu praticables, & pour ceste raison, de petite efficace pour les pilotes.'[34]

The Englishman Robert Hues, in his *Tractatus de Globis et eorum usu*, London 1594,[35] quotes Nunes and expounds with clear understanding his concept of the loxodromic curve, criticizing him on one small point: 'And therefore I cannot assent to Pedro Nonius, who would have the Rhumbes to consist of portions of great circles. For seeing that the portion of a great circle, being intercepted between diverse Meridians, though never so little distant from each other, maketh unequal angles with the same, a Rhumb cannot consist of them by the precedent proposition.[36] But this inequality of Angles is not perceived (saith he) by the sense unlesse it bee in Meridians somewhat farre removed from one another. Be it so. Notwithstanding, the error is discoverable by art and demonstration. Neither doth it become so great a Mathematician to examine rules of art by the judgement of the sense.'[37]

Hues was of course right in this criticism expressed from the point of view of a pure mathematician. But Nunes' solution was an approach and his suggested difference of one degree, though too large from a mathematical point of view, was negligible for the practice of navigation. Hues' reproach that he used the 'judgement of the sense' is quite unfair.

Another Englishman, Edward Wright, in his *Certaine Errors of Navigation*, published in 1599, was heavily influenced by Nunes' *Tratado em defensam da carta de marear*, copying from it most of the first chapter.[38] The part he copied concerned the exaggeration on the chart of the real distances between Lisbon and the Azores. Following Nunes, Wright explains the advantages of sailing a great circle course. Like Nunes, Wright defended the 'mariner's chart with aequidistant meridians and streight-lined rhumbes'. 'And although the globe', he declared, 'be commended by some as the most absolute & perfect for all courses & climates whatsoever: yet for changeableness thereof, troublesome carriage, stowage and tedious usage for the most part in navigation [...] it will for the most part be found unmeete and combersome, and nothing so fit and ready for the mariner's common use at sea as the nautical planisphere truely made'.[39]

Wright's 'nautical planisphere truely made' is based on Mercator's projection with increasing latitudes and Wright is the first to give a table for calculating the Mercator projection.[40]

Last of all we come to the influence of Nunes on Mercator, a subject on which heated polemic took place between Joaquim Bensaude and Hermann Wagner.[41]

Wagner, in his reply to Bensaude, recognized Nunes' priority over Mercator regarding the discovery of the loxodromic curve and of the former's influence over the latter. Wagner, however, insisted that Mercator had produced his globe (1541) with loxodromes shown on it (though not carried to the pole) before Nunes had published the method for accomplishing this in his Latin text. Wagner recognized that Nunes, in his work in Portuguese of 1537, was the first to draw, on a chart framework, having an equidistant polar projection, two loxodromic arcs from the equator to the pole. But these drawings show appreciable errors and so Mercator, according to Wagner, is the first to have accurately drawn loxodromes on his globe in 1541. Since Mercator gives no explanation of how he went about this, Pedro Nunes can be said to be the first to describe, in his Latin text of 1566, a method for drawing loxodromes on a globe using his own mathematical definition.[42]

Mercator's fame, as is well known, lies in his chart of 1569, whereby he eliminated for seamen the problems of the loxodromic curve, by spacing his parallels at rising intervals from the equator and from each other, towards the poles (whilst keeping the meridians parallel). He did this through the use of the method known as that of 'increasing latitudes'. His chart is based on a congruent projection; that is, the angles are constant so that the projection makes all the loxodromes appear as straight lines. The chart thus becomes a matchless instrument for navigation since it enables sailors to solve graphically and simply all the problems relating to loxodromic navigation.

Mercator did not give any explanation of his methods, whether graphic or mathematical, and it was left for his successors to work them out. Charts drawn according to his projection were never used by seamen before the early part of the seventeenth century.

The device of increasing latitudes appears, however, to have been thought of long before by Erhardt Etzlaub, who drew a map of Europe and Africa as far as the equator with increasing latitudes, in the lids of two sundials made in Nuremberg as early as 1511 and 1513.[43]

Nunes' theoretical contributions to navigation came too much in advance of his time. The difficulty in applying them lay mainly in the then insufficient accuracy of the instruments available, in the limitations of a sailing ship, and in the fact that the mathematics of his time were too cumbersome and laborious to be used at sea.

It was Nunes' tragedy to have been born in a country which could have benefited greatly from his outstanding talents had it been possible to put them to profitable use in ships.

ACKNOWLEDGEMENT. Commander D. W. Waters and Professor John

Kemp have been kind enough to review and comment on this paper. My deepest gratitude is here expressed for their thorough detailed criticism and suggestions, of which I have tried, to the best of my ability, to take account and to incorporate into the text. I take full responsibility for what errors and obscurities still remain.

NOTES AND REFERENCES

[1] Cf. W. G. L. Randles, 'From the Mediterranean portulan chart to the marine world chart of the Great Discoveries: the crisis in cartography in the sixteenth century', in *Imago Mundi*, vol. 40 (1988), pp. 111–114.

[2] In Pedro Nunes, *Obras*, Lisbon, 1940, vol. 1, pp. 159–174.

[3] Cf. A. Fontoura da Costa, *A Marinharia dos Descobrimentos*, Lisbon, 2nd ed. 1939, p. 219.

[4] Pedro Nunes, *op. cit.* p. 159.

[5] Pero Lopes de Sousa, *Diário de Navegação de Pero Lopes de Sousa (1530–32)*, Prefácio do Commandante A. Teixeira da Mota, Leitura do Doutor Jorge Morais Barbosa. Ed. Agência Geral do Ultramar, Lisbon, 1968.

[6] Rodolfo Guimarães, *Sur la vie et l'œuvre de Pedro Nunes*, Coimbra, 1915, p. 26 (original in French). In a letter dated 8 April 1985, Professor Luís de Albuquerque wrote to us that it was obvious to him that Martim Afonso de Sousa could not have been using a contemporary chart and that it was probable that he was carrying a globe in spite of their use being prohibited by the King.

[7] J. Ramos Coelho (ed.), *Alguns Documentos da Torre do Tombo*, Lisbon, 1892, pp. 138–139. ('...defendemos que não façam nenhuns mestres das cartas de marear, nem outros alguns oficiais nenhumas pomas grandes nem pequenas, de pouco, nem muito, porque não queremos que se façam em maneira alguma...')

[8] The term 'quadratic chart' is used by Pedro Nunes, *Tratado...em defensam da carta de marear*, Lisbon, 1537, in Idem, *Obras*, vol. 1, pp. 176–177. Cf. Luís de Albuquerque's very pertinent remarks that Portuguese nautical charts of the sixteenth century were not in fact true quadratic charts. Luís de Albuquerque, *Ciência e experiência nos Descobrimentos Portugueses*, Biblioteca Breve, Lisbon, 1983, pp. 15–19. and Idem. 'Considerações sobre a carta portulano', in *Revista da Universidade de Coimbra, vol.* 31 (1984), pp. 19–22.

[9] Cf. the analysis of this problem in W. G. L. Randles, *art. cit.* (n. 1), pp. 112–113.

[10] Martin Fernandes de Enciso, *Suma de Geographia*, Seville, 1519, sign. b.ix (v°).

[11] João de Lisboa, *Livro de Marhinaria*, ed. Brito Rebelo, Lisbon, 1903, pp. 197–198. Marcel Destombes has drawn attention to a chart in the Top Kapu Sarayi Museum in Istanbul which he attributes to Pedro Reinel and which he dates to between 1522 and 1524 and which was probably presented at the Badajoz Conference in 1524. It shows the southern hemisphere with converging meridians, just as João de Lisboa declared should ideally be the case. The chart was, however, probably prepared for diplomatic purposes rather than for navigational use. Cf. Marcel Destombes, 'L'Hémisphère austral en 1524; une carte de Pedro Reinel à Istanbul', in Marcel Destombes, *Selected Contributions to the History of Cartography and Scientific Instruments* (eds) Gunter Schilder, Pieter van der Krogt and Steven de Clerq, Utrecht/Paris, 1987, pp. 175–184, (with illustration of the chart).

[12] Left out of consideration here is the phenomenon of magnetic variation, a separate problem never touched on by Nunes in his works. In the sixteenth century it was recognized but not clearly understood.

[13] Nunes appears to be imagining a ship heading home from the east coast of Brazil (Martim Afonso de Sousa) on a north-easterly course to cross the equator.

[14] Pedro Nunes, 'Tratado [...] sobre certas duvidas de navegação', in *op. cit. ed. cit.* vol. 1, p. 161. 'Mas posto que o circulo grande sobredito nos encaminhe ao oriente equinocial e se represente pello leste da [a]gulha: e quem pello tal cirulo for: va ter ao dito oriente equinocial: não auemos porem de cuidar: que quem por elle for yra a leste. porque tanto que por elle andar achara que o leste da agulha não vay na proa do seu navio. e andando espaço de caminho: em que esta deferença possa sentir achara que vay ja per outro rumo. E por tanto o que gouerna sem

entender o porque o faz: emenda logo de principio sua navegação: se quer jr em huma mesma altura. E he isto per tal maneyra que se gouernassemos a leste: e atassemos o gouernalho de sorte: que nenhuma mudança fezesse: e o mar fosse tam tranquillo: que nenhuma cousa nos embargasse nossa navegação: e per cima de tudo isto ho vento nos fauorecesse: como quisessemos: e corresse pera aquel parte onde vay endereçado o leste da agulha: todauia se assi andasemos notauel espaço de caminho e oulhasemos a agulha: achariamos que imos fora de leste.'

[15] In his Portuguese text, Nunes believed that the pole would be reached, but in his Latin text of 1566, he wrote that the ship would get closer and closer, but never touch it. Pedro Nunes, 'Tratado... em defensam da carta de marear' in *Obras*, vol. **1**, p. 184, and Pedro Nunes, *Opera*, Basel, 1566, p. 173.

[16] Cf. the exhaustive treatment of the mathematical definition by Raymond d'Hollander, 'Historique de la Loxodromie', In *Mare Liberum*, no. 1, Lisbon, 1990, pp. 29–69.

[17] Cf. the remark of Joaquim de Carvalho in his introduction to Pedro Nunes, [*Defensão do Tratado da Rumação do Globo para a Arte de Navegar*, *s.d.*], Coimbra, 1952, p. xx.

[18] Pedro Nunes, 'Tratado... em defensam da carta de marear', in *Obras*, vol. **1**, p. 183.

[19] Pedro Nunes, *Obras*, vol. **1**, pp. 167–68.

[20] Idem, *Obras*, vol. **1**, p. 183.

[21] Cf. Bk IV (1608), pp. 85–87. Cf. also Hermann Wagner, 'Gerhard Mercator und die ersten Loxodromen auf Karten' in *Annalen der Hydrographie und Maritimen Meteorologie*, Bd. 43, Heft VII–IX, Berlin, 1915, p. 302.

[22] Pedro Nunes, 'De regulis & instrumentis ad varia rerum tam maritimarum quam cœlestium apparentias deprehendas, ex Mathematicis disciplinis, Liber II', in *Opera*, Basel, 1566.

[23] This is a reference to Book I of his Latin text, which is a translation of his earlier treatises in Portuguese, 'Certas dúvidas da navegação' and 'Tratado... em defensam da carta de marear', both published in Lisbon in 1537.

[24] The late Professor Pierre Costabel emphasized to us the importance of the sequential nature of the two movements.

[25] Pedro Nunes, 'De iis quae praemitti debent ad ducendum eas lineas in globo quas nautae rumbos appellant', in ch. 2 of Book II of Pedro Nunes, *Opera*, Basel, 1566, p. 158.

[26] Pedro Nunes, 'Tratado sobre certas duvidas da navegação' in *Obras*, vol. **1**, p. 167.

[27] Geber (Jabir ibn Aflah) (1st half of the 12th century), *De astronomia libri IX*, Latin translation by Gerard de Cremona, 2nd half of the 12th century. First published by Peter Apian, Nuremberg, 1535. Cf. Bk I, Prop. XIII (p. 11), ('I declare that in a triangle formed by the arcs of great circles, the proportion of the sine of each of the sides to the sine of the arc of the angle subtended by that side is constant').

[28] Diogo de Sá, *De Navigatione*, Paris, 1549, f. 80 r°-v°. The late Professor Luís de Albuquerque had prepared a Portuguese translation of Diogo de Sá's book which I believe has remained unpublished. He was kind enough to make available to me a photocopy of the typescript. The above translation is, however, made directly from the Latin and I take full responsibility for it.

[29] One of the scholastic arguments (not however invoked by Diogo de Sá) against the existence of the loxodromic curve, was that the force of gravity (Aristotelian gravity) acting from the centre of the Earth, would pull on the ship to keep it on a great circle course (!).

[30] Pedro Nunes [*Defensão do Tratado da Rumação do Globo para a Arte de Navegar*], Subsídios para a História da Filosofia e da Ciência em Portugal, IV, publicados por Joaquim de Carvalho, Coimbra, 1952.

[31] Pedro Nunes, *op. cit.* p. 31.

[32] Pedro Nunes, *Traité de Navigation*, n. d., Ms Fr 1338 (Fonds Colbert), Bibliothéque Nationale, Paris.

[33] Michel Coignet, *Instruction Nouvelle des poincts plus excellents et nécessaires touchant l'art de naviguer*, Antwerp, 1581, ch. IIII, 'Des cartes marines et de ce qu'en dépend', pp. 16–26.

[34] Idem, p. 26.

[35] Cf. the English edition: *A Learned Treatise of Globes, both celestial and terrestrial*, London,

1638, re-edited by Clements Markham, Hakluyt Society, 1st series, vol. **79**, London, 1889. A French edition appeared in Paris in 1618.

[36] The 'precedent proposition' stated that: 'A great circle drawne through the vertex of any place that is in any whit distant from the Equator cannot cut diverse meridians at equal Angles'.

[37] Robert Hues, *A Learned Treatise of Globes...*, Hakluyt Society edition, p. 130.

[38] Edward Wright, *Certaine errors of Navigation*, London, 1599, reprint Amsterdam, 1974, preface.

[39] E. Wright, *op. cit.* ch. 1.

[40] Idem, ch. II.

[41] Hermann Wagner, 'Gerhard Marcator und die ersten Loxodromen auf Karten', in *Annalen der Hydrographie und Maritimen Meteorologie*, vol. **43** (1915), pp. 299–311; Joaquim Bensaude, *Histoire de la Science Nautique Portugaise*, Geneva, 1917, pp. 78–85; Hermann Wagner, 'Die loxodromische Kurve bei G. Mercator. Eine abwehr gegenüber Senhor Joaquim Bensaude', in *Nachrichten von der Königlichen Gesellschaft der Wissenschaften zu Göttingen*, Phil.-Hist. Classe Berlin, 1917, pp. 254–264, and Joaquim Bensaude, *Les Légendes Allemandes sur l'histoire des Découvertes Maritimes Portugaises*, Part I, Geneva, 1917–20, pp. 19–44; Part II, Coimbra, 1925–7.

[42] Pedro Nunes, *Opera*, Basel, 1566, Livre II, ch. 26. 'Proposition globum rumbis delineare'.

[43] The map, in the Germanisches National Museum, Nuremberg, is reproduced in Wolfgang Köberer, *Das rechte Fundament der Seefahrt*, Hoffmann und Campe, Hamburg, 1982, Plate 28. Cf. also Joseph Drecker, 'Ein instrument, eine Karte und eine Schrift des Nürnberger Kartographen und Compostmachers Erhardt Etzlaub', in *Annalen der Hydrographie und Maritimen Meteorologie*, vol. **45** (1917), pp. 217–224 and Ernst Hammer, 'Die Mercator projection und Erhardt Etzlaub', in *Petermann's Mittheilungen*, vol. **63** (1917), pp. 303–304.

KEY WORDS

1. History. Loxodromes. 3. Rhumb lines.

XV

Le Nouveau Monde, l'Autre Monde et la Pluralité des Mondes

«Nostre Monde vient d'en découvrir un autre et qui nous respond si c'est le dernier de ses frères». — MONTAIGNE: *Essais,* Liv. III, cap. VI

«Novos mundos ao mundo irão mostrando» — CAMÕES: *Os Lusíadas,* II, 45.

Il est souvent admis que le terme «Nouveau Monde» apparait pour la première fois dans la fameuse lettre, le *Mundus Novus,* attribuée à Americ Vespuce et dont la première édition est datée de Paris, 1503 (1).

(1) Pour la date de l'édition *princeps* nous prions le lecteur de se reporter au *Times Literary Supplement* du 25 Déc. 1955, dernière page, «Notes on the Sales». Il y eut 22 éditions du *Mundus Novus* avant 1506. La popularité de cette lettre éclipsa complètement celle de la lettre analogue de Colomb, dont il n'y eut que 9 éditions: 7 en 1493, 1 en 1494 et 1 en 1497 (V. Harrisse: *Bibliotheca Americana Vetustissima,* Paris, 1886).

S'il est vrai que la publication de cette lettre lui a donné un retentissement considérable, le terme n'est nullement de l'invention de Vespuce, comme nous allons le voir.

Pour certains, l'idée de considérer la découverte de l'Amérique comme celle d'un Nouveau Monde est d'origine vespucienne et portugaise ([2]), alors que l'idée de l'associer à l'Inde d'Outre-Gange est d'origine colombienne et espagnole.

À vrai dire, c'est dans une lettre de Pierre Martyr d'Anghiera ([3]), adressée au Cardinal Ascanio et datée de Barcelone, 1er Novembre 1493, que le terme «Nouveau Monde» se trouve appliqué à l'Amérique pour la première fois, à notre connaissance. On y trouve la phrase suivante:

> «...Colomb, cet illustre découvreur du Nouveau Monde...» ([4]).

([2]) Tel qu'il fut imprimé, le *Mundus Novus* n'est pas de la main de Vespuce. Il était calqué sur un autre texte aujourd'hui perdu. Le voyage décrit dans le *Mundus Novus* est bien celui qui se fit en 1501 sur les côtes du Brésil, sur l'ordre du Roi D. Manuel, et auquel prit part Vespuce; la preuve en est fournie par l'Acte Notarié de Valentim Fernandes, du 20 mai 1503 (publié par A. Fontoura da Costa, *Cartas das Ilhas do Cabo Verde de Valentim Fernandes, 1506-1508,* Lisbonne, 1939, p. 91).

([3]) Plusieurs érudits ont prétendu que l'*Opus Epistolarum,* de Pierre Martyr d'Anghiera, était d'une authenticité douteuse. (V. ceux que cite Harrisse, *Bibl. Am. Vet.* pp. 281-282). Les accusations portent surtout sur les dates des lettres, qui ne sont pas toujours les mêmes dans les diverses éditions. Pour les lettres citées, nous avons comparé les deux éditions de Alcala, 1530 (Bibl. Nat. Paris, Fol. Rés. Z 499) et d'Amsterdam, 1670 (Bibl. Nat. Paris, Rés. Z 500) et nous n'avons trouvé aucune différence, ni de date, ni de contenu. Dans la traduction française des lettres d'Anghiera, parue dans la *Revue de Géographie,* T. XV, 1884, p. 466 et suiv., et basée sur l'édition de 1670, la deuxième des lettres citées (du 13 Novembre 1494) porte la date du 20 Octobre 1494. Nous ignorons si l'édition utilisée par le traducteur de la *Revue de Géographie* est du même éditeur que celle que nous avons consultée, et qui porte la cote Rés. Z 500. Anghiera n'a pas été sans défenseurs. Mariéjol (*Un lettré italien à la cour d'Espagne, Pierre Martyr,* Paris, 1887, p. 170) dit que: «La meilleure réfutation [des charges d'authenticité douteuse] est dans une lecture attentive».

([4]) Pierre Martyr d'Anghiera, *Opus Epistolarum,* Alcalá, 1530, f. xxxiii, (r).

Dans une autre lettre à Jean Boromée, datée de Alcalá, 13 Novembre 1494, il dit:

> «...Chaque jour on apprend des choses étonnantes sur le Nouveau Monde découvert par Colomb...» ([5]).

La présente étude a pour objet de démontrer que l'idée du Nouveau Monde se trouve étroitement liée à deux problèmes qui n'avaient cessé, depuis l'Antiquité et à travers tout le Moyen Age, de passionner philosophes, théologiens et hommes de lettres: l'existence d'un *autre monde* austral peuplé d'hommes antipodiens et la *pluralité des mondes.*

La croyance en l'existence d'hommes antipodiens habitant l'hémisphère austral avait certes été déclarée hérétique par St. Augustin ([6]), mais elle avait néanmoins survécu durant tout le Moyen Age et trouvé des défenseurs, en la personne, par exemple, d'Albert le Grand et de Roger Bacon. Il est d'ailleurs intéressant de noter que ces controverses sur la question des antipodes naissaient en général dans l'église même et non pas entre chrétiens et libertins.

Le grand retentissement du *Mundus Novus* est vraisemblablement dû en partie au fait qu'il affirme catégoriquement l'existence d'hommes antipodiens, pour constater ensuite que les Anciens l'ignoraient ou la niaient.

Mais ce n'est pas dans le *Mundus Novus* que l'on voit pour la première fois établir un rapport entre la découverte de l'Amérique et les antipodes.

Voici ce que disait Anghiera, le 14 Mai 1493, dans une lettre à Jean Borromée:

> «Quelques jours se sont écoulés depuis qu'est venu des Antipodes occidentaux un certain Christophe Colomb...» ([7]).

([5]) *Ibid.*, f. xxxiiii (r).

([6]) St. Augustin, *La Cité de Dieu,* XVI, 9. En 1496, Zaccaria Lilio écrivait encore que le Christ «...n'est pas allé aux Antipodes, n'y a envoyé aucun de ses Apôtres, qui n'ont pas prêché dans une seule partie du monde, ni chez un seul peuple, mais partout, l'Evangile du Christ». (cf. «Contra Antipodes», in *De origine et Laudibus Scientiarum,* Florence, 1496, b. f. iiii (v).

([7]) Pierre Martyr d'Anghiera, *Opus Epistolarum,* Alcalá, 1530, f. xxxi, (v).

Le 1er Octobre 1493, il écrit de Barcelone à l'archevêque de Braga:

«Un certain Colomb a navigué jusqu'aux Antipodes de l'Occident et aux rivages de l'Inde; du moins on le suppose...» (8).

Il est assez curieux de voir Anghiera placer les Antipodes à l'ouest plutôt qu'au sud, comme on le fait d'habitude (9). Selon la tradition du Moyen Age, c'était la zone torride, jugée infranchissable pour l'homme, qui séparait les Antipodes de l'*oikoumenê,* situé dans l'hémisphère septentrional (10).

Dans les doctrines traditionnelles de l'Antiquité en matière d'astronomie, il n'est pas rare de trouver toute une «géographie» complexe de la voûte céleste (11). L'image cosmographique de la terre se reflétait dans le ciel nocturne. Une partie du zodiaque (la Balance) fut appelée le Styx (12) Macrobe (Ve siècle) nous apprend que:

«...les physiciens ont, sous le nom de Vénus, honoré cet hémisphère supérieur du monde, dont nous occupons une partie; l'hémisphère inférieur, ils l'ont appelé Proserpine» (13).

Il rapporte aussi la tradition suivante des Assyriens:

«Donc, chez les Assyriens ou les Phéniciens, on représente la déesse [Venus] en deuil, lorsque, dans sa course annuelle, qui le fait passer par les douze signes du zodiaque, le soleil entre aussi dans la partie d'entre eux qui appar-

(8) *Ibid.* f. xxxii, (v). La comparaison de ces deux lettres dans les deux éditions n'a pas révélé de différences.

(9) Martian Capelle (Ve siècle) les plaçait, lui aussi, à l'ouest (cf. J. K. Wright, *Geographical Lore of the Time of the Crusades,* New York, 1925, p. 160).

(10) Sur les idées de l'Antiquité et du Moyen Age au sujet des Antipodes, v. J. K. Wright, op. cit.

(11) Cf. Franz Boll, *Sphera,* Leipzig, 1903, pp. 246-251.

(12) Ibid., p. 247, citant Firmicus.

(13) Macrobe, *Saturnales,* Livre I, XXI, 1. (Édition Garnier, Vol. I, p. 217).

tient à l'hémisphère inférieur; car, sur les douze signes du zodiaque, six sont réputés supérieurs, six inférieurs. Et lorsque le soleil est dans les signes inférieurs, et que, par suite, les jours sont moins longs, on croit que la déesse est en deuil, comme si le soleil était perdu, enlevé par une mort temporaire, et retenu par Proserpine, déesse qui, nous l'avons dit, est celle de l'hémisphère inférieur et des antipodes. [...] Mais lorsque le soleil est sorti des parties inférieures de la terre, qu'il franchit les limites de l'équinoxe du printemps et prolonge la durée du jour, alors Vénus est gaie et belle; les blés verdissent dans les champs, l'herbe dans les prés, les feuilles sur les arbres. Voilà pourquoi nos ancêtres ont consacré le mois d'avril à Vénus» ([14]).

D'après ce texte, Macrobe assimile manifestement les Enfers à l'hémisphère austral et aux Antipodes.

Gervase de Tilbury (fl. 1211) nous rapporte une curieuse histoire concernant l'autre monde des antipodes. Ce qui frappe dans cette anecdote, c'est son caractère plus folklorique qu'astronomique ou cosmographique. Voici le texte:

«*Des antipodes et de leur pays*
(De la forteresse que les Angles appellent Pech)

Il y a en Grande Bretagne une ville fortifiée située au milieu des montagnes, à laquelle on a donné le nom de Pech. Ses murs sont difficiles à prendre et, dans la montagne, se trouve l'orifice d'une caverne d'où souffle parfois un vent violent, comme d'un tuyau ([15]). Les gens se sont demandé d'où venait tant de vent, et entre autres sujets d'étonnement, je tiens d'un homme très pieux, Robert Prior de Renildwrta ([16]), originaire de là, l'histoire

---

([14]) Ibid., Livre I, XXI, 2 et 6 (Ed. cit. Vol. I, pp. 217-218).

([15]) En réalité, le château de Pech, dont parle Gervase de Tilbury dans ce chapitre, portait, en raison de sa position sur le Mont Peak, le nom de «Castle in the Peak», et se trouvait près de Burgh (Castleton, Derbyshire). La caverne qui se trouve à proximité s'appelle «the Devil's Arse». (Cf. Félix Liebrecht, ed. Gervasius von Tilbury *Otia Imperialia,* Hanovre, 1856, p. 117 (Anmerkung 44).

([16]) Peut-être Kenilworth ?

que voici: en ce temps-là, un homme noble et puissant, riche en animaux divers, et qui s'appelait Guillerme Peverel, tenait ladite ville et la baronie attenante; un beau jour son porcher, négligent dans le travail qui lui avait été assigné, perdit une truie pleine, d'une excellente race. Craignant donc pour cette faute une sévère réprimande de la part du régisseur du maître, il pensa qu'il retrouverait peut-être sa truie, si par hasard elle avait pénétré dans la fameuse caverne Pech, inexplorée jusqu'alors. Il décida de se faire l'explorateur de ce lieu caché. Il entre dans la caverne, à ce moment calme et libre de tout vent, et quand il eut parcouru un long chemin, il sortit enfin des ténèbres pour aboutir en un lieu éclairé donnant sur une vaste plaine couverte de champs. Ayant largement pénétré dans les terres cultivées, il trouva des paysans qui récoltaient les fruits mûrs, et entre les épis mûrs, il reconnut la truie, qui s'était multipliée et avait mis bas des gorets. Alors le porcher, étonné et heureux de voir sa faute ainsi réparée — le récit a été fait avec la description de cette terre et de tout ce qui s'est alors passé — reprit la truie, et partant avec joie la reconduisit au troupeau de porcs. Chose étrange: venant des moissons souterraines, il voit que les froids de l'hiver persistaient dans notre hémisphère» ([17]).

Dans la littérature arthurienne, il est aussi souvent question d'un autre monde, parfaitement réel. La légende place parfois le séjour d'Arthur sous terre ou aux Antipodes ([18]). Selon Jean Marx,

«...dans la conception celtique, l'Autre Monde n'est pas séparé du Monde Terrestre par une différence et une hétérogénéité totales de nature. Caesar nous attestait déjà que les Celtes se considéraient comme les fils de Dispater, c'est-à-dire le Roi de l'Autre Monde» ([19]).

([17]) Gervase de Tilbury, *Otia Imperialia*, ed. Felix Liebrecht, Hanovre 1856, p. 24.

([18]) Jean Marx, *La légende arthurienne et le Graal*, Paris, 1952, p. 64.

([19]) *Ibid.*, p. 82. La littérature arthurienne fut connue au Portugal, comme en témoigne le texte *A Demanda do Santo Graal*, qui date du

L'italien Alvise da Cadamosto, qui a visité les îles du Cap Vert en 1456, écrit dans la relation de ses voyages que,

> «...en vérité, si l'on compare notre manière de vivre, nos costumes et nos maisons à toutes les choses que j'ai vues et connues, on peut bien parler d'un Autre Monde» ([20]).

Et pourtant, les îles du Cap Vert se trouvent encore dans l'hémisphère nord.

Les Portugais franchirent l'équateur vers 1472-1474; ils parvinrent à l'embouchure du Congo en 1484.

En 1489, dans un discours prononcé devant le Roi de Portugal, D. João II, au cours d'une cérémonie où Pedro de Meneses reçut le titre de Marquis de Vila Real, l'orateur royal João Teixeira disait:

> «Mais notre Roi, grâce à la divine Providence et aussi à ses grandes vertus, lança tant de navigations qu'il découvrit pour ainsi dire ([21]) un autre monde (...) Ce fut une chose comme on n'en avait jamais vu, car nos marins passèrent le cercle équinoxial et virent le soleil droit au-dessus de leurs têtes...» ([22]).

---

premier quart du XVe siècle (cf. ed. de A. Magne, Rio de Janeiro, 1944). Eugène Beauvois a réuni de nombreux textes, tant de l'Antiquité classique que de tradition celtique, où il est question d'un autre monde d'outre--Atlantique (Eugène Beauvois, «L'Élysée transatlantique et l'éden occidental», In *Revue de l'Histoire des Religions,* t. VII, 1883, pp. 273-318 et t. VIII, 1883, pp. 672-727, V. aussi Howard Rollin Patch, *The Other World according to description in Medieval Literature,* Cambridge, Mass. 1950.

([20]) Alvise da Cadamosto, *Voyages,* texte italien du MSS ital. cl. 6, N° 454 de la Bibliothèque Marciana, édité par l'Académie Portugaise d'Histoire, Lisbonne, 1948, p. 3. Le texte cité est du plus vieux manuscrit connu, qui date du XVe siècle.

([21]) On peut se demander si cette réserve introduite par l'auteur correspond franchement à du scepticisme de sa part, ou si, n'étant pas personnellement conservateur, il se rendait compte que la notion de l'existence d'un autre monde pouvait soulever des difficultés d'ordre théologique. Nous évoquerons plus loin les problèmes théologiques suscités par l'idée du Nouveau Monde.

([22]) João Teixeira, *Oratio habita insigni viro Ioanne Teixeira...* Coimbra, 1562, (sans pagination). Bibliothèque Nationale, Lisbonne, Rés. 80 v. Il n'en existe que deux exemplaires connus.

Dans le *Liber Chronicarum* de Hartmann Schedel, publié à Nuremberg en 1493, nous trouvons le texte suivant, qui se rapporte au voyage de Diogo Cão:

> «En l'année 1483, Jean II, Roi de Portugal, homme à l'âme élevée, envoya des galées (galeas) bien armées faire un voyage de découverte vers le sud, au-delà des colonnes d'Hercule, vers l'Éthiopie. Il leur donna deux capitaines: Jacobus Canus (Diogo Cão), un portugais, et Martinus Bohemus ([23]), un allemand né à Nuremberg d'une bonne famille originaire de Bohême, qui avait une connaissance approfondie des pays du monde et une grande expérience de la mer. Ayant navigué pendant de longues années, il connaissait aussi parfaitement les longitudes et les latitudes de Ptolémée vers l'ouest. Tous deux, grâce à la faveur des dieux, naviguèrent assez près de la côte vers le sud, et, ayant franchi la ligne équinoxiale, pénétrèrent dans un autre monde (alterum orbem) où, lorsqu'ils étaient tournés vers l'est, leur ombre tombait vers le sud et à droite. Ainsi, grâce à leurs efforts, ils avaient découvert un autre monde (alium orbem) jusqu'ici inconnu de nous, et cherché en vain pendant de longues années par les Gênois» ([24]).

Dans l'Acte Notarié de Valentim Fernandes, daté du 20 Mai 1503, il est aussi question d'un autre monde, cette fois le Brésil, découvert par Pedro Álvares Cabral, et où le Roi D. Manuel envoya de nouveau une expédition en 1501. C'est ce dernier voyage qui devait fournir à Vespuce la base de son *Mundus Novus*.

> «...Partie du très riche port de commerce de Lisbonne en direction de l'Inde, une flotte de 13 grandes nefs du très puissant D. Manuel Ier, roi de Portugal et des Algarves, du Continent et d'Outre-Mer en Afrique, seigneur de la Guinée ainsi que de la Conquête, de la Navigation et du Commerce d'Éthiopie, d'Arabie, de Perse et de l'Inde, découvrit au-delà du Gange, dans une mer inconnue,

---

([23]) Martin Behaim n'a pas accompagné Diogo Cão.

([24]) Reproduit par António Brásio, *Monumenta Missionária Africana*, vol. I, Lisbonne 1952, p. 6.

au-dessous de la ligne équinoxiale, un autre monde que la Divine Providence avait celé à toutes les autres puissances. Ce fut en l'an du Christ de 1500 et le dernier jour du mois d'Avril. Elle était commandée par le valeureux chevalier Pedro Álvares Cabral. Les habitants de ce monde-là n'ont ni foi, ni religion, ni idolâtrie et ne savent rien de leur Créateur; ils ne sont pas soumis à des lois, ni à aucune domination, mais uniquement au conseil des anciens; rien ne leur appartient en propre, mais tout leur est commun, sauf les femmes; tous vivent complètement nus et ni les hommes, ni les femmes ne se couvrent les parties honteuses [...] Deux ans plus tard, une autre flotte ([25]) du même roi très chrétien, envoyée dans ce but, après avoir suivi le littoral de cette terre sur près de 760 lieues, découvrit que ces peuples parlaient une seule langue; elle baptisa beaucoup de gens et, avançant vers le sud, parvint à la hauteur du pôle antarctique, à 53 degrés; ayant trouvé alors de grands froids en mer, elle prit le chemin du retour vers sa patrie. [...] Et moi, Valentim Fernandes de Moravia, tabellion public par ordre de ce même roi de Portugal, j'ai lu le présent acte devant sa royale Majesté, ses barons, ses grands capitaines et pilotes ou gouverneurs de ses navires de la dite terre des antipodes, qui porte le nouveau nom de terre de Santa Cruz, et tous l'ont confirmé à l'unanimité et j'ai recueilli tout cela d'un livre écrit par moi...» ([26]).

([25]) Il s'agit évidemment de celle de 1501. L'expression «deux ans plus tard» exige cependant une explication.

([26]) Reproduit par A. Fontoura da Costa, *Cartas das ilhas de Cabo Verde de Valentim Fernandes 1506-1508,* Lisbonne, 1939, pp. 91-93. On ignore le sort du livre que Valentim Fernandes affirme avoir écrit. Au XIXe siècle, Harrisse a cherché en vain à la Biblioteca Colombina, à Séville, un ouvrage qui portait, dans le catalogue, le titre: *Nouvelles de l'autre monde in versu gallico* (N° 19727 de la Bibl. Colombina) (V. Harrisse, *Christophe Colomb,* Paris, 1884, Tome II, p. 37). Mais il a trouvé, dans la *Crónica de Aragon,* de Gaubert Fabricius Vagad, parue à Constance en 1499, le curieux texte suivant: «Oyd pues otra novedad no menos grande que nueva y estraña. Oyd un gran fecho y tan digno de oyerse q̃ del se maravillan todas las mares: tiemblan y se espantan las yslas tan de nuevo: mas tan nunca fasta agora falladas, y fasta el mundo del otro mundo queda vencido y maravillado: que allende nuestro emisperio dizen que estã en la buelta y del otro cabo del mar se han

Il est assez difficile de savoir exactement ce qu'on voulait dire en appelant les terres qu'on venait de découvrir *autre monde*. Peut-être faut-il voir ici un écho de la légende d'Alexandre le Grand ([27]) qui, si l'on en croit la tradition, aurait pleuré en apprenant qu'il existait plusieurs mondes car, disait-il, il n'avait encore pu se rendre maître d'un seul ([28]).

Dans le *Libro de Alexandre,* texte espagnol du Roman d'Alexandre très connu au Moyen Age, nous trouvons les strophes suivantes:

«Disen las escripturas yo ley el tratado
que siete son los mundos que Dios avo dado
de los siete el uno apenas es domado
por esto yo non conto que nada he ganado.

Quanto avemos visto antes nos lo sabiemos
si al non anpendremos ([29]) *en balde nos viuiemos*
por Dario e por Poro que vencido avemos
yo por esto non cuydo que grant cosa fisiemos.

Enbionos por esto Dios en partidas
por descobrir las cosas que jasen sofondidas ([30])
cosas sabran por nos que non serian sabidas
seran las nuestras nuevas en coronicas metidas.

---

de nuevo fallado estas islas: que nuevamente descubre la gente animosa y valiente de españa. Y poco pensans que estan espantadas las gentes del otro cabo del mundo del tanto poderio magestad y grandeza tã desygual de coraçon y effuerço de nuestras hespañoles: que han osado prender lo que nunca las cesares los osaron poner...» (Gaubert Fabricius Vagad, *Crónica de Aragon,* Constance, 1499, Prologo, fl. Biiij (r) cité par Harrisse, *Bibliotheca Americana Vetustissima,* Add. ed. 1872, pp. 10-11).

([27]) Il n'est pas rare de voir comparer les grands promoteurs des découvertes maritimes à Alexandre le Grand. Dès le XVe siècle, Gomes Eannes de Zurara rapportait «...qu'on disait couramment que l'Infant (Henri) était bien un nouvel Alexandre...» (Gomes Eannes de Zurara, *Crónica dos feitos de Guiné,* ed. de l'Agence Générale des Colonies, Lisbonne, 1949, p. 99).

([28]) Voir la tradition rapportée d'après Plutarque par Francisco López de Gómara, citée plus loin.

([29]) Variante: «has apresiessemos».

([30]) Variante: «escondidas».

Los oms que non saben prescio prender
esto prender por gloria en balde se jaser
mas diselo el maestro mandalo rretener
qui prodesa ([31]) *quiere afan deve prender*

Con todos vos a una queriendome seguir
buscare los antipades querolos conquerir
estos estan de jus ([32]) *la tierra commo oyemos desir*
mas yo non lo afirmo que cuydo de mentir» ([33]).

Que cet *autre monde* pouvait parfois être pris dans le sens de *au-delà,* il suffit, pour s'en convaincre, de lire le texte suivant, qui est une partie du discours de João Teixeira cité plus haut:

«O Roi, notre Maître, le plus grand et l'unique, tu es comparable à cette Reine de Saba, venue, dit-on, de cette Arabie que tu explores, jusqu'à Jerusalem, afin de connaître la sagesse de Salomon, toi qui as aussi découvert beaucoup de pays, rencontré des peuples nouveaux et traversé des mers. De même Pythagore a pénétré les secrets des devins de Memphis, et Platon a parcouru à grande peine l'Égypte, connu Archytas de Tarente et ce rivage d'Italie, qu'on appelait autrefois la Grande Grèce. Et d'après le récit de Callimaque, Cleombrotus d'Ambracie, ayant lu le livre de Platon sur l'immortalité de l'âme, se précipita d'un point élevé, afin d'apprendre de sa propre expérience si tout était dans l'autre monde tel qu'il l'avait lu dans ce livre» ([34]).

En confrontant les deux extraits du discours de Teixeira que nous avons cités, on peut aboutir à la conclusion suivante: L'autre monde constitué par l'hémisphère austral semble

---

([31]) Variante: «proeza».
([32]) Variante: «so».
([33]) *El Libro de Alexandre,* ed. par Raymond S. Willis Jr., Elliot Monographs, N° 32, Princeton, 1934, lignes 2289-2293. Le texte est de la Bibliotèque Nationale de Paris (M. S. Espagnol, 488). Les variantes sont du MS de la Biblioteca Nacional, Madrid (V-5-N° 10).
([34]) João Teixeira, *ibid.*

s'identifier ici à l'*autre monde* de l'*au-delà*, et les navigations ordonnées par le Roi de Portugal dans l'Atlantique sud et l'Océan Indien sont présentées comme une quête de la *sagesse*, cette même sagesse qu'on pouvait espérer trouver dans le monde de l'*au-delà*.

En 1501, alors que les Portugais avaient déjà atteint l'Inde, Pietro Pasqualici, ambassadeur de Venise auprès du Roi de Portugal, lui présentait ainsi ses félicitations, dans un discours prononcé devant lui:

> «Voilà une chose étonnante et remarquable, une chose qu'on n'avait encore jamais vue, une chose mémorable et digne de passer dans l'histoire (...) toi, roi invincible jusque dans l'hémisphère souterrain, et qui peux te vanter d'avoir créé un empire aux antipodes, et ce qui est plus grand, plus mémorable encore, tu as réuni des peuples séparés par la force de la nature et établi des relations entre deux mondes différents de l'univers...» ([35]).

C'est chez Valentim Fernandes, l'imprimeur allemand qui s'est fixé au Portugal, que nous trouvons pour la première fois l'expression «Nouveau Monde» appliquée aux découvertes portugaises dans l'Océan Indien.

Voici ce qu'il dit dans le prologue de son édition des voyages de Marco Polo, publiée à Lisbonne en 1502, *une année avant* la première édition du *Mundus Novus:*

> «Quelle chose merveilleuse nous avons vue aujourd'hui! Comment le Roi Jean II, de glorieuse mémoire, votre ancêtre [de D. Manuel] a lutté de toutes ses forces pour pénétrer dans cette terre promise, qui vous était destinée, à vous et à vos successeurs. Il en fut de lui comme de Moïse, qui avait lutté tant d'années pour pénétrer dans la [terre promise] et put enfin la contempler du haut du Mont Nébo. De même, ce Roi Jean II put contempler cette autre [terre promise] du Cap de Bonne Espérance. Qu'il est noble ce Cap, et quelle belle renommée d'espoir que ce

---

([35]) Pietro Paschalici Veneti, *Oratoris ad Hemanuelem Lusitanae Regem Oratio,* Venise, 1501. (Bibliothèque Nationale, Paris, Rés. K. 1292).

nom, qui lui fut accordé non par l'intelligence humaine mais par la Divine Providence. C'est elle qui choisit Bartolomeu Dias, de votre maison (lui dont la devise évoque l'«espérance»), pour être digne comme Josué d'entrer dans ce nouveau monde que nous pouvons bien appeler la terre promise» ([36]).

Lorsque Lourenço d'Almeida parvint à Ceylan en 1505, croyant reconnaître en cette île la Taprobana de l'Antiquité ([37]), les Portugais se souvinrent que celle-ci était considérée par les auteurs classiques comme se trouvant dans l'autre monde ([38]).

Dans une lettre au Pape Jules II, datée d'Abrantes, le 25 Septembre 1507, et publiée à Paris la même année, le Roi D. Manuel disait:

> «... [Lourenço de Almeida] parvint à cette fameuse île de Taprobane, parfois considérée comme un autre monde: également appelée Zoyla par les indigènes» ([39]).

([36]) Valentim Fernandes, «A Epistola sobre a tralladaçam», In *O Livro de Marco Polo,* Lisbonne, 5 Février 1502, f. Aij (v), Réimprimé par Francisco Maria Esteves Pereira, Lisbonne, 1922.

([37]) On a pu constater une certaine confusion en ce qui concerne l'identification de Taprobane avec Ceylan, car on a affirmé aussi qu'il s'agissait de Sumatra. Dans les Lusíadas, Camões identifie Taprobane et Ceylan (cf. *Os Lusíadas,* X, 107), mais dans l'Ode qu'il écrivit pour les *Colloques* de Garcia de Orta, il l'identifie à Sumatra. C'est aussi l'opinion de Pedro Nunes dans son *Tratado em defensam da Carta de marear,* Lisbonne, 1537, et celle de Nicolas Caverio et d'Albert Cantino, dans leurs célèbres planisphères, datés de 1502. Castanheda (*Historia do Descobrimento e Conquista da India,* Coimbra, 1551, Liv. II cap. xxii) dit: «[Ceylan] est celle que l'on nommait autrefois Taprobana...», mais il dira ailleurs, Liv. II, cap. cxi, «[Sumatra] est celle-là même, croit-on, que les cosmographes anciens appelèrent Taprobana».

([38]) Pline disait: «Taprobanem alterum orbem esse diu existimatum est, Antichthonum appellatione», (Pline, *Histoire Naturelle,* Livre VI, 22). Sur Taprobane, Pomponius Mela citait Hipparque: «...quae Hippacho prima pars orbis alterius dicitur». Les citations sont tirées de A. Humboldt, *Examen critique...* Paris, 1836, Vol. I-II, pp. 121-122.

([39]) *Epistola serenissimi Regis Portugaliae de victoria cõtre infideles habita. Ad Julium papam Secũduz...,* ed. fac-similé de Eugenio do Canto, Ponta Delgada, 1905, Bibl. Acad. Ciências, Lisbonne (Rés. 37, 3/29).

Voici comment Giovanni Poggio chante, dans son *Emmanuelis Portugaliae Regis Elogium,* écrit en 1514, l'arrivée des Portugais à Taprobane (Ceylan):

«Tout dernièrement, envahissant cette île de Taprobane, la première du monde, mais plus justement désignée par les Géographes comme un autre monde, ou encore, selon l'opinion de Pline, située en dehors du monde (40), ta flotte [celle du Roi D. Manuel] a dispersé avec des effectifs réduits les forces innombrables des ennemis et fait du roi de l'île ton sujet. Chose merveilleuse ! Chose inouïe ! Chose incroyable ! Qui eût jamais cru que les Portugais s'aventureraient jusqu'aux rivages de l'Inde» (41).

Le fait que cette île ait été identifiée à l'autre monde explique, dans une certaine mesure, que Camões ait voulu en parler dans la première strophe de son grand poème épique:

«As armas e os barões assinalados
Que da Occidental praia Lusitana
Por mares nunca de antes navegados
Passaram ainda além de Taprobana» (42).

Les géographes de l'Antiquité attribuaient à l'île de Taprobana des dimensions plus considérables que n'en ont aujourd'hui Ceylan ou Sumatra. João de Barros en donne une explication:

«D'après ce qu'en ont écrit les géographes, il semble que dans ces temps très anciens, [Ceylan] était aussi grande que le disent les indigènes, lorsqu'ils affirment qu'elle avait plus de sept cents lieues de tour et que la mer l'a rongée peu à peu; c'est pourquoi sans doute (si

(40) Jacopo Filippo Foresti, consacrant un paragraphe à Colomb dans son *Supplementum Chronicarum* (1503), l'intitulait: «De quattuor p maximis insulis in Indis extra orbem nuper inventis», J. Jacopo Filippo Foresti, *Supplementum Chronicarum,* Venise, 1503, ff. 441 (v) — 442 (v), Bibliothèque Nationale, Paris (fol. Z. Don. 94 (13).

(41) Cité par Fidelino de Figueiredo, *A Épica Portuguesa do século XVI,* São Paulo, 1950, pp. 112-113.

(42) Luis de Camões, *Os Lusiadas,* I, 1.

nous voulons citer Ptolémée) il dit qu'elle s'étend dans la longueur deux degrés et demi au-delà de la ligne Equinoxiale vers le sud» (43).

C'est sans doute justement parce que l'on supposait que l'île s'étendait sous l'Équateur, que la croyance qu'elle était dans l'autre monde s'est trouvée renforcée.

Si nous n'avons pas pu apporter de documents prouvant incontestablement que l'expression «Nouveau Monde» découle de «Autre Monde», nous croyons néanmoins avoir montré qu'il existait un rapport étroit entre les deux idées.

Par ailleurs, nous l'avons vu, loin d'être exclusivement appliquée à l'Amérique, l'expression «Nouveau Monde» désignait également le monde au-delà du Cap de Bonne Espérance.

Il nous semble qu'elle représentait surtout, et en premier lieu, le monde austral, car ce qui était vraiment «nouveau» pour les navigateurs européens, c'était le ciel et les étoiles de l'hémisphère sud (44).

La découverte d'un nouveau ciel apportait un démenti catégorique aux affirmations d'Aristote, dont l'autorité restait énorme (45).

---

(43) João de Barros, Dec. III, Liv. II, Cap. I.

(44) Pedro Nunes disait: «Les Portugais ont osé affronter la grande Mer Océane. Ils s'y sont lancés sans aucune crainte. Ils ont découvert de nouvelles îles, de nouvelles terres, de nouvelles mers, de nouveaux peuples, et qui plus est: un nouveau ciel et de nouvelles étoiles». (Pedro Nunes, *Tratado da Sphaera,* Lisbonne, 1537, fac-similé ed. Munich, 1915, p. 117).

Et Camões: «Lá no novo hemisfério, nova estrela — Não vista de outra gente» (*Os Lusiadas,* V, 14). Il est intéressant de comparer ces affirmations avec les prophéties de la Sainte Écriture: «Or nous attendons, selon sa promesse, de nouveaux cieux et une nouvelle terre, où la justice habite». (2 *Pierre,* 3,13) voir aussi, *Esaie,* 65, 17, et *Apoc.* 21,1.

Dans une lettre écrite vers la fin de 1500, Colomb affirmait que «Del nuevo cielo y tierra que decia nuestro Señor por S. Juan en el Apocalipse, despues de dicho por boca de Isaías, me hizo dello mensagiero y amostró en cual parte. En todos hobo incredulidad». Cf. R. M. Major, *Select Letters of Christopher Colombus,* Hakluyt Society, London, 1870, p. 153.

(45) Dans son traité sur le Ciel, Aristote disait: «...il n'y a pas maintenant plusieurs cieux; il n'y en a jamais eu plusieurs; il n'y a qu'un seul ciel, unique et complet, celui que nous connaissons (Aristote, *De Caelo,* Lib. I, Cap. IX, trad. française de J. Barthélémy Saint-Hilaire, Paris, 1866, p. 80).

Selon Alexandre Koyré, «La grande ennemie de la Renaissance, du point de vue philosophique et scientifique, a été la synthèse aristotélicienne et l'on peut dire que sa grande oeuvre a été la destruction de cette synthèse» ([46]).

Lorsque l'expression «Mundus Novus» fut appliquée au Brésil, on ne se doutait pas encore que cette terre tenait aux autres terres qui devaient être appelées plus tard Amérique du Nord. On ignorait, en effet, que la côte du nouveau continent s'étendait dans le sens nord-sud et traversait l'Équateur. On s'attendait, au contraire, à ce qu'elle s'étendît dans le sens est-ouest. Le terme «Nouveau Monde» date d'une époque où régnait encore la notion d'une opposition «symétrique» entre l'*oikoumenê* (ou ancien monde) et l'autre monde austral, la barrière entre les deux étant constituée par la zone torride, jugée infranchissable dans l'Antiquité.

La vieille opposition «horizontale» entre le «monde» et «l'hémisphère inférieur» devait subir un déplacement de 90° pour aboutir à une opposition «verticale» entre l'Europe et l'Amérique. La «barrière» de la zone torride se trouva remplacée par l'Océan Atlantique.

Le texte de Macrobe déjà cité, où l'on trouve établie l'opposition entre le «monde» et l'hémisphère inférieur (ou les Enfers), suggère une antithèse affective, qui découle de l'antithèse archétypale: *Ciel-Enfer* ou *Joie-Douleur*.

On retrouve cette même idée dans la nouvelle opposition Europe-Amérique, née des découvertes, et sa première manifestation est le mythe du bon sauvage et d'une vie heureuse et idyllique outre-atlantique, chez les primitifs que la civilisation n'avait pas encore corrompus. Il arrive aussi que cette polarisation affective se trouve inversée, l'Amérique n'étant plus présentée comme un paradis, mais comme un monde «défavorisé». Dans un livre fort intéressant ([47]), Antonello Gerbi a montré que Buffon et l'Abbé Corneille de Pauw, par exemple, considèrent l'homme et la faune américains comme inférieurs et débiles.

([46]) Alexandre Koyré, «L'apport scientifique de la Renaissance», In *Revue de Synthèse,* Jan.-Juin 1950, p. 31.

([47]) Antonello Gerbi, *La Disputa del Nuovo Mundo, Storia di Una Polemica, 1750-1900,* Milan et Naples, 1957.

Et au fond, la polémique née de cette polarisation affective entre l'Europe et le Nouveau Monde reste aujourd'hui aussi vivante qu'elle le fut au XVIe, au XVIIe et au XVIIIe siècles.

★

Si l'Église acceptait sans trop protester le «fait» de l'existence d'hommes antipodiens habitant l'hémisphère austral, il n'en était pas de même en ce qui concerne la thèse de la pluralité des mondes, qu'elle repoussait avec intransigeance.

Et cette question de la pluralité des mondes fut souvent discutée, au XVIe siècle, à propos du Nouveau Monde.

On trouve ce que l'on pourrait appeler les premières références à la question de la pluralité des mondes au sujet des découvertes, dans le *Discours d'Obédience* prononcé par Vasco Fernandes de Lucena devant Innocent VIII en 1485, imprimé à Rome la même année, et dans la lettre écrite par l'humaniste italien Angelo Poliziano à D. João II en 1489.

Vasco Fernandes de Lucena louait D. João II d'avoir amené à l'Église Romaine:

> «...novas provincias, nova regna, novas insulas et quasi novos et incognitos orbes...» ([48]).

De même que dans le texte de João Teixeira cité plus haut, on remarquera la prudente réserve exprimée par «quasi».

Voici ce qu'écrivait Angelo Poliziano:

> «Car qu'as-tu fait d'autre, je le demande, ô notre roi, que de découvrir d'autres terres, une autre mer, d'autres mondes et enfin d'autres astres, que de les faire sortir des ténèbres éternelles, et je dirais presque de l'ancien chaos, pour les ramener à cette lumière commune?» ([49])

([48]) Texte latin reproduit par A. Fontoura da Costa, *Às Portas da India em 1484,* Lisbonne, 1936, p. 77.

([49]) Texte latin reproduit par António Caetano de Sousa, *Provas da Historia Genealogica da Casa Real Portuguesa,* Lisbonne, 1742, Tome II, p. 164. La première édition de cette lettre fut imprimée à Venise, en 1498, par Aldo Manucio, avec les oeuvres complètes de Poliziano. On connaît 13 éditions de la lettre au XVIe siècle (V. *Bibliografia Geral Portuguesa,* Lisbonne, 1941, Vol. I, pp. 305-323).

La thèse de la pluralité des mondes avait été repoussée de façon catégorique par Aristote ([50]), pour qui la coexistence de deux univers était une absurdité.

Cette opinion concorde d'ailleurs avec celle qu'il énonçait dans la *Métaphysique,* où il n'attribuait à Dieu aucun pouvoir créateur. «Mais», comme le note Duhem, «n'était-elle pas [cette opinion] en contradiction avec la notion chrétienne de Dieu ? Si Dieu est le Tout-Puissant, capable de faire de rien le Ciel et la Terre, osera-t-on prétendre que sa puissance créatrice est épuisée par la formation d'un monde unique ? Affirmera-t-on qu'il ne saurait, en dehors de ce monde, en produire plusieurs autres?» ([51])

C'étaient là des questions que se posaient au Moyen Âge les théologiens, surtout quand ils étaient de tendance nominaliste.

Le 7 Mars 1277, «les maîtres en théologie et d'autres prud'hommes» de l'Université de Paris se réunirent, sous la présidence de l'évêque Etienne Tempier, pour extirper de leur enseignement certaines erreurs péripatéticiennes et averroïstes. Parmi les propositions condamnées se trouvait la suivante:

> «Quod prima causa non posset plures mundos facere» ([52]).

«De ce jour», selon Duhem, «on compta, en l'école de Paris, de nombreux maîtres qui ne considérèrent plus la pluralité des mondes comme une absurdité» ([53]). Parmi ceux qui admettaient que Dieu avait pu créer une multitude pratiquement infinie de mondes distincts, se trouvait Jean de Bassols. Au début du XVIe siècle, sa doctrine fut défendue par Maître Jean Majoris, qui soutenait que l'on ne saurait trouver une

---

([50]) Dans son *Traité sur le Ciel,* Aristote disait: «...il ne résulte... aucune nécessité qu'il y ait plusieurs mondes, et il ne peut même pas y en avoir plusieurs...» (Aristote, *De Caelo,* Lib. I Cap. IX, trad. de Barthélemy Saint-Hilaire, Paris, 1866, p. 76).

([51]) P. Duhem, *Études sur Leonardo da Vinci,* seconde série, Paris 1909, p. 72.

([52]) Ibid. p. 75 et P. Duhem, *Études sur Leonardo da Vinci,* troisième série, Paris, 1913, p. 233.

([53]) Ibid. loc. cit.

contradiction en cette hypothèse: Il existe une infinité de mondes ([54]).

Cependant la doctrine de la pluralité des mondes ne fut pas acceptée par Jean de Celaya et Louis Coronel, tous deux maîtres à l'Université de Paris au début du XVIe siècle.

> «Actuellement», écrivait Celaya, «il n'existe qu'un seul monde. Cette conclusion se prouve par ceci: La foi catholique ne nous fournit aucune autorité d'où découle l'existence de plusieurs mondes, et, en outre, il n'existe aucune raison capable de nous contraindre à supposer la pluralité des mondes; bien plus, quelques-unes des raisons que le philosophe [Aristote] invoque contre cette pluralité ont une certaine apparence de vérité. Actuellement, donc, nous n'avons pas à supposer l'existence de plusieurs mondes...» ([55]).

En 1511, Louis Coronel, espagnol originaire de Segovie, écrivait:

> «La pluralité des mondes ne doit jamais être supposée sans nécessité; nous ne devons donc pas admettre l'existence de plusieurs mondes, car rien de ce que l'expérience nous enseigne ne requiert la réalité d'un autre monde...» ([56]).

Ni Celaya, ni Coronel, ne semblent néanmoins esclaves de la tradition péripatéticienne. S'ils sont finalement parvenus à une conclusion analogue à celle d'Aristote, ce n'est pas en invoquant les mêmes raisons.

Ces deux savants avaient-ils entendu parler de la découverte d'un Nouveau Monde outre-Atlantique ? Très probablement. Mais peut-être n'admettaient-ils pas que ce phénomène eût sa place dans le cadre de leurs spéculations, peut-être estimaient-ils qu'il s'agissait là d'un domaine différent.

([54]) Ibid. loc. cit.

([55]) Jean de Celaya, *Expositio in quatuor libros de coelo et mundo Aristotelis: cum questionibus ejusdem,* Paris, 1518, cité par P. Duhem, *Études sur Leonardo da Vinci,* troisième série, Paris, 1913, p. 234.

([56]) Louis Coronel, *Physicae perscrutationes,* lib. VIII, pars II, cité par P. Duhem, Ibid., p. 236.

On pourrait d'ailleurs objecter que le débat sur la pluralité des mondes se situait sur le plan de l'astronomie et non sur celui de la géographie. Il est certain que c'est dans le cadre de l'astronomie que la question a retenu l'attention des savants au XVIIe siècle et après, mais ce n'était pas toujours le cas au XVIe, nous allons le voir.

C'est Francisco López de Gómara qui, le premier, en 1552, soulève la question de la pluralité des mondes à propos de la découverte de l'Amérique:

«El Mundo es uno, y no muchos, como algunos filósofos pensaron

Opinion y tema fué de muchos y grandes filósofos, hombres en su tiempo tenidos por mui sabios, que habia muchos mundos. Leucipe, Democrito, Epicuro, Anaxmandro y los otros, porfiados en que todas las cosas se engendran y crian del *tamo* y átomos, que son unos pedacitos de nada como los que vemos al rayo del sol, dijeron que habia muchos mundos; y que asi como de solas viente y tantas letras se componen infinitos libros, asi, ni mas ni menos, de aquellos pocos y chicos átomos y menudencias se hacen muchos y diversos mundos. Esto afirmaban, creyendo que todo era infinito. Y asi a Metrodoro le parecia cosa fea y desproporcionada no haber en esto infinito mas de un solo mundo, como seria si en una muy gran viña no hubiese sino una cepa, ó en una gran pieza una sola espiga.

Orfeo tuvo que cada estrella era un mundo, à lo que Galeno escribe de historia filosofica. Y lo mesmo dijeron Heraclides y otros pitagóricos, segun refiere Teodorito, *De materia y mundo.* Seleuco, filósofo, segun escribe Plutarco, no se contentó con decir que habia infinitos mundos, sino que tambien dijo ser el mundo infinible, como quien dejese que no puedo tener cabo donde fenezca su fin. Creo que de aqui le tomó ansia al gran Alejandre de conquistar el universo; pues claramente à lo que Plutarco cuenta, lloró oyendo un dia disputar esta quistion à Anaxarco. El cual, preguntada la causa de lágrimas tan fuera de tiempo, respondio que lloraba con justa y gran razon, pues habiendo tantos mundos como Anaxarco decia, no era él aun señor de ninguno. Y asi, después, cuando emprendió la conquista deste nuestro mundo, imaginaba otros muchos y pretendia señorearlos todos. Mas atojóle la muerte los pasos antes que pudiese sujetar medio.

Tambien dice Plinio: «Creer que hay infinitos mundos procedió de querer medir el mundo à pies», lo cual tiene por atrevimiento; aunque dice llevar tan sotil y buena cuenta, que seria vergüenza no creerlo. De la opinion destos filósofos salió el refran que cuando uno se halla nuevo en alguna cosa dice que le paresce estar en otro mundo.

Poco estimáramos el dicho destos gentiles, pues como dice sant Augustin, se revolcaron por infinitos [mundos ?] con su vano pensamiento; ni el de los herejes dichos ofios ([57]) ni el de los talmudistas, que afirman decinueve mil mundos, pues escriben contra los Evangelios, si no hubiese teólogos que hagan mencion de mas mundos. Baruch habló de siete mundos, como dice Origines; y Clemente, discipulo de los apóstolos, dijo en una su epístola, segun Origines lo acota en el *Periarcon:* «No es navigable el mar Oceano; y aquellos mundos que detrás de él están, se gobiernan por providencia del mismo Dios».

Tambien sant Jerónimo alega esta misma autoridad sobre la epistola de sant Pablo à los efesios, donde dice: «Todo el mundo está puesto en malignidad». En muchas partes del Testamento Nuevo está hecha mencion de otro mundo; y Cristo, que es la mesma verdade, dijo que su reino no era deste mundo, llamo al diabo principe deste mundo. Diciendo este, paresce que hay otros, à lo menos otro; y por eso erraron los herejes ofios, que no entendiendo bien la Escriptura Sagrada, inferian ser innumerables los mundos; y quien creyese que hay muchos mundos como el nuestro, erraria malamente como ellos.

Mundo es todo lo que Dios crio: cielo, tierra, agua, y las cosas visibles, y que como dice sant Augustin contra los académicos, nos mantienen; lo cual afirman todos los filósofos cristianos, y aun los gentiles, sinos es Aristoteles con sus discipulos, que hace el cielo diferente del mundo, en el tratado que dellos compuso.

([57]) Secte apparentée aux Gnostiques.

Este pues es el mundo que Dios hizo, segun lo certifican sant Juan Evangelista, y mas largamente Moisen; que si hubiera mas mundos como él, no los callaron.

El reino de Cristo, que no era deste mundo, porque respondamos à ellos, es espiritual, y no material; y asi, decimos el otro mundo, como la otra vida y como el otro siglo; lo cual declara mui bien Esdras, diciendo: «Hizo el Altisimo este siglo para muchos; y el otro, que es la gloria, para pocos»; y sant Bernardo llama inferior à este mundo en respecto del cielo. Cuanto à los mundos que pone Clemente detrás del Océano, digo que se han de entender y tomar por orbes y partes de la tierra; que asi llama Plinio y otros escritores à Scandinavia, tierra de Godos; y à la Taprobana, que agora dicen Zamotra.

Y Epicuro, segun Plutarco refiere, tenia por mundos à semejantes orbes e bolas de tierras apartadas de la Tierra Firme como islas. Y por ventura estos pedazos de tierra son el orbe y redondez que la Escriptura llama de tierras, y la que llama de tierra ser todo el mundo terrenal. Yo, aunque creo que no hay mas de un solo mundo, nombraré muchas veces dos aqui en esta mi obra, por variar de vocabulos en una mesma cosa, y por entenderme mejor llamando nuevo mundo à las Indias, de las cuales escribimos» ([58]).

Il suffit de considérer le nombre d'auteurs cités dans ce texte, d'ailleurs assez confus, pour se persuader que la question de la pluralité des mondes a dû beaucoup préoccuper les esprits à l'époque où écrivait Gómara. On notera d'ailleurs qu'il importait plus à Gómara de concilier la découverte de l'Amérique avec la théologie qu'avec la science Aristotélicienne. Quant à l'autre monde, il veut qu'il soit «espiritual y no material». Les «mondes» supposés au-delà de l'Océan ne sont plus, pour Gómara, que «orbes y partes de la tierra». Il ne précise pas quelle différence il suppose entre les termes «monde», «orbe» et «tierra». D'après les documents que nous avons présentés, il est bien évident que le mot «monde» avait

([58]) Francisco López de Gómara, *Historia General de las Indias,* Saragoza, 1552, Primera Parte (Ed. de la Biblioteca de Autores Españoles, Tome XXII, p. 156).

alors un sens très différent de celui qu'il a aujourd'hui et qu'il n'est nullement l'équivalent de «globe».

Mais si Gómara affiche une orthodoxie irréprochable sur le plan religieux, il ne consent pas à renoncer au beau nom ronflant qui a été donné au continent dont il se consacra à écrire l'histoire. Et lorsqu'il explique qu'il a voulu appeler «nuevo mundo à las Indias» pour «variar de vocabulos», cet argument ne nous semble pas dépourvu d'humour.

Jerónimo Osório, l'évêque de Silves, au Portugal, s'est également penché sur la question de la pluralité des mondes. Ses pensées, dont le sens n'est guère différent de l'opinion de Gomara, se trouvent exposées dans un petit traité, dont on ne connaît que la traduction française, publiée à Dijon en 1666 ([59]) (Osório est mort en 1580).

Voici ce qu'il dit:

«De l'unité du monde et de la pluralité imaginée par certains philosophes

Question IV

Ouy Metrodore, Anaximandre, Anaximẽ: Archelaüs, Aristarque, Xenophanes, Diogenes, Leucipe, Epicure, et Anaxarque, ont cru qu'il y avait plusieurs Mondes, et Plutarque rapporte que ce dernier Philosophe discourant en la présence d'Alexandre de l'infinité des Mondes, ce Prince cõquérant se prit à pleurer, et comme ce Philosophe luy eût demandé, s'il auait quelque déplaisir, ou si quelque accident luy estoit arriué, il répondit qu'il auoit bien suiet de pleurer, puisqu'il entendoit dire qu'il y auoit une infinité de Mondes, et qu'il ne s'estoit pas encore rendu le Maistre d'un seul.

La raison qu'apportoit Metrodore, pour prouuer cet infinité de Mondes, est bien plus agréable, que solide: cõme ce seroit chose ridicule (disait ce Philosophe) que dans un grand champ il ny eût qu'un seul épy, aussi ne

([59]) Jerónimo Osório, *Théologie curieuse contenant la naissance du Monde,* Dijon, 1666, Trad. par le Chevalier de Iant. (Bibliothèque Nationale, Paris, R. 13678). Ce traité ne se trouve ni dans les *Opera Omnia,* Rome, 1592, ni dans les *Obras Ineditas,* Lisbonne, 1818.

seroit-il pas biẽ seant, que dans l'enceinte d'un espace infiny qui peut contenir une infinité de Monde, il n'y eût qu'un seul.

Pour moy ie ne puis me persuader, que ces Philosophes ayent entendu autre chose, par cette infinité de Mondes, sinon qu'il y auoit une grande quantité de contrées esloignées, et inconnuës différentes en moeurs, en religion, en couleur de corps, en langues, en habits, et autres choses semblables: ce qui toutesfois n'a pu obliger Platon à soustenir cette pluralité de Mondes. Ce Philosophe ayant enseigné au contraire qu'à fin de pouuoir dire que le Monde est parfait il falloit comprendre sous son nom, generalement toutes les Créatures; autrement il n'y auroit point de Monde qui pût estre appellé parfait: il alléguoit encore cette autre raison de cette unité, qu'il estoit conuenable que le Monde répondit à son Prototype, et à son exẽplaire intelligent, et que comme il n'y a qu'un Dieu et une seule première cause, il n'y eut aussi qu'un seul Univers.

Nous pouuons dire encore qu'il n'y a qu'un seul Univers, tant à cause de la subordination, qui est entre les causes, que l'on appelle efficientes, en vertu de laquelle les choses sublunaires sont soumises et dépendent des corps Célestes, ceux cy des intellegences qui les gouvernent, et les intelligences de Dieu qu'à cause de l'accord et de l'harmonie auec laquelle toutes choses conspirent au bien commun de l'Univers, et enfin à cause de l'ordre qui est entre toutes les Créatures par un effet duquel, d'un cõmun concert, comme des soldats d'une mesme armée, elles sont voüées au service, et à la gloire de leur Chef, qui est Dieu, auquel elles tendent chacune en leur manière, comme à la dernière fin de toute la nature et au souverain bien» ([60]).

([60]) Ibid. p. 62 — Dans le Prologue de ses *Décades,* João de Barros semble accueillir favorablement la doctrine de la pluralité des mondes «...vos [de D. João III] armes et vos trophées de victoire se sont imposés, non seulement sur toutes les côtes d'Afrique et d'Asie, mais encore en d'autres mondes plus grands qu'Alexandre déplorait de n'avoir pas connus...» Dec. I, Prólogo. Ailleurs, cependant, il en parle comme d'une hypothèse à laquelle il ne se ralliait pas: «C'est assurément une chose

Lorsque, pour expliquer pourquoi les philosophes de l'Antiquité avaient envisagé «une infinité de Mondes», Osório suggère que c'était à cause de la «...grande quantité de contrées esloignées et inconnües, différentes en moeurs, en religion, en couleur de corps, en langues, en habits et autres choses semblables», il évoque là une des révélations apportées par les grandes découvertes, et qui devait ébranler plus que toute autre l'image du Monde, que le Moyen Age avait si laborieusement synthétisée ([61]).

---

grave et pitoyable de constater qu'une nation à qui Dieu a donné tant de courage que, s'il avait créé d'autres mondes, elle y aurait planté les trophées d'autres victoires, néglige à ce point de perpétuer sa gloire, comme s'il n'y avait pas autant de mérite à la répandre par la plume qu'à la conquête avec la lance» (Déc. I, Liv. IV, Cap. 11). Il se sert pourtant de l'expression «nouveau monde» sans l'appliquer spécifiquement à l'Amérique. Il parle «doutro novo mundo de infieis» (Dec. I, Liv. V, Cap. I) et «deste novo mundo, tantas centenas de anos encoberto» (Dec. I, Liv. IV, Cap. II).

([61]) L'effet bouleversant des découvertes sur l'imagination européenne se trouve exprimé dans ce texte de João de Barros: «E ainda a muitos, vendo sòmente na carta de marear ũa tam grande costa de terra pintada, e tantas voltas de rumos que parecia rodearem as nossas naus duas vezes o mundo sabido, por entrar no caminho doutro novo que queriamos descobrir, fazia neles esta pintura ũa tam espantosa imaginação, que lhe assombrava o juizo. E se esta pintura fazia nojo à vista, ao modo que faz ver sobre os ombros de Hércules o Mundo que lhe os poetas puseram, que quási a nossa natureza se move com afectos a se condoer dos ombros daquela imagem pintada, como se não condoeria um prudente homem em sua consideração, ver este reino (de que ele era membro) tomar sobre os ombros de sua obrigação um Mundo, não pintado, mas verdadeiro, que às vezes o podia fazer acurvar com o grã peso da terra, do mar, do vento e ardor do sol que em si continha; e o que era muito mais grave e pesado que estes elementos — a variedade de tantas gentes como nêle habitavam ?...» (João de Barros, Dec. I, Liv. VI, Cap. 1).

Dans ce passage, Barros semble ébaucher ce qu'on appellera plus tard: «The White Man's Burden».

On trouve aussi chez Montaigne des réflexions sur la signification des découvertes: «Quand tout ce qui est venu, par rapport, du passé jusques à nous, seroit vray, et seroit sceu par quelqu'un, ce seroit moins que rien, au prix de ce qui est ignoré. Et de cette mesme image du monde qui coule pendant que nous y sommes, combien chestifve et raccourcie est la cognoissance des plus curieux ? Non seulement des évènements particuliers, que fortune rend souvent exemplaires et poisants, mais de l'estat des grandes polices et nations, il nous en eschappe cent fois plus

Ce n'est que vers le milieu du XVIe siècle, quelques décades après les principaux voyages de découverte, que le terme «monde» fut définitivement assimilé à «globe». Si l'esprit européen avait pu faire entrer tout le globe terrestre dans son idée de l'*oikoumenê*, qui coïncidait, jusqu'à la fin du Moyen Age tout au moins, avec la *Christianitas* médiévale, il lui avait bien fallu admettre que les autres «mondes» ou «contrées esloignées» qu'on venait de découvrir faisaient partie intégrante du Monde. Or il était indéniable que certaines de ces «contrées» n'avaient vraisemblablement pas connu la révélation chrétienne, et qu'en tout cas elles ne faisaient certes pas partie de l'Église. L'unité chrétienne du globe se trouvait donc gravement mise en question.

De ce fait, et aussi en raison du démenti apporté à la synthèse aristotélicienne par la découverte du ciel austral, la conception médiévale du monde subit un bouleversement.

Obligé par la voix impérieuse de l'*expérience* à admettre la multiple variété de la Nature entourant l'homme, Osório recourt à l'*autorité* de Platon pour soutenir l'unicité du monde.

Le passage que nous avons cité est d'ailleurs le seul où Osório fasse allusion aux problèmes théologiques posés par les découvertes.

L'argument de la «subordination», auquel recourt Osório et que l'on retrouvera chez les Jésuites, se présente comme l'un des instruments les plus importants de l'école néo-scolastique. On remarquera qu'Osório parle de Dieu comme du «Chef» d'une armée dont les soldats [les créatures de Dieu] sont voués à son service et vers lequel ils tendent comme «à la dernière fin de toute nature et au souverain bien». Il n'eût pas été sans intérêt de savoir par quel artifice dialectique il aurait expliqué Dieu comme «la dernière fin de toute nature».

---

qu'il n'en vient à nostre science: nous nous escrions du miracle de l'invention de nostre artillerie, de nostre impression; d'aultres hommes, un aultre bout du monde à la Chine en jouissoient mille ans auparavant. Si nous veoyions autant du monde comme nous n'en veoyons pas, nous appercevrions, comme il est à croire, une perpétuelle multiplication et vicissitude de formes. Il n'y a rien de seul et de rare, eu égard à nature, ouy bien eu egard à nostre cognoissance, qui est un misérable fondement de nos règles, et qui représente volontiers une tresfaulse image des choses». (Michel de Montaigne, *Essais,* Livre III, Chap. VI — Des Coches).

À Coimbra, au XVIe siècle, les Jésuites de l'Université s'acharnèrent à reconstruire l'ancienne synthèse, aristotélicienne et théologique. Dans ce but, ils préparèrent une série de manuels de philosophie, de physique et de géographie, auxquels ils donnèrent la forme de commentaires d'Aristote, et qui se répandirent dans toute l'Europe.

Dans le commentaire du *De Caelo,* ils traitent du problème de l'unité du monde. Ils passent en revue tous les auteurs classiques et prouvent, en s'appuyant sur l'autorité des anciens et sur la dialectique, qu'en effet le monde est unique. Ils apportent ensuite des preuves tirées de la Bible et aboutissent à la conclusion suivante:

> «De ces citations et de quelques autres, on déduit que tout le genre humain a été engendré et propagé dans notre monde et qu'il y est contenu. À moins qu'il ne se trouve quelqu'un pour dire que Dieu a créé un autre monde où il n'y a pas d'hommes; ou que quelques-uns ont été transportés par Dieu de ce monde-ci dans l'autre comme dans une colonie. Ces théories sont absurdes l'une et l'autre et n'ont aucune vraisemblance» ([62]).

Cependant, de même qu'Osório, les Jésuites se virent obligés de reconnaître ce fait écrasant qu'il existait d'énormes différences entre les différentes parties de la terre:

> «*Pourquoi dit-on que le monde constitue un tout unique alors qu'il comporte des natures diverses*
>
> Si quelqu'un se demande: pourquoi tout cet univers, qui réunit une telle diversité de natures différentes, est-il

---

([62]) *Commentarii Collegii Conimbricensis societas iesu, in quatuor libros de coelo,* Cologne, 1600, Lib. I, Cap. VIII, Quaest I, Art. II Col. 110 (Bibliothèque Nationale, Paris, R. 1871). À peu près à la même époque où écrivaient ainsi les Jésuites, Giordano Bruno, qui fut brûlé à Rome comme hérétique en 1600, clamait avec un «impétueux dogmatisme» les propos suivants: «Nous disons que l'Univers est une substance infinie dans un espace infini, c'est-à-dire un infini à la fois vide et plein. L'Univers est un, mais les mondes sont innombrables; chacun des corps du Monde, en effet, est de grandeur finie, mais pris en leur ensemble, ils sont numériquement infinis». (Giordano Bruno, «Camoeracenis acrotismus», Art. IX, cité par P. Duhem, *Études sur Léonardo da Vinci,* troisième série, Paris, 1913, p. 229).

considéré comme un tout ? Après avoir exposé l'opinion de ceux qui donnaient à toutes choses un seul principe vital, dont toutes les parties sont reliées entre elles par un rapport, un lien, de même que les membres d'un animal, et qui ont conçu ainsi l'idée d'un monde unique, nous dirons que c'est là une erreur que nous allons réfuter ailleurs et à laquelle il faut répondre que l'univers est considéré comme un tout unique en raison de la subordination des causes efficaces, en vertu de laquelle les corps inférieurs sont soumis aux corps célestes, ceux-ci aux intelligences, et ces dernières à Dieu, le meilleur et le plus grand. Au sujet de cette subordination, v. Aristote, lib. 2, Métaph. cap. 2 à text. 6, D. Thomas lib. 3 contra gent. cap. 17. En outre l'univers est un à cause de l'harmonie par laquelle tout concourt au bien commun de l'univers tout entier: Aristote l'enseigne, 12 Métaph. cap. 10 text. 52. Et aussi à cause de l'ordre qui régit l'univers, de même que les soldats d'une même armée, disciplinés aussi bien entre eux qu'à l'égard du seul chef et seigneur, Dieu; c'est vers lui que tout tend comme vers la fin suprême de toutes choses et le bien suprême et dernier, comme nous l'avons exposé ailleurs conformément à la doctrine d'Aristote, citée non loin d'ici par D. Dionysii 4 c. de dium. rom., celle de Trismegiste dans Pimandro, de Platon dans le Timée, et de bien d'autres» ([63]).

Dans ce texte, nous voyons avancer deux idées, la «hiérarchisation et l'organicisme». Il est intéressant de noter que les Jésuites approuvent la première et repoussent la seconde, alors que celle-ci a trouvé un défenseur convaincu en la personne de Francisco de Vitoria, dans sa *Relectione «De Indis»* ([64]). L'interprétation du concept du droit naturel présentée alors par Vitoria est, à notre avis, incompatible avec le concept de «subordination».

([63]) *Ibid.* Lib. I, Cap. VIII, Quaest. I, Art. III. col. 111.

([64]) Francisco de Vitoria, *Relectione «De Indis»*, prononcée à l'Université de Salamanque en 1539, imprimée à Lyon en 1557.

Les Jésuites de Coimbra avaient beau insister sur l'idée de la «subordination» et d'une hiérarchie rigide de la Création, les triomphes des découvertes n'en avaient pas moins amené l'homme européen à se croire moins «créature» et plus «dieu». Dans un texte où il évoque l'herbe magique, le Pantagruélion, Rabelais commente ironiquement ce désir de l'homme de briser l'ordre spirituel, comme il avait bouleversé l'ordre physique (la synthèse aristotélicienne):

«Icelle moyennant, sont les nations que Nature sembloit tenir absconces, imperméables et incongneues, à nous venues, nous à elles: choses que ne feroient les oyseaulx, quelque légièreté de pennaige qu'ilz ayent et quelque liberté de nager en l'aër que leur soit baillée par Nature. Taprobana a veu Lappia; Java a veu les mons Riphées; Phébol voyra Thélême; les islandoys et Engronelands voyront Euphrates. Par elle Boreas a veu le manoir de Auster; Eurus a visité Zéphire.

De mode que les Intelligences célestes, les Dieux, tant marins que terrestres, en ont esté tous effrayez, voyans par l'usage de cestuy benedict Pantagruelion, les peuples Arctiques, en plein aspect des Antarcticques, franchir la mer Athlantique, passer les deux Tropicques, volter sous la zone torride, mesurer tout le Zodiaque, s'esbattre sous l'AEquinoctial, avoir l'un et l'autre Pole en veue à fleur de leur Orizon. Les Dieux Olympicques ont en pareil effroy dict: Pantagruel nous a mis en pensement nouveau et tedieux, plus que oncques ne feirent les Aloïdes, par l'usage et vertus de son herbe. Il sera brief marié. De sa femme aura enfans. A ceste destinée ne pouvons nous contrevenir; car elle est passée par les mains et fuseaulx des soeurs fatales, filles de Nécessité. Par ses enfans (peut-estre) sera inventée herbe de semblable énergie, moyenant laquelle pourront les humains visiter les sources des gresles, les bondes des pluyes et l'officine des fouldres; pourront envahir les régions de la Lune, entrer le territoire des signes célestes, et là, prendre logis, les uns à l'Aigle d'or, les aultres au Mouton, les aultres à la Couronne, les aultres à la Herpe, les aultres au Lion d'Argent, s'asseoir à table avecques nous et nos Déesses prendre à femmes, qui les seulz moyens d'êstre déifiez» ([65]).

L'herbe pantagruelion, vraisemblablement une sorte de mandragore, devait ici représenter la «technique» de l'homme moderne, le «truc» qui permettait de dominer la Nature. Cette technique, c'était alors la science nautique (calcul des latitudes par des observations astronomiques) et la construction navale (développement de la caravelle).

Aujourd'hui, alors que nous nous trouvons au seuil de la conquête d'«autres mondes» dans l'espace cosmique, ce texte de Rabelais sur les perspectives qu'ouvre la conquête de la terre prend un aspect prophétique.

APPENDICE

Cette communication avait déjà été présentée au Congrès, lorsque nous avons pris connaissance de deux textes, l'un de Pétrarque, l'autre d'António Luís, qui nous ont paru susceptibles d'éclairer encore notre thème.

Nous remercions ici le Secrétaire du Congrès de nous avoir donné la possibilité de les ajouter ici sous forme d'appendice.

Voici le texte de Pétrarque (1346):

«...Puis [après Thulé et l'Irlande] je vais aux îles Fortunées, qui sont à l'extrême pointe de l'Occident et tout aussi éloignées des Indes que de l'Arctique, mais pour nous plus voisines et plus familières, terre [multorum (?)], mais surtout fameuse par le poème lyrique d'Horace [Flacci], dont la très ancienne gloire est toujours vivante. Si ce que racontent nos pères est vrai, une flotte armée de Gênois ([66]) y pénétra, et récemment Clément VI donna

---

([65]) François Rabelais, Tiers Livre, Chapitre LI (Ed. de Jean Plattard, Paris, 1948).

([66]) Malgré les divergences d'opinions entre les historiens qui se sont intéressés aux voyages aux Canaries vers cette époque, il nous semble bien qu'il s'agit ici de l'expédition de Lanzarote Malocello (appelé aussi Llansrote de Framqua) c. 1336.

à ce pays un prince, chez qui se mêle le sang des rois d'Espagne et de France et vraiment courageux entre tous. En effet, tu te souviens qu'en ce jour où il s'avançait en cortège à travers la ville, portant la couronne et le sceptre, il se mit tout à coup à pleuvoir si fort, qu'il rentra chez lui trempé, ce qui était signe que ce pays où il était arrivé était surtout pluvieux et humide. J'ignore par qui cette épithète [fortunée] a été attribuée à cette région située en dehors du monde [extra orbem]: mais je sais, car on a écrit et rapporté beaucoup de choses, que la fortune ne correspond pas toujours à ce surnom de terres fortunées. Mais tu dis que ces gens aiment la solitude plus que presque tous les mortels, qu'ils ont des moeurs rudes et qu'ils sont si peu différents des animaux que, agissant plus par instinct que par choix, non seulement ils vivent seuls, mais ils errent dans des endroits isolés, tantôt avec les bêtes sauvages, tantôt avec leurs troupeaux. Mais j'ai assez parcouru en long et en large, par curiosité, ces coins à l'écart du monde [séparés du monde-disiunctos mundi], dont je rapporte des choses lues et entendues, qui ne sont pas affirmées par moi mais par des auteurs très dignes de foi. Mais ceci dit, je passerai à des choses plus claires et mieux connues de nous» (67).

Il est intéressant de rapprocher cette idée de Pétrarque que les Canaries se situent «en dehors du monde» avec celle de Jacopo Foresti (1503), qui intitule sa description de la découverte par Colomb des îles des Antilles: «De quatuor maximis insulis in Indis extra orbem nuper inventis» (68):

Si les Canaries ne représentaient pas pour Pétrarque «un autre monde», comme les îles du Cap Vert pour Cadamosto, il les situait tout de même «en dehors du monde».

L'Église a toujours eu une attitude, sinon équivoque, du moins assez mouvante, devant la thèse de l'existence d'un autre

Cf. Ch. Verlinden, «Lanzarote Malocello et la découverte portugaise des Canaries», In *Rev. belge de Philolog. et d'Hist.*, Bruxelles, t. xxxvi, (1958), N.° 4, pp. 1173-1209.

(67) Pétrarque, «De Vita Solitaria», lib. II, sect. VI, Ch. 3, In *Opera quae extant omnia,* Bâle, 1581, p. 277.

(68) V. plus haut note 40.

monde ou de la pluralité des mondes. La tradition augustinienne l'écartait résolument. Si un homme comme Albert le Grand citait Lucain, parlant des Arabes venus pour aider Pompée:

«Ignotus vobis Arabes venistes in orbem» ([69]), le Pape Zacharie déclarait (748) au sujet d'un prêtre nommé Virgile:

> «En ce qui concerne la doctrine perverse et criminelle qu'il [Virgile] a proférée contre Dieu et contre son âme — si la chose est dûment établie, cet individu [Virgile] aurait proclamé qu'ils existerait un autre monde et d'autres hommes sous la terre, [...] après avoir réuni le conseil des évêques, chasse-le de l'Église, en le dépouillant des honneurs du sacerdoce» ([70]).

À propos de la doctrine de la pluralité des mondes, Isidore de Séville (c. 570-636) écrivait:

> «...d'autres affirment qu'il y a d'innombrables mondes [...] Ce sont des hérésies opposées à la foi catholique et d'avance condamnées par les Apôtres et par les saints Pères ou les Conciles» ([71]).

Gratien (XIIe siècle) le reproduit textuellement ([72]).

En revanche, au XVe siècle, le cardinal Nicolas de Cues, évêque de Brixen, affirmait l'existence d'une pluralité des mondes habités ([73]).

([69]) Lucain, *Pharsalia*, iii, 247-248, cité par Albert le Grand, «Lib. de Natura Locorum», tract 1, In *Opera Omnia*, ed. Auguste Borgnet, Paris, 1890, t. IX, p. 544.

([70]) Transcrit par Youssouf Kamal, *Monumenta Cartographica Africae et Aegypti*, Le Caire, Tome III, fasc. I, f. 505.

([71]) Isidore de Séville, *Etymologiarum*, Lib. VIII, v.

([72]) Gratien, *Decretum*, Causa 24, quaest. 3, In Migne, *Pat. Lat.* Vol. 187, col. 1318.

([73]) Nicolas de Cues, *Op. laud.* II, cap. XII, cité par P. Duhem, *Le Système du Monde*, Paris, 1959, Tome X, p. 324.

«Peut-on — demande Duhem — souhaiter preuve plus manifeste de l'extrême liberté que l'Église catholique, au déclin du Moyen Age, laissait à la méditation du philosophe et aux tentatives du physicien?» ([74]).

«À la fin du XVe siècle, — toujours selon Duhem — le problème de la pluralité des mondes donnait lieu, dans les écoles, à des débats passionnés...» ([75]). «Au XVIe siècle, Humanistes catholiques et Humanistes réformés étaient [...] fort peu disposés à souscrire aux thèses de Giordano Bruno sur l'infinie grandeur de l'univers ou sur la pluralité des mondes; leurs sentiments à l'égard de ces propositions ne diffèrent guère sans doute de l'opinion des Averroïstes les plus endurcis; seuls les Scottistes et les Nominalistes devaient écouter ces affirmations sans effroi et les discuter sans parti pris» ([76]).

C'est dans le cadre de cette atmosphère turbulente du XVIe siècle où s'entrecroisent et se heurtent de puissantes idées sur la théologie et la physique, que se situe le texte que nous présentons maintenant.

C'est en 1539 qu'un médecin de Lisbonne, António Luís, aussi réputé pour ses connaissances professionnelles que pour son érudition classique, adresse au Roi D. João III un panégyrique en latin. Il semble peu probable, comme on l'a prétendu, qu'il l'ait écrit dans le but de s'assurer la protection du Roi en raison de certains ennuis avec l'Inquisition. Le panégyrique était déjà rédigé depuis plusieurs années lorsqu'il fut publié ([77]), et l'Inquisition fut introduite au Portugal en 1536. D'ailleurs, rien dans le texte ne laisse transparaître la moindre crainte de l'Inquisition.

L'ouvrage d'António Luís, qui est autant une évocation de la *geste* héroïque des découvertes portugaises qu'un éloge du Roi, nous semble un exemple exceptionnel de la floraison éphémère mais brillante de l'esprit humaniste au Portugal, là

([74]) Ibid. loc. cit.

([75]) P. Duhem, *Études sur Leonardo da Vinci,* 2e série, Paris, 1909, p. 93.

([76]) P. Duhem, *Études sur Leonardo da Vinci,* 3e série, Paris, 1913, p. 253.

([77]) A. da Rocha Brito, «Duas Assinaturas de António Luís», in *O Lar do Médico,* Porto, N.º 40 (Nov. 1946).

surtout où il proclame la supériorité des modernes sur les anciens.

Voici le passage où l'auteur évoque la pluralité des mondes:

«Alors ils [les Grecs] auraient plus justement admiré le courage d'Alexandre, jamais assez loué pour ses prédictions, lui qui, enfermé dans ce seul monde [orbis] étroit, insuffisant pour un esprit aussi grand, aspirait à d'autres mondes [alios orbes] dont son intelligence prophétique lui révélait l'existence. Lui qui ne pouvait voir sans envie les exploits réalisés par d'autres (il avait même l'habitude de dénigrer les prouesses de son père Philippe et de s'attribuer à lui-même ses hauts faits), je crois qu'il aurait pleuré en voyant possédés par toi [D. João III] ces mondes [mundos] qu'il avait rêvés (et dont, dit Anaxarque, il souffrait de n'en pas posséder un seul entre tant), et il aurait eu raison d'en éprouver de la douleur et de la honte et de se lamenter. Il t'envierait certes, mais non pas, je crois, parce que c'est à toi que Dieu a donné tant de gloire et confié cette mission; en vérité, il t'aurait envié pour la grandeur et l'excellence des prouesses réalisées, à tout jamais admirables, et qu'on ne croirait pas possibles sans l'aide de Dieu: tu as triomphé des ténêbres par ta splendeur, tu as convaincu les hommes, obtenu leur accord et leur approbation.

Maintenant les Abdéritains diraient que Démocrite est sain d'esprit, lui qui pensait qu'il y avait beaucoup d'infinités de mondes [mundorum], et qui engageait à contempler toute l'étendue de la nature. Il n'aurait plus été nécessaire de faire chercher Hippocrate à Cos pour lui administrer une purge [veratro], et ce très sage philosophe n'aurait plus été soupçonné de folie. S'ils étaient nés de ton temps, Métrodore, le disciple d'Épicure, et Épicure lui-même, ainsi que son maître Leucippe, n'auraient pas été considérés comme les créateurs d'une philosophie étrange et lointaine, eux qui estimaient qu'il y avait d'innombrables mondes [mũdos] à l'infini et que tout était compris dans l'immensité. Et Thales l'Egyptien et d'autres avec lui, qui ont jugé que plusieurs mondes [mũdos], des sphères [orbes] et des globes [globos] s'étaient rapprochés et unis les uns aux autres pour former un tout, quelles louanges ne se seraient-ils pas attirées, s'il avait été possible de prolonger leur vie jusqu'en ce moment où tu règnes, car ils auraient certainement conçu les grandes théories probables et conformes à la vérité des choses de la nature. Sois glorifié César invaincu, toi qui peux te vanter d'avoir enfanté un autre monde [alium orbem], toi qui as ajouté des terres à ton empire et un plus grand élément à tous les autres. N'aurais-tu pas plus de raisons de te dire César ? — ou il faut être bien hostile au nom lusitanien pour te refuser ce qui est ton droit. Quand César, qui inventa le nom d'empereur envahit en protecteur la Germanie, qui n'est qu'une toute petite partie de la terre, il écrivit qu'il avait découvert un autre globe terrestre [alium orbem terrarum] — il l'avait jugée très grande parce qu'elle n'était pas entourée par l'Océan, mais semblait l'entourer — et pourtant, dans cette expédition, il n'avait rien accompli d'aussi grand, et de loin, que s'il avait franchi l'Océan. Que dire, alors, de toi, ô Roi, qui as découvert d'autres mondes [alios mundos], de nouveaux soleils, de nouveaux astres, grâce aux heureuses navigations de tes lusitaniens, et les as (dit-on) révélés au grand jour» ([78]).

Il est intéressant de comparer la dernière phrase de ce texte aux citations d'Angelo Poliziano et de Pedro Nunes, que nous avons données plus haut.

À ce propos, nous nous permettons de revenir une fois de plus sur le problème que posent les expressions «nouveau monde» et «autre monde»: y a-t-il entre elles une identité, ou au contraire une différence ?

Il semble tout d'abord que les hommes du XVIe siècle aient éprouvé de façon poignante le besoin de comprendre les

---

([78]) António Luis, *Panagyrica oratio elegantissima plurima rerum & historiarum copia referta Ioanni huius nominis tertio inuictissimo Lusitaniarum regi nuncupata,* Ulysbonae apud Logdouicũ Rotorigiũ, 1539, ff. xxii (r) — xxiij (r), Bibl. Nacional, Lisbonne, Rés. 76V & 77V.

Il en existe aussi un exemplaire au British Museum. Une traduction libre et résumée, en portugais, fut publiée avec une introduction par A. da Rocha Brito, «Duas Assinaturas de António Luís», in *O Lar do Médico,* Porto, N.os 40, 41, 42, 43, 45, 47, 48, 49, 50, 51 (1946-1947).

découvertes qu'ils venaient de faire, surtout quand elles s'annonçaient comme «sensationnelles». Pour aussi «inédite» que puisse se révéler une découverte, elle ne saurait être «comprise» qu'en fonction de structures mentales déjà existantes. Du moment où l'on admettait que la découverte d'une terre signifiait davantage que la découverte de quelques nouvelles îles, on en arrivait fatalement à se demander si elle correspondait à la découverte d'un «monde». Admettre l'existence de mondes successifs dans la dimension de l'espace entraînait inévitablement à des spéculations sur l'existence de mondes successifs dans la dimension du temps (ou dans l'Histoire). Si (comme nous l'avons vu dans le cas du texte de Macrobe) la mort représentait une rupture de niveau ontologique ([79]) entre notre monde et l'autre monde — celui des Enfers —, le Jugement Dernier équivalait à une rupture de niveau ontologique entre notre monde et le «nouveau monde» — le Royaume du Ciel à venir ([80]). Cette dernière correspond à la délivrance promise par le Christ.

Il n'est pas étonnant qu'en présence d'un fait aussi extraordinaire que la découverte de l'Amérique, des esprits «exaltés», comme Colomb, soient allés jusqu'à «comprendre» cette découverte comme l'accomplissement de la promesse de l'Apocalypse «d'un ciel nouveau et d'une terre nouvelle» ([81]).

L'erreur de ces esprits «exaltés» consista à confondre l'«Histoire» et la «Géographie», et à croire que s'ils suivaient à travers l'Océan, la voie ouverte par une *apparente* rupture de niveau ontologique dans l'espace, ils pourraient, en modifiant leur situation géographique, échapper à leur condition d'homme et à leur destin terrestre dans le Temps.

Combien de ceux qui accompagnaient, pleins d'espoir, les grandes vagues d'émigration vers le Nouveau Monde au XIXe siècle n'étaient-ils pas hantés par cette idée ?

([79]) L'expression est de Mircea Eliade.

([80]) Pour Valentim Fernandes (v. plus haut), ce «Royaume du Ciel» devait prendre la forme de la «Terre Promise».

([81]) Apocalypse, 21, 1, 2. «Je vis ensuite un ciel nouveau et une terre nouvelle; car le premier ciel et la première terre étaient passés et la mer n'était plus. Et moi, Jean, je vis la sainte cité, la nouvelle Jérusalem, qui descendit du ciel...»

RELATÓRIO

PELO

P.[e] DOMINGOS MAURÍCIO

William Graham Lister Randles deu à sua comunicação a seguinte epígrafe: *O novo mundo, o outro mundo e a pluralidade dos mundos.*

Supõe-se, geralmente, que a expressão *Novo mundo* aparece, pela primeira vez, na famosa carta *Mundus Novus,* atribuída a Américo Vespúcio e impressa em Paris, em 1503.

A expressão não foi inventada por Vespúcio, embora a impressão da célebre carta lhe desse larga ressonância. Quem primeiro a emprega, referida à América, é Pedro Mártir d'Anghiera, na sua carta de 1 de Novembro de 1493. A ideia de *Novo mundo* encontra-se ligada, ìntimamente, a dois problemas que, desde a Antiguidade, através da Idade Média, apaixonaram filósofos, teólogos e homens de letras: a existência dum outro mundo austral, povoado por homens antípodas; e a pluralidade dos mundos.

S. Agostinho declarou herética, a crença na existência de homens antípodas, habitantes do hemisfério austral. Alberto o Magno e Rogério Bacon, porém, defenderam-na. A repercussão do impresso famoso *Mundus Novos* proveio, verosìmilmente, de nessa carta se afirmar, categòricamente, a existência de antípodas. Mas, também, não é nela que, pela primeira vez, se relaciona a descoberta da América com tal facto. Já Anghiera o fizera, em 1493, colocando os antípodas, não ao Sul mas a Oeste. Para a Idade Média, era a zona tórrida que os separava da *oikouméne,* situada no hemisfério setentrional. A Antiguidade compraz-se numa geografia complexa da abóbada celeste. E a imagem cosmográfica da terra reflectia-se no céu nocturno. As regiões austrais e os antípodas correspondem ao *Outro mundo:* os Infernos.

Em Macróbio (séc. V), como em Gervásio de Tilbury e na literatura do ciclo arturiano, há referências a esse mundo diferente do nosso. Cadamosto, ao tratar de Cabo Verde, em 1456, fala, embora referindo-se a coisas situadas no hemisfério norte, desse *Outro mundo.*

Os Portugueses atravessam o Equador, em 1472-1474, e chegam à embocadura do Congo, em 1484. João Teixeira, em 1489, interpreta essas viagens como descobertas de *Outro mundo,* o mesmo fazendo Hartmann Schedel no *Liber Chronicarum* de Nuremberg, em 1493, e Valentim Fernandes, em 1503.

Que pretendiam significar, estes autores, com a expressão *Outro mundo,* referida a terras recém-descobertas ? É difícil determiná-lo.

O *Outro mundo* parece ser o mundo de além. Valentim Fernandes é o primeiro a usar a expressão *Novo mundo,* aplicada aos descobrimentos portugueses, no Oceano Índico. Lourenço de Almeida, chegando a Ceilão, em 1505, identificava esta ilha com a Taprobana dos antigos que a supunham existente no *Outro mundo.* E João Poggio, em 1514, adoptava a mesma interpretação dessa ilha, supondo-a constituir um *Outro mundo*, opinião reforçada por Barros, que pretende justificá-la com a posição dela para além da linha equinocial.

Entre as duas expressões *Outro mundo* e *Mundo novo* — parece haver, assim, estreita relação, aplicando-se não só à América, mas ao mundo situado para além do Cabo da Boa Esperança, representando, sobretudo e em primeiro lugar, o mundo austral, pois para os navegadores europeus, verdadeiramente novo era o céu com as estrelas do hemisfério sul. A descoberta desse céu vinha deitar por terra a concepção cosmográfica aristotélica.

Quando a expressão *Mundus Novus* se aplicou ao Brasil, ninguém pensava que este fizesse um todo continental com a América do Norte. Supunha-se, antes, que se estendesse no sentido Este-Oeste. Ela data da época, em que reina a noção duma oposição simétrica entre a *oikouméne* (mundo antigo) e o mundo austral, separados pela zona tórrida, julgada intransponível.

A antiga oposição «horizontal» entre o «mundo» e o «hemisfério inferior» sofria um desvio de 90°, para dar lugar a uma oposição vertical entre a Europa e a América, passando a barreira da zona tórrida a ser substituída pelo Oceano Atlântico.

A oposição geográfica sugere uma antítese afectiva: o mito do bom selvagem e da vida feliz e idílica de além Atlântico, contraposto ao civilizado e à vida complicada da Europa ou, pelo contrário, a noção do homem e da fauna americana, inferiores ao homem e à fauna do velho mundo. A polémica nascida desta polarização afectiva mantém-se, hoje, ainda viva...

Conexo com a noção do *Mundo novo* é o problema da *pluralidade dos mundos*. As primeiras referências que, a propósito dos descobrimentos, se lhe fazem, encontram-se no *Discurso de Obediencia*, pronunciado por Vasco Fernandes de Lucena, diante de Inocêncio VIII, em 1485, e na carta de André Poliziano a D. João II, em 1489.

Aristóteles repudiara a pluralidade dos mundos como absurda.

Na Idade Média, enunciou-se o problema, admitindo a possibilidade dessa pluralidade. Em 1277, a Sorbona de Paris condenou a doutrina de Aristóteles e Averrois. As opiniões dividem-se. No século XVI, João de Celaya e Luis Coronel não admitiam a doutrina da pluralidade dos mundos.

A questão podia considerar-se, sob o ponto de vista astronómico e geográfico. Francisco Lopez de Gómara é o primeiro, em 1522, a levantar a questão da pluralidade dos mundos, a propósito da descoberta da América. Não a admite e aplica a expressão *Novo mundo* às Índias só para variar os vocábulos. Jerónimo Osório, bispo de Silves, também abordou o problema e segue as pisadas de Gómara. A «pluralidade de mundos» dos antigos teria um sentido figurado. As descobertas portuguesas rectificaram a noção medieval do mundo. É nos meados do século XVI que, após os grandes descobrimentos, o termo *mundo* passa a corresponder definitivamente a *globo*. A experiência põe em foco a múltipla variedade da natureza que circunda o homem. Para sustentar a unicidade do mundo, Osório recorre à autoridade de Platão. Os Conimbricenses retomam o argumento da «subordinação» para apoiar a tese da unicidade do mundo. No *De coelo,* tratam do problema da unidade do universo e pronunciam-se pela sua unicidade. Esta unicidade não exclui a diversidade de naturezas. Os Jesuítas, segundo o Autor, admitem a hierarquização delas, mas repudiam o seu organicismo. Francisco de Vitória admitiu-o. Mas o seu conceito de direito natural parece incompatível com o conceito de «subordinação».

As descobertas concorreram, talvez, para o repúdio deste conceito. Rabelais comenta, irònicamente, esta pretensão humana de romper a ordem espiritual do mundo como subvertera a ordem física nos quadros da síntese aristotélica. Quando o homem de hoje se encontra no limiar da conquista de *outros mundos,* através do espaço cósmico, o texto de Rabelais parece assumir um sentido profético.

XVI

LA DIFFUSION DANS L'EUROPE DU XVIe SIECLE DES CONNAISSANCES GEOGRAPHIQUES DUES AUX DECOUVERTES PORTUGAISES

Les sources imprimées, par lesquelles l'Europe prit connaissance des découvertes portugaises, peuvent se diviser en trois grandes catégories. D'abord les opuscules publiés à la fin du XVe siècle et jusque vers la première moitié du XVIe, des «Discours d'Obédience» adressés au Pape, des plaquettes d'actualité, de brefs récits de voyage et des allusions aux découvertes dans des ouvrages consacrés à des thèmes plus généraux. Ensuite viennent, au milieu du XVIe siècle, les grandes histoires nationales relatant les hauts faits des conquêtes en Afrique et en Orient. Ce sont les œuvres de Fernão Lopes de Castanheda, de João de Barros et de Jerónimo Osório.

La dernière catégorie englobe les compendiums géographiques qui paraissent à partir du dernier quart du XVIe siècle. Les auteurs de ces compendiums procèdent au dépouillement des récits des premiers voyages, puis ils extraient des histoires nationales tout le savoir géographique qui s'y trouve. En réorganisant ces connaissances géographiques et en suivant systématiquement les côtes des continents, ils cherchent à dresser des inventaires géographiques de tous les pays du monde.

Sans nullement prétendre à être exhaustif, examinons de plus près la première catégorie. Elle est importante, car plus que dans la suivante, on y perçoit la surprise et le choc que provoqua la découverte de terres dont aucun auteur de l'antiquité n'avait prévu l'existence.

Le *Discours d'Obédience* prononcé devant le Pape en 1484 par Vasco Fernandes de Lucena, imprimé à Rome c. 1484 [1] et à nouveau

---

1. *Valasci ferdinandi utriusque iuris consulti, Illustrissimi regis Portugalliae oratoris, ad Innocentium VIII, pontificem maximum de obediencia Oratio*, Rome, c. 1484. Reproduit en fac-similé par A. Fontoura da Costa, *As Portas da India em 1484*, Lisbonne, 1936.

en 1491, est la première notice imprimée concernant les découvertes portugaises: elle fournit des informations importantes sur le progrès des voyages le long de la côte occidentale de l'Afrique au sud de l'Equateur. L'affirmation, fondée sur le résultat apparent du premier voyage de Diogo Cão, que les Portugais étaient parvenus dès cette date à quelques jours de voyage de l'Océan Indien, fut plus tard cruellement démentie lors du second voyage de celui-ci, en 1486.

Le Discours adressé au Roi D. Manuel par l'ambassadeur italien Pietro Pasqualigo en 1501, et imprimé à Venise la même année, est une louange dans le style ampoulé de l'époque: le souverain y est qualifié d'«invincible jusque dans l'hémisphère souterrain», on déclare qu'il a «créé un empire aux antipodes» et «réuni des peuples séparés par la force de la nature et établi des relations entre deux mondes différents de l'univers» [2]. C'est dans ce discours que, pour la première fois, les régions découvertes par les Portugais sont traitées comme un «monde» distinct du «monde» connu de l'Occident médiéval.

L'idée sera reprise par Valentim Fernandes dans sa préface à la traduction portugaise des voyages de Marco Polo, publiée à Lisbonne en 1502. Il emploie l'expression *mundo novo* pour désigner toutes les régions orientales au-delà du Cap de Bonne Espérance [3].

On retrouve l'expression dans la célèbre plaquette *Mundus Novus*, qui décrit le voyage portugais de 1501 le long de la côte est du Brésil. Ce petit ouvrage de quatre feuillets a connu, au XVIe siècle, entre 1503 et 1529, au moins soixante éditions en latin, français, allemand, flamand et tchèque [4].

Bien plus que la lettre analogue de Colomb, publiée en 1493, ce «best seller» du XVIe siècle a frappé les imaginations pour la raison suivante: l'auteur (qui n'est pas Améric Vespuce, comme on l'a dit, mais un anonyme) y affirme à propos de ces régions nouvelles que «... nos ancêtres n'en avaient aucune connaissance [...] Cela va plus loin que l'opinion de nos anciens auteurs, dont la plupart disent qu'au-delà de l'équateur [...] il n'y a aucun continent, mais seulement une mer. [...] Cette opinion est fausse et totalement contraire à la vérité [...]

---

2. *Pietri Paschalici Veneti Oratoris ad Hemanuelem Lusitaniae Regem Oratio*, Venise, 1501. Edition avec une traduction anglaise par Donald Weinstein, Minneapolis, 1960.

3. Valentim Fernandes, «Epistola sobre a tralladaçam», in *O Livro de Marco Polo*, Lisbonne, 1502. Rééd. F. M. Esteves Pereira, Lisbonne, 1922, sign. Aij (r.).

4. Cf. Rudolf Hirsch, «Printed Reports on the early Discoveries and their reception», in Fredi Chiapelli (éd.) *First Images of America*, University of California Press, Los Angeles, 1975, vol. II, p. 540.

car on a trouvé un continent plus densément peuplé [...] que notre Europe, ou même l'Asie ou l'Afrique ...» [5].

C'est la première fois que, à propos des régions nouvelles découvertes par les Portugais, on voit le terme «continent», car au Moyen Age le seul «continent» c'est l'*œcumène*, composé de trois parties: l'Europe, l'Asie et l'Afrique.

Dans ce contexte, le terme «monde» ou «continent» renvoie au mot latin *orbis*. Les auteurs de l'antiquité, Pline et Pomponius Mela avaient parlé de Taprobane (l'île de Ceylan) comme d'un autre *orbis* [6].

Ayant sans doute ces références présentes à l'esprit, le roi D. Manuel adresse au Pape Jules II, en 1507, une lettre où il annonce l'arrivée à Ceylan, en 1505, de Lourenço de Almeida, «... à cette fameuse île de Taprobane, parfois considérée comme un autre monde (*alterum aliquando orbem*)». La lettre est imprimée à Paris la même année [7], ce qui témoigne de l'importance que revêtent, dans l'esprit du public, les lieux qu'une culture humaniste avaient rendus célèbres.

Cette lettre inspire à l'Italien Giovanni Francesco Poggio, fils du célèbre humaniste Poggio Bracciolini, un éloge de D. Manuel qui sera publié l'année suivante. On y lit: «... tout dernièrement, envahissant cette île de Taprobane, la première du monde, mais plus justement désignée par les Géographes comme un autre monde, ou encore, selon l'opinion de Pline, située hors du monde, ta flotte a dispersé avec des effectifs réduits les forces innombrables des ennemis [...] Chose merveilleuse! Chose inouïe! Chose incroyable! Qui eût jamais cru que les Portugais s'aventureraient jusqu'aux rivages de l'Inde?» [8]. On voit bien pourquoi Camões, dans la première strophe des *Lusiades* a écrit ... *passaram ainda além da Taprobana.* La comparaison avec Alexandre le Grand annonce la Querelle des Anciens et des Modernes, thème qui va traverser tout le XVIe siècle.

---

5. *Mundus Novus*, éd. Augsbourg, 1504, in Henri Vignaud, *Améric Vespuce, 1451-1512*, Paris, 1917, p. 305.

6. Pline, *Hist. Nat.*, VI, 81: *Taprobanem alterum orbem terrarum esse diu existimatum est.* Pomponius Mela, *De Chorographia libri tres*, éd. G. Ranstrand, Gotheborg, 1971, III, 70, p. 59: *Taprobane aut grandis admodum insula aut prima pars orbis alterius, id Hipparcus dicitur, sed quia habitatur nec quisquam circum eam isse tradıtur prope verum est.*

7. D. Manuel, *Epistola serenissime Regis Portugalliae de victoria contra infideles habita ad Julium papam secundum*, Paris, 1507. Rééd. en fac-similé de José V. de Pina Martins, Fondation Gulbenkian, Centre Culturel Portugais, Paris, 1978.

8. Giovanni Francesco Poggio, *Emmanuelis Portugaliae Regis Elogium*, Florence, 1514. Cité par Fidelino de Figueiredo, *A Epica Portuguesa do século XVI*, São Paulo, 1950, pp. 112-113.

Dès les premières années du siècle, on voit paraître trois petites plaquettes. La première s'intitule *Calcoen*, elle a été publiée à Anvers en 1504 [9]. Ecrite par un Flamand en flamand, elle décrit le second voyage de Vasco da Gama. La deuxième, un texte apocryphe, a pour titre: *Copia de una lettera del Re de Portogallo mandata al Re de Castella del viaggio & successo de India*, Rome et Milan, 1505 [10]. Le «roi de Castille» est évidemment Ferdinand d'Aragon. Le titre de la troisième est *Den rechten weg ausz zu faren von Lissbona gen Kallakuth*, Nuremberg, 1505 [11].

Aucune de ces plaquettes n'apporte d'informations très importantes, mais elles témoignent de la curiosité avide du reste de l'Europe pour tout ce qui concerne les découvertes des peuples ibériques.

Le récit du voyage en Inde de D. Francisco de Almeida en 1505: *Gesta proxime per Portugalenses in India Ethiopiam & aliis orientalibus terris*, Rome et Cologne, 1506, et Nuremberg, 1507, est déjà plus détaillé et plus consistant. Mais c'est le récit d'un Allemand, Balthasar Sprenger, *Die Merfart* [12], Munich, 1509, à propos du même voyage, qui aura un plus grand retentissement, parce qu'on y trouve des gravures sur bois de Hans Burkmair représentant les peuples d'Afrique et de l'Inde, les premières qu'ait connues la Renaissance; surtout, elles donnent l'impression d'avoir été croquées sur le vif. Une traduction flamande et une anglaise parurent à Anvers en 1508 et en 1521-22. La traduction anglaise est le premier livre imprimé en langue anglaise, où il est question de l'Afrique noire et de l'Inde, révélées par les découvertes portugaises [13].

Il ne faut pas oublier, pour ces premières années du XVIe siècle, les quelques pages sur les découvertes portugaises figurant dans un ouvrage consacré aux villes du monde, et qui a connu neuf éditions au XVIe siècle. L'auteur y apporte des informations intéressantes, qui ne seront pas reprises par d'autres plus tard. Il s'agit du *Commentariorum urbanorum libri XXXVIII* de Raffaello Maffei da Volterra, publié

---

9. *Calcoen*, Anvers, 1504. Réédition moderne avec trad. angl. par J. Ph. Bergeau, Londres, 1874.

10. Rééd. moderne avec fac-similé de l'original et trad. angl. par Sergio J. Pacifici, Université de Minnesota, Minneapolis, 1955.

11. Rééd. moderne avec trad. angl. par Alvin E. Prottengeier et John Parker, Université de Minnesota, Minneapolis, 1956.

12. Texte latin in *Voyage littéraire de deux religieux bénédictins de la Congrégation de St. Maur*, Paris, 1724, Tome II, p. 361. Cf. aussi F. Schulze, *Balthasar Springers Indienfahrt*, Strasbourg, 1902.

13. Réimpression de la traduction anglaise, *Of the newe landes and of the peoples found by the messengers of the Kynge of Portugal named Emanuel*, in Edward Arber, *The First Three English Books on America 1511-1555*, Londres, 1885, Reprint Kraus, 1971.

à Rome en 1506 [14]. On y trouve des remarques sur la culture du sucre à Madère, sur celle du coton aux îles du Cap Vert et sur le commerce des esclaves échangés contre des coquillages; puis sur le royaume du Congo, la population de sa capitale et sur la pratique du cannibalisme; ainsi qu'une notice sur le voyage de Cabral de 1500, et sur le voyage portugais de 1501 au Brésil. A propos de ce dernier, il est affirmé que les Portugais n'admettent pas que l'on soit parvenu jusqu'à 50° de latitude, comme l'avait déclaré l'auteur du *Mundus Novus*. On y découvre enfin quelques renseignements sur l'Inde et sur le royaume de Narsinga. C'est un Portugais, le Cardinal Jorge da Costa, qui, à Rome même, avait fourni à l'auteur ses informations [15].

En 1516 et en 1517 parviennent en Europe deux lettres, datées de 1515 et de 1516, envoyées d'Asie par un Italien parti avec les flottes des Portugais. Il s'agit d'Andrea Corsali. Ces lettres seront publiées la même année à Florence, puis rééditées dans la collection de Ramusio. Corsali donne d'importantes informations sur la géographie de l'Orient, notamment sur l'île de Ceylan, sur Malacca, sur la Chine et sur les Moluques. Il évoque, à l'instar de la représentation sur la carte de Lopo Homem, de 1519 [16], la terre australe qui relierait l'Asie du sud-est au Brésil, aux Antilles et à la «terre ferme du roi de Castille» [17]. Corsali compare la géographie de l'Orient, telle que la figuraient les Portugais, avec celle de Ptolémée, dont il relève les erreurs. Dans sa lettre de 1515, il fait remarquer que Ptolémée n'avait pas correctement situé l'île de Taprobane (Ceylan), «comme on peut en juger d'après «la carta de navigare», carte certainement d'origine portugaise. L'île de Taprobane, dit Corsali, est bien plus petite que ne le dit Ptolémée, et les longitudes que lui attribue l'Alexandrin sont fausses. Il annonce son intention de rester en Orient afin de refaire la géographie de la région, d'étudier l'intérieur des terres et de comparer les latitudes et les toponymes donnés par Ptolémée à «ceux d'aujourd'hui». Dans ce but, il affirme avoir emporté un astrolabe et beaucoup d'autres instruments nécessaires [18].

---

14. Raffaello, Maffei da Volterra, *Commentariorum Urbanorum libri XXXVIII*, Rome, 1506, «Loca nuper reperta», ff. 138 (v.) à 140 (r.).

15. Cf. Vsevolod Slessarev, «Raphael Maffei's contribution to the history of the Portuguese Discoveries», in *Congresso Internacional de História dos Descobrimentos*, Lisbonne, 1961, vol. III, pp. 551-574.

16. Reproduite in Armando Cortesão & A. Teixeira da Mota, *Portugaliæ Monumenta Cartographica*, Coimbra, 1958, vol. I, Pl. 16.

17. Andrea Corsali, lettre de 1515, in G. B. Ramusio, *Navigationi et Viaggi*, Venise, 1550, ff. 192 (v.) à 195 (v.).

18. Id., lettre de 1515, in *op. cit.*, f. 195 (r.-v.).

Au XVIe siècle, l'Europe va beaucoup s'intéresser à l'Ethiopie, surtout à travers les ouvrages de Damião de Góis (1502-1574), dont le premier, renfermant des informations fournies par l'ambassadeur éthiopien Mathieu, venu à Lisbonne, s'intitule: *Legatio Magni Indorum Imperatoris Presbyteri Ioannis ad Emanuelem Lusitaniae Regem anno Domini MDXIII*, et fut publié à Anvers en 1532. Une traduction anglaise parut en 1533.

Plus ambitieux et très diffusé en Europe sera l'ouvrage attribué à Paul Giovio, *Legatio David Aethiopiae Regis ad Clementem Papa VII*, Bologne, 1533, dont on fit la même année des éditions à Anvers, Paris, Bâle, ainsi que des traductions: allemande à Dresde en 1533, et italienne à Bologne en 1533 et 1535. Il s'agit d'un ensemble de lettres échangées entre le Négus, le Pape et le roi D. Manuel, auxquelles sont ajoutées des remarques sur l'expédition en Ethiopie de Francisco Alvares et une première tentative de classification des ethnies en Afrique. Toujours de Damião de Góis, le *Fides Religio moresque Aethiopum sub Imperio Pretiosi Ioannis* ... ne connaîtra pas moins de quatorze éditions à travers l'Europe entre 1540 et 1591 [19]. On y trouve, exposés par un prêtre éthiopien, les rites du christianisme dans son pays.

En 1540 paraît à Lisbonne le premier témoignage à parvenir en Europe de la vraie réalité de l'Ethiopie. Il donne de ce pays une image totalement opposée à l'image mythique de l'Ethiopie fabuleuse du Prêtre Jean, qui n'avait cessé depuis le XIIe siècle de fasciner les esprits. Par son style prosaïque et par la banalité de ce qu'il décrit, l'ouvrage de Francisco Alvares, *A Verdadeira Informação das Terras do Preste Joham*, met fin à tout jamais au mythe d'une Ethiopie opulente et puissante. L'historien João de Barros exposera, dans sa troisième décade, l'effondrement du mythe et la déception qu'il provoqua [20]. La relation de Francisco Alvares a connu trois éditions en italien dans la collection de Ramusio, deux en français en 1588, une en allemand (1566) et trois en espagnol (1561, 1577, 1588).

La première d'une longue série de collections de voyages, qui auront une si grande vogue au XVIe siècle et encore longtemps après, est celle de Fracanzano da Montalboddo, *Paesi novamente ritrovati* [21], publiée à Vicence en 1507 et dont on connaît cinq autres éditions en italien, six en français, deux en allemand, une en latin et une en espagnol.

---

19. Sur l'œuvre de Góis, cf. Francisco Leite Faria, *Estudos bibliográficos sobre Damião de Góis e sua época*, Lisbonne, 1977. Il existe une traduction moderne de *Fides Religio* ... etc., in Damião de Góis, *Opúsculos Históricos*, Porto, 1945, pp. 125-201.

20. João de Barros, *Terceira Década de João de Barros*, Lisbonne, 1563, Livro IV, Cap. II.

21. Réédition en fac-similé de l'édition de Milan de 1508, Princeton, U.S.A., 1916.

On y trouve notamment les relations des voyages de Cadamosto sur la côte occidentale de l'Afrique, du pilote anonyme de Cabral, une lettre de Girolamo Sernigi sur le voyage de Vasco da Gama, et une autre de Giovanni Matteo Cretico sur le voyage de Cabral. Une autre collection de voyages est celle de Simon Gryneus, *Novus Mundus*, parue à Bâle en 1532, puis à nouveau en 1534, 1537 et 1555, à Strasbourg en allemand en 1534 et en néerlandais en 1563. Elle contient les relations des voyages de Cadamosto et de Cabral, mais l'importance de cet ouvrage tient à sa préface: une longue réflexion sur la portée historique des découvertes. Et, en fait de préface, n'oublions pas celle, dédiée à D. João III, que le Français Jean Fernel écrivit pour sa *Cosmotheoria*, Paris, 1528. Fernel y évoque en termes élogieux l'Infant Henri, et Diogo Cão et Bartolomeu Dias pour leurs voyages sur les côtes de l'Afrique.

La première véritable collection de voyages est celle de Giovanni Baptista Ramusio, *Navigationi et Viaggi*, Venise, 1550. Elle connut un énorme succès et eut de nombreuses rééditions au XVIe siècle. Le premier volume est consacré à l'Afrique et à l'Orient; on y trouve les récits des premiers voyages des Portugais autour de l'Afrique, ainsi que l'importante relation de Duarte Barbosa (c. 1518) sur l'Afrique orientale et sur l'Orient. Dans la seconde édition du premier volume de sa collection, Ramusio ajoute des extraits de l'*Ásia* de João de Barros.

Em 1559, sont imprimés à Poitiers les *Voyages Aventureux* de Jean Alphonse, qui se dit «natif de Saintonge». Il y en aura dix éditions entre 1559 et 1605. Il s'agit d'un routier des deux rivages de l'Atlantique et de la route autour de l'Afrique jusqu'en Extrême-Orient. Soit calqué sur un routier portugais, soit rédigé d'après une expérience personnelle, l'ouvrage contient des informations que l'on ne rencontre pas dans les sources en langue portugaise. La nationalité d'Alphonse est depuis longtemps l'objet de controverses, certains le disant Français, d'autres Portugais. C'est vers cette dernière hypothèse que penche le consensus actuel.

A partir du milieu du siècle, commencent à paraître les grandes histoires nationales des découvertes portugaises.

La première est celle de Fernão Lopes de Castanheda (c. 1500-1559), *História do Descobrimento e Conquista da Índia.* L'édition originale, sortie à Lisbonne en 1551, est aujourd'hui très rare; le livre fut traduit en français par Nicolas Grouchy et cette traduction publiée à Paris en 1553. Une traduction allemande de l'édition française a vu le jour en 1565. Deux éditions d'une traduction anglaise parurent à Londres en 1582 et 1587 et deux éditions d'une traduction italienne furent imprimées en 1557 et 1578. Il y eut une traduction espagnole à Anvers en 1544.

Une seconde édition du premier livre sortit à Lisbonne en 1554; en la comparant à la première, on y constate des suppressions significatives, notamment en ce qui concerne l'Ethiopie et le voyage de Pero de Covilhã. A la différence de Barros, que nous étudierons plus loin, Castanheda attache peu d'importance à la géographie des régions où passaient les Portugais. Son style est assez plat et son mérite tient à la relative impartialité de ses jugements [22].

Passons maintenant aux *Décadas da Ásia* de João de Barros, vaste somme de connaissances sur les découvertes et véritable chef-d'œuvre littéraire. Les deux premières Décades sortirent à Lisbonne en 1552 et 1553. Une traduction italienne par Affonso Ulloa parut à Venise en 1562. Il existe à la Bibliothèque Nationale de Paris une traduction française, manuscrite et inédite, des deux premières décades, mais la traduction n'est pas très bonne. Les autres Décades de Barros — la troisième, parue en 1563, et la quatrième, remaniée par Lavanha et publiée à Madrid en 1615 — n'ont jamais été traduites en d'autres langues [23].

Si Barros a voulu, en tout premier lieu, retracer l'histoire des conquêtes des Portugais en Orient, on trouve aussi partout, dans son *Ásia*, de longues digressions d'un très grand intérêt sur la géographie, la société et l'organisation politique des pays d'Afrique et d'Orient. Dans la première Décade, notamment, Barros écrit de longs passages sur le Congo, le Monomotapa et sur les Naires du Malabar, leur système de parenté matrilinéaire et leur polyandrie. Des descriptions analogues sur Ceylan, le sud-est asiatique et la Chine, dans la troisième Décade, n'ont pas été traduites au XVIe siècle et n'ont donc pas connu une aussi large diffusion. Mais il est tout à fait certain que c'est sur ces passages, traduits ou non, que se jeta le public lettré de l'Europe de la Renaissance, avide de connaissances concernant les pays d'Afrique et d'Orient.

D'une qualité nettement inférieure pour ses informations et d'un style prétentieux, le *De Rebus Emanuelis*, de Jerónimo Osório, Lisbonne, 1571, a connu, parce que rédigé en latin, une diffusion exceptionnelle en Europe. Réédité six fois à Cologne enre 1574 et 1597 et une fois à Rome en 1592, cet ouvrage fut, pour l'Europe savante, la source principale et directe de connaissances sur les découvertes portugaises. Douze

---

22. Cf. Georges Le Gentil, *La littérature portugaise*, Paris, 1935, pp. 62-63.

23. Sur João de Barros, cf. I. S. Révah, «João de Barros», in *Revista do Livro*, São Paulo, n.º 9 (1958), pp. 61-71; António José Saraiva, «A Concepção Planetária da História em João de Barros», in *Idem, Para a História da Cultura em Portugal*, Lisbonne, 1961, vol. II, pp. 329-355; C. R. Boxer, *João de Barros, Portuguese Humanist and Historian of Asia*, Xavier Centre of Historical Research, n.º 1, Concept Publishing Co., New Delhi, 1981; et António Alberto Banha de Andrade, *João de Barros, Historiador do pensamento humanista português de quinhentos*, Lisbonne, 1980.

des vingt livres furent traduits en français par Simon Goulart et réédités trois fois à Paris en 1580, 1581 et 1597, et une fois à Genève en 1610 [24].

A partir du dernier quart du XVIᵉ siècle vont apparaître les grands compendiums géographiques. Dans l'élaboration de ces géographies systématiques du globe, les auteurs suivent les côtes des continents, en réunissant autant d'informations sur l'intérieur qu'ils en peuvent trouver. Les deux premiers ouvrages d'auteurs français, les *Cosmographies* de François Belleforest et d'André Thevet, ont vu le jour la même année: 1575. Belleforest connaît bien les historiens portugais, et son livre est utile et sérieux. Ce n'est pas le cas de Thevet, esprit médiocre et confus, mais chez qui l'on trouve des informations qui ne figurent nulle part ailleurs. Il est souvent impossible de savoir avec certitude si elles sont vraies, ou transformées par une imagination fertile.

En Italie, Giovanni Botero fera paraître en 1592, à Venise, son *Delle Relazioni Universali.* Il y en aura trois autres éditions de 1595, 1602 et 1605, deux traductions espagnoles de 1603 et 1605, une allemande de 1596 et une anglaise de 1599. L'ouvrage résume utilement les connaissances géographiques des Portugais.

L'un des meilleurs compendiums du XVIᵉ siècle est sans doute le *Geografia dell'Africa* de Livio Sanuto, paru à Venise en 1588, qui est aussi un atlas. Il rassemble, sur l'Afrique, tout ce que les historiens portugais avaient publié jusqu'à cette date.

Pour finir, nous évoquerons un auteur, publié au-delà de notre limite chronologique, mais qui apporte des éléments assez détaillés sur les découvertes portugaises: Pierre du Jarric, *De l'Histoire des choses les plus mémorables advenues tant es Indes Orientales que autres païs de la découverte des portugais,* Bordeaux, 1608-1613, en trois volumes. Le volume II renferme des passages sur l'Afrique, la Chine et le Brésil.

L'Europe s'appuiera longtemps sur les connaissances acquises sur l'intérieur des continents par les Portugais du XVIᵉ siècle. On les reprendra jusqu'au XVIIIᵉ siècle et même au XIXᵉ, époque des grands explorateurs de ces régions.

---

24. Il existe une traduction portugaise de l'édition latine par Francisco Manuel do Nascimento (pseudonyme «Filinto Elíseo»). Lisbonne, 1804.

XVII

The Artilleries and Land Fortifications of the Portuguese and of their Adversaries in the Early Period of the Discoveries*

Artillery technology ('real barrel guns'), first appeared in use in China from 1288.[1] It was diffused westward, either through Islamic societies, or directly to Europe with great rapidity. In the 15th and 16th centuries, the technology, after being considerably improved by Europeans, was diffused back to the East where, in spite of partial transfers to their adversaries, the superiority of the former was, by and large, never put in question and used to great advantage by them.

Artillery was first used in European expansion outside Western Europe by the Portuguese at the beginning of the 15th century in their bid to conquer Morocco, where they found it also in use by their adversaries.

In the description of the Portuguese conquest of the Moroccan town of Ceuta in 1416, the records make no explicit reference to the use by either side of artillery, though it is known that numerous pieces (bombards and trons) figured in the preparations for the Portuguese expedition.[2] Three years later, in a bid to re-take the city, the Moroccans, aided by the kings of Fez, Granada and Tunis unsuccessfully attacked it with trebuchets and artillery (*muytos engenhos e artilharias*).[3] When these had been set in place, the Portuguese aimed two trebuchets at the adversary's bombards and when one was about to be fired, a heavy load of stones from one of the trebuchets shattered it and killed the gunner.[4] Deprived of this specialist, '... they either knew not how, nor dared to fire the other one'.[5]

Artillery in this context seems to have been better adapted to battering fortifications, than being capable of defending them.

---

* First published in *Ars Nautica, Actas da VIII Reunião Internacional de História da Náutica e da Hidrografia,* Viana do Castelo, Portugal, 19–24 September 1994, (Cascais, 1998), pp. 329–340.

1 Joseph Needham, *Science and Civilisation in China*, Vol. V, Pt. 7, (Cambridge, 1986), p. 404.

2 Gomes Eanes de Zurara, *Crónica da tomada de Ceuta,* Introdução e notas de Reis Brasil, (Lisbon, 1992), Cap. XXX, p. 118. (*bombardas e troons, as quais eram muitas e grandes...*).

3 Idem. *Crónica dos feitos de Guiné*, Cap. II, ed. A.J. Dias Dinis, (Lisbon, 1949), Vol. II, p. 25.

4 Idem. 'Crónica de D. Pedro Meneses', in *Monumenta Henricina*, Vol. II, (Coimbra, 1960), p. 333.

5 *Ibid. loc. cit.*

In 1437 the Portuguese suffered a severe defeat in their attempt to take Tangiers. They had taken with them a considerable quantity of artillery and trebuchets, (... *muyta artilharia e boõs aparelhos* ...).[6] Finding resistance, Prince Henry, who commanded the operation, sent to Ceuta for two large bombards, stone and gunpowder, but the bombards turned out to be too small to have any effect on the wall of the town.[7] The Moroccans proceeded to use their own artillery of which they had many trons and a bombard, wounding many Portuguese and killing several.[8] Forced to withdraw on account of the solidity of the fortifications and the superiority of the adversary's artillery, the Portuguese were compelled to surrender the Infante D. Fernando as a hostage and to abandon all their arms.[9]

In 1458, when the Portuguese attacked the Moroccan town of Alcacer Ceguer, Prince Henry, whom the chronicler describes as being an old hand at artillery, had a large bombard fired which caused no less damage than it did astonishment and which immediately brought the inhabitants to surrender.[10] The historian Damião de Góis, repeating the story, describes how Prince Henry ordered the artillery officer to, ' ... load the bombard well, promising him a recompense if it battered in the city wall'. When a large part of the wall collapsed at the first shot, ' ... the enemy saw that against the fury of this bombard there was no resistance (*contra a furia daquela bombarda nam havia resistencia* ...)' and surrendered.[11] The King of Fez, having mobilized considerable forces with large pieces of artillery, (*bombardas grossas*), found he was unable to re-take the town,[12] in spite of attacks on succeeding days in which the attackers used not only their own artillery, but also one of the large bombards abandoned by the Portuguese at Tangiers which fired stones of four *quintals* in weight (= 58 kgs). Finding that even with this weapon, with which they fired '810 large stones', the walls remained in sound condition, the Moroccans abandoned their attempt.[13]

In the course of the 15th century, if we can go by the above evidence, the artillery of the Portuguese and in particular their expertise in the use of it steadily improved. At the end of the century, the German traveller in Portugal, Jerome Münzer refers to the presence in the country and in Morocco of German artillery

---

6 Rui de Pina, *Chronica do Senhor Rey D. Duarte,* Cap. XXI, ed. M. Lopes de Almeida, (Oporto, 1977), p. 538.

7 *Ibid.* Cap. XXVI, p. 544.

8 *Ibid.* Cap. XXVIII, p. 548.

9 *Ibid.* Cap. XXXIII, p. 558.

10 Rui de Pina, *Chronica do Senhor Rey D. Afonso V*, Cap. CXXXVIII, ed. M. Lopes de Almeida, (Oporto, 1977), p. 777.

11 Damião de Góis, 'Crónica do Príncipe D. João', Cap. XIII, in Vitorino Magalhães Godinho, *Documentos sobre a Expansáo Portuguesa*, Tome, III, (Lisbon, 1956), pp. 251–52.

12 Rui de Pina, *Crónica do Senhor Rey D. Afonso V*, Cap. CXXXIX, *ed. cit.* p. 779.

13 *Ibid.* Cap. CXL, pp. 780–82.

specialists. In the town of Alcacer Ceguer, noted Münzer, the King of Portugal ' ... has highly expert bombardiers, among them a certain Jacobus Swevus from Würtemberg'.[14] Münzer recounted meeting in Lisbon several other German bombardiers, some of whom had been overseas in Portuguese service. In Lisbon he was impressed by an ' ... enormous workshop with many furnaces' (*maximam officinam cum multis fornacibus*), where anchors, cannons etc. are made'. Later he saw ' ... four buildings with numerous very large and fine guns, [...] mortars and hand guns. Nuremberg's ordnance (*instrumenta Nurembergen*)', he wrote, 'is nothing in comparison'.[15]

Although in the first half of the fifteenth century, Portugal seems to have been dependant to a certain degree on arms and munitions imported from Flanders,[16] by the end of the century such was the degree of self-sufficiency that artillery was apparently being occasionally exported to the same region.[17]

Was this change achieved with or without foreign technological tutelage in the production of artillery and what exactly were, during their reigns, the respective rôles of King D. Afonso V (1438–1481) and of King D. João II (1481–1495) in bringing about this development of the production of artillery in Portugal? These are questions to which for the moment no satisfactory answers are forthcoming.

One of the key innovations in Portuguese artillery in the 15th century was the introduction of breech-loading guns referred to in a document of 1418 containing a list of objects carried by the São Cristovão, a ship returning from Ceuta in that year. The list includes 'a tron with three chambers'.[18] Mention of trons and culverins carried on a voyage along the African coast in 1456 to the neighbourhood of Cape Blanco (20° Lat. N.), provides clear evidence, at least as far as the culverins are concerned, of breech-loading guns mounted on a Portuguese ship; the guns, it is said, were only used for salutes.[19]

The major innovation in artillery technology in Portugal came in the reign of King D. João II (reigned 1481–1495). Instead of the traditional built-up iron guns (strips of wrought iron welded together with rings to hold them in place, shrunk

---

[14] Jerome Münzer, *Itinerário,* [1495], Latin text and Portuguese translation by Basílio de Vasconcelos, (Coimbra, 1931), p. 33.

[15] *Ibid.* p. 27.

[16] 'Carta de Quitação ao feitor del-Rei em Flandres Pedro Eanes', 23/6/1443, in João Martins da Silva Marques, *Descobrimentos Portugueses,* Vol. I, (1147–1469), (Lisbon, 1944), facsimilé ed. (Lisbon, 1988), Vol. I, p. 431.

[17] Quirino da Fonseca, *Os Portugueses no Mar,* 2nd ed. (Lisbon, 1989), p. 144, (*A Nau Bretoa – [1495] – algumas bombardas*).

[18] Document dated 28 & 29 April, 1416, in J.M. da Silva Marques, *Descobrimentos Portugueses, ed. cit.* facsimilé ed. (Lisbon, 1988), Vol. I, p. 243, (*hum troo com tres camaras*).

[19] Gomes Eanes de Zurara, *Crónica dos feitos de Guiné*, Cap. LIIII, *ed. cit.* Vol. II, p. 234, (*troõns e collobretas*).

on), guns now appeared cast in bronze, thus enabling them to support higher charges of gunpowder with resulting higher pressures within the barrel and a higher corresponding muzzle velocity.

Evidence that guns were being cast in bronze in King D. João II's reign has been brought together by Colonel Valdez dos Santos. In letters written in 1484 to local authorities (*vereadores*), the King reserved for the Crown a monopoly of the export of cork to buy copper in Flanders, France and England for the founding of artillery: ('two thousand *quintals* [=120,000 Kgs] of copper for certain pieces of artillery which we are having made').[20] Further direct evidence that bronze cannon were being made in Portugal at this time appears in another letter from King D. João II dated 24/10/1488 in which he granted privileges to João Goterrez 'Our master founder of bronze cannon (*tiros de metal*)'.[21]

In view of this evidence it may be necessary to revise Padfield's statement that ' ... while the Portuguese made themselves masters of the Eastern seas' with 'wrought iron bombards, gun-founders in Germany and in the Low Countries developed [he does not say exactly when] the cast bronze muzzle-loading gun.[22] Padfield also notes that bronze breech-loaders were exceptions to the general rule.[23]

We do not have textual evidence that the bronze guns cast in King D. João II's reign were breech-loaders. But proof that they were in the reign of his successor King D. Manuel I (reigned 1495–1521), comes from a letter dated 22/3/1513 addressed to the King by Estevão Pais. Pais refers to the 'bronze chambers' of a *berço* made by Joham Goterrez,[24] who, as we have seen, was 'the master founder of bronze cannon'.

King D. Manuel I is further described as being the 'inventor of a method of casting a *berço* (breech-loader) in bronze and of making it fire heavy cannon-balls designed for the much larger *selvagem* (presumably an iron gun). The *berço* founded by Joham Goterrez, had, in a test, fired a cannon-ball from Lisbon to Almada, a distance of some 2 and 1/4 kilometers. It is implied in Pais' letter that King D. Manuel's 'invention' consisted of designing for use on caravels, a breech-loading gun of relatively moderate size, capable, because it was of bronze, of firing

---

20 See Col. Nuno Valdez dos Santos, 'Canhões de Portugal em terras de Maroccos', in *Revista de Artilharia*, N°s 737–740, Março-Abril, 1987, pp. 497–98 and pp. 507–508.

21 Sousa Viterbo, 'Fundidores de Artilharia', in *Revista Militar*, Anno LIII, (1901), p. 343. (*...Joham Goterrez, nosso fundidor e mestre de tiros de metall...*), *Metall* here refers to bronze.

22 Peter Padfield, *Guns at Sea*, (London, 1973), p. 29.

23 *Ibid*. p. 14.

24 Letter from Estevão Pais, 22/3/1513 in Sousa Viterbo, 'Armarias e Arsenaes Portuguezes no século XVI', in *Revista Militar*, Anno XXXIX, No 13, (1887), pp. 394–95, (*...as camaras são de muito metal*).

a heavy projectile.[25] This bronze *berço* was, noted Pais, 'the best form of artillery possible for ships and capable of producing results' (*os mais prestes tiros de artilharia que pera naoos se pode fazer e averem bençam*). '*Selvagems*, added Pais, are dangerous for caravels because of their weight and size.'[26]

How were guns actually used on a ship in the early period of the Discoveries? The Portuguese historian Garcia de Resende describes King D. João II as 'clever in the practical arts (*officios*) and as knowing a lot about artillery'. After carrying out many experiments at Setubal across the river from Lisbon, he invented and had put (*achou e ordenou*) on small caravels, large bombards (*muyto grandes bombardas*) which were arranged to fire so low that the shots almost touched the water (*tam rasteiros que hiam tocando na agoa*).[27] King D. João II was, according to Resende, the first to have invented this method, which enabled small caravels to take on large ships. The caravels, so armed, were for a time, greatly feared until the method came to be generally known and adopted by others.[28]

It should perhaps be noted that D. João II's 'invention' for firing cannon horizontally amounts to the only possible way of using artillery on a ship to attack another with any hope of success, since in the period, as Padfield has noted, in the absence of effective range-finders and accurate methods of setting the elevation of the gun, any attempt to angle the gun upward to achieve a range beyond point-blank range was waste of time.[29]

Frequent references appear in the works of Portuguese historians to the use in the East of the technique of firing horizontally at the level of the water and of its success in sinking enemy ships, (*tiros grossos ao lume d'agoa*), (*artelharia meuda tã rasteira, que meterã muytos bárcos no fundo), (tiro grosso [...] por baixo da amûra ao lume d'agoa & passoulhe ambos os costados*), (*bombarda grosa a huma das naaos dos imigos pelo lume d'agoa que llogou se foy ao fundo*).[30] To enable such firing, and given the low profile of the caravel, the guns, it has been claimed,

---

25 Col. Nuno Valdez dos Santos, *A Artilharia Naval e os Canhões do Galeão 'Santiago'*, Academia de Marinha, (Lisbon, 1986), p. 35.

26 Estevão Pais, *loc. cit.*

27 Garcia de Resende, *Crónica de Dom João II*, (Evora, 1545), ed. (Lisbon, 1973), Cap. CLXXXI, pp. 255–56.

28 Idem. *loc.cit.*

29 Peter Padfield, *op. cit.* p. 37.

30 Gaspar Correia, *Lendas da Índia*, ed. (Oporto, 1978), Vol. I, Cap. XIII, p. 332. (Vasco da Gama – 1502); João de Barros, *Ásia*, Déc. II, Liv. II, Cap. III, (Lisbon, 1553), ed. (Lisbon, 1988), p. 58, (Albuquerque at Ormuz – 1507); F. Lopes de Castanheda, *História do Descobrimento [...] da Índia*, Liv. II, Cap. C, ed. (Oporto, 1979), Vol. I, p. 348; Anon. *Crónica do Descobrimento e Conquista da Índia pelos Portugueses;* Introdução e notas de Luís de Albuquerque, Leitura de Adélia Lobato, Agrupamento de Estudos de Cartografia Antiga, Vol. LXXXVI, (Coimbra, 1974), p. 185; João de Barros, *op. cit.* Déc. II, Liv. III, Cap. VI, *ed. cit.* p. 130; G. Correia, *op. cit. ed. cit.* Vol. I, p. 940, (Almeida at Diu – 1509).

must have been placed low down amidships, to fire broadside, yet no explicit reference is known to us that this was so done. On Vasco da Gama's caravels at Calicut in 1502, the smaller guns were placed to fire forward *(polas perchas, dez berços)*, the heavier ones aft (*dous falcões tiravão por popa*).[31]

It has been claimed by Colonel Nuno José Varela Rubim that in Garcia de Resende's reference to shots being fired low over the water, (... *hiam tocando na agoa* ...) implies that they were deliberately made to ricochet off the water in order to extend the firing range.[32] There appear in Gaspar Correia's *Lendas da India* several references to shots being made to ricochet off the water (*chapeletas*) in naval encounters in the East, but there is no clear evidence that this was done to extend the firing range.[33]

When the Portuguese reached India at the turn of the 15th century,[34] all the societies round the Indian Ocean possessed artillery, but it was on the whole much inferior to that of the Portuguese. The Samory of Calicut had, 'wretched bombards,' (... *bombardas rruims* ...) (João de Nova's voyage 1501).[35] Portuguese guns were 'much more powerful than his' [the Samory's] (... *muy possantes mais que os seus* ...), (Vasco da Gama's voyage - 1502).[36] The Samory's gunners fired 'five or six bombards, which were not very big, but their firing was as good as ours,' (... *5 ou 6 bombarde le quale non erano molto grosse ma tiravano tanto bene como le nostro* ...) (Vasco da Gama's voyage – 1502).[37]

Malacca, according to a report made in 1510 to Afonso de Albuquerque by nineteen Portuguese captives held there before its seizure by the Portuguese, was very poorly equipped with ' ... bombards, most of which resembled large hand guns' (*as bombardas* [...] *a maior parrte delas são como spingardões*), others were muzzle loaders (*outras como as que soya aver em Calicut que tiram pelouros atochados na boca*).[38]

---

[31] G. Correia, *op. cit. ed. cit.* Vol. I, Cap. XIII, p. 330.

[32] Col. Nuno José Varela Rubim, 'D. João II e o Artilhamento das Caravelas de Guarda-Costas. O tiro de ricochete naval', in *Revista de Artilharia,* (1990), pp. 18–21.

[33] G. Correia, *op. cit. ed. cit.* Vol. I, Cap. XIV, p. 49, (Vasco da Gama at Melinde – 1497) (*...alguns pelouros dos tiros grossos [...] que forão dando chapeletas pelo mar...*), Idem. *op. cit. ed. cit.* Vol. I, Cap. IX, p. 599, (Francisco de Almeida – 1505) (*...os pelouros das caravellas [...] fazendo chapeletas alcançavão muytos dos paraos..* = Moorish ships).

[34] General accounts, though unsatisfactory, of this period can be found in J. Cortesão, *Os Descobrimentos Portugueses,* Livros Horizonte, 3rd ed. (Lisbon, 1981), Vol. V, Pt VI, Chapters II & III and R.S. Whiteway, *The Rise of Portuguese Power in India*, (London, 1899), Reprint, (New York, 1969).

[35] Anon. *Crónica do Descobrimento e Conquista da Índia pelos Portugueses, ed. cit.* pp. 27–28.

[36] G. Correia, *op. cit.* Vol. I, Cap. XIII, p. 331.

[37] Mateo Begnino, letter, 30/6/1503, in F. Hummerich, *Vasco da Gama und die Entdeckung des Seewegs nach Osten,* (Munich, 1898), reprint, Georg Olms, (Hildersheim & New York, 1977), p. 198.

[38] R.A. de Bulhão Pato, ed. *Cartas de Affonso de Albuquerque*, (Lisbon, 1893), Vol. III, p. 5.

When Fernão Peres approached Canton in 1517, a Chinese coastguard vessel fired at him ' ... a few weak shots from iron bombards' (... *alguns tiros de bombarda de ferro fracos* ...).[39] During Martim Afonso de Melo Coutinho's naval encounter with a Chinese fleet in 1523, 'very well armed with light artillery' (... *muj bem armadas d'artilharia mjuda* ...),[40] which ended disastrously, the Chinese nevertheless showed 'great fear of our artillery since it was much more powerful than theirs' (... *grande medo da [...] nossa artilharia por ser muyto mais furiosa que a sua* ...).[41] The Japanese never had any artillery at this time.[42]

The ships of the western Indian Ocean were unable to resist the effects on them of Portuguese artillery. An Italian Florentine supercargo on Vasco da Gama's voyage (1497) described them as,'very frail without arms or artillery' (*molto deboli & non portano arme ne artegliaria* ...)[43] and the Italian traveller Chá Masser [c. 1506] described them as, 'weak and fragile without artillery, though some now carry it' (... *debele e flache, senza artellaria, benché al presente alcuna de quelle ne portano* ...).[44] By 1507 in Gujerat and in 1510 at Goa, there were already local attempts to imitate Portuguese ship construction.[45]

The Malayan junk of the eastern waters of the Indian Ocean, showed quite different characteristics from the ships of the western part of the Ocean. The hull, built over with successive layers of sheathing (from four to seven), made it invulnerable to Portuguese guns.[46] Afonso de Albuquerque attacked one in 1511 for two days, unable to make any impression on it or make it surrender till he shot away the rudders.[47]

Rapid firing and the ability to re-load quickly was one of the crucial factors in securing an advantage at sea. Vasco da Gama at Calicut in 1502 used a type of proto-cartridge in the form of little sacks of powder prepared beforehand containing

---

39 João de Barros, *op. cit.* Déc. III, Liv. II, Cap. VIII, (Lisbon, 1563) ed. (Lisbon, 1992), Vol. II, f.48^{v}.

40 Carta de Martim Afonso de Melo Coutinho para o rei, 25/10/1523', reproduced in João Paulo Oliveira e Costa,'Do sonho manuelino ao realismo joanino. Novos documentos sobre as relações luso-chineses na terceira década do século XVI', in *Studia,* Vol. 50, (1991), p. 152.

41 F. Lopes de Castanheda, *op. cit.* Liv. VI, Cap. XIV, *ed. cit.* Vol. II, p. 178.

42 Luís Fróis, *História do Japão,* ed. J.Wicki, S.J. (Lisbon, 1984), Vol. I, p. 315.

43 See G.B. Ramusio, *Navigationi et Viaggi,* (Venice, 1550), 3rd ed. (Venice, 1563), f.120^{r}.

44 Lunardo da Chá Masser,'Relazione', [c.1506], in *Memórias da Academia Real das Sciencias de Lisboa,* Classe de Sc. Mor. Pol. e Bellas Lettras, Nova Série, Tome VI, Pt. I, (Lisbon, 1895), p. 84.

45 G. Correia, *op. cit.* Vol. I, Cap. XIII, p. 750. (*um mouro [...]* fez *um galeão muy forte á feição de nao nossa)* and Vol. II. Cap. VIII, p. 60, (*...dezaseis náos feitas á nossa feição, mas por dentro estroncadas e de pouca liação*).

46 F. Lopes de Castanheda, *op. cit.* Liv. II, Cap. CXII, *ed. cit.* Vol. I, p. 461, (*amuradas tão grossas que as não passa hum camelo* [*camelo* =a relatively heavy gun.])

47 G. Correia, *op. cit. ed. cit.* Vol. II, Cap. XXVI, p. 217.

the right measure.[48] There was a limit to the extent to which repeated and continual firing was possible because of the damage it caused to the ships. When Cabral in 1500 wanted to bombard Calicut for several days in succession, the captains of the ships refused to let him *... polo mal que o tirar fazia ás naos ...* [49]

In the East the Portuguese used their artillery as much against land fortifications as in naval encounters. With the exception of the Islamic cities of the northern and western part of the Indian Ocean, some of which had stone gates and towers,[50] the cities along the Malabar coast and in Malacca were only protected by wooden stockades with an earth fill.[51] Such fortifications made them extremely vulnerable to Portuguese artillery; Aden remaining, with its high fortified walls,[52] the sole exception when Albuquerque abandoned his attempt to take it in 1513.[53]

Chinese cities were completely different, being surrounded by thick walls, invulnerable to any artillery. When Fernão Peres visited Canton in 1517, it had a wall of the thickness of five *braças* (=5 x 2.2m = 11 metres) and was the size of Evora (Portugal), with a large wall, faced with dressed stone and with battlements and loop holes.[54] Japanese fortifications were on the other hand impossible to defend against artillery, as Luís Fróis noted: ' ... with four pieces of artillery the whole would be destroyed in half a day.' (*... com quatro peças de artilharia em meio dia se destruiria toda ...*).[55]

The arrival of the Portuguese in the Indian Ocean with their manifest naval and mil-

---

48 G. Correia, *Ibid.* Vol. I, Cap. XIII, p. 330, (*...e tanta pressa davão a tornar a carregar que carregavão as peças com saccos de polvora que traziam pera isso feitos da medida...*)

49 *Ibid.* Vol.I, Cap. X, p. 207.

50 Anon. *Crónica do Descobrimento e Conquista da Índia, ed. cit.* pp. 182–3, (Chaul): *portas lavradas de cantaria;* (Diu): (*muros e torres ameadas e allvas*); João de Barros, *op. cit.* Déc. I, Liv. VIII, Cap. IV, (Lisbon, 1552), ed. (Coimbra, 1932, facsimile reprint, 1988), p. 305: (Kilwa): (*casas de pedra e cal*).

51 Anon. *Crónica do Descobrimento..., ed. cit.* p. 180, (Dabul): (*barreira de madeira [...] entulhada toda d'area);* João de Barros, *op. cit.* Déc. II, Liv. VI, Cap. II, *ed. cit.* p. 262, (Malacca): (*povoação quasi toda da madeira [...] tambem [...] outras torres, muros e arquitecturas de melhor parecer e defensão*). A report on Malacca in 1510 by the nineteen Portuguese captives held there before its seizure by the Portuguese stated that of the 10,000 buildings of the place 500 were of *adobe* (*terradas*), the rest were of 'straw like those of India' (*sam de palha como as da Índia*), See R.A. de Bulhão Pato, *Cartas de Affonso de Albuquerque,* (Lisbon, 1893), Vol. III, p. 5.

52 Duarte Barbosa, *Livro em que dá relação do que viu e ouviu no Oriente,* [c.1518], ed. A. Reis Machado, (Lisbon, 1946), p. 40, (Aden): (*cidade de mui altas casas cercada de muros, torres e cubelos com suas ameias à nossa maneira*).

53 G. Correia, *op. cit. ed. cit.* Vol. II, Cap. XLI, p. 343 and F. Lopes de Castanheda, *op. cit.* Liv. III, Cap. CVI, *ed. cit.* Vol. I, pp. 750–754.

54 G. Correia. *op. cit. ed. cit.* Vol. II, Cap. XIII, p. 526 and F. Lopes de Castanheda, *op. cit.* Liv. IV, Cap. XXX, *ed. cit.* Vol. I, p. 918.

55 Luís Fróis, *op. cit.* Vol. V, p. 315.

itary superiority, set in motion throughout the whole region, an immediate and intense challenge to their artillery. In the Islamic societies it took the form, first of attempts to acquire, with Turkish aid, the technical assistance of Christian renegades to found and serve locally made artillery, then later to solicit the direct intervention of Turkish armed fleets in the Indian Ocean. In the non-Islamic societies (Ethiopia, China and Japan) attempts were made either to solicit the friendly help of Portuguese mercenaries, together with their arms, or to learn and copy the making and serving of European artillery. The challenge was taken up so quickly that as early as 1513 Afonso de Albuquerque wrote to the King that on the west coast of India, ' ... artillery, arms and fortresses have now all been transformed to our way of using them'.[56]

Throughout the 16th century, the Portuguese were to build and continually re-enforce their forts and fortresses of dressed stone set at strategic spots around the Indian Ocean: Cananor and Cochin - 1507; Mozambique - 1507; Ormuz - 1508; Goa - 1510; Malacca - 1511; Diu - 1535. The Portuguese did not in the 16th century build fortified bases in China or in Japan. One was planned for China as early as 1522, but never built till much later.[57]

As early as 1503 the Samory of Calicut was able to enlist the services of two Milanese who had defected from a Portuguese ship on arrival in the East. They founded for him some 180 pieces of artillery, most of them of bronze,[58] with which he attacked Duarte Pacheco Pereira. But the guns founded were only small light pieces (*não erão de mais grandura de berços*)[59] and the projectiles 'no greater than an orange, knocking a man down and leaving him with a mere bruise on the shoulder'.[60] In 1505 a ship arrived at Calicut from Mecca with four Venetians, 'masters of artillery', whom the Samory had asked the Sultan of Cairo to be sent to him,[61] but it is not known what they achieved, if anything.

The artillery with which the Turks armed their fleets sent into the Indian Ocean to attack the Portuguese, far outweighed that which the occasional European renegades had been able to provide the local rulers with and in their naval victory over the Portuguese at Chaul, it proved more lethal than anything hitherto seen by the latter.[62]

---

56 Afonso de Albuquerque, letter to the King, 30 Nov. 1513, in *ed. cit.* Vol. I, p. 127, (*artelharia e armas e fortelezas he já tudo tornado a nosa usança*).

57 Carta de nomeação de Martim Afonso de Melo Coutinho, Lisbon, 7/3/1521, Reproduced by João Paulo Oliveira Costa, *art. cit.* p. 146.

58 Anon. 'Conquista da Índia per humas e outras armas reaes e evangelistas' Ms. Egerton 1646, British Library, [16th cent.] reproduced in *Documentação Ultramarina Portuguesa*, (Lisbon, 1960), Vol. I, p. 314.

59 G. Correia, *op.cit. ed. cit.* Vol. I, Cap. XVII, p. 455.

60 Anon. 'Conquista da Índia' in *op. cit. loc. cit.* Cf. also F. Lopes de Castanheda, *op. cit.* Liv. I, Cap. LXVIII, *ed. cit.* Vol. I. p,147.

61 F. Lopes de Castanheda, *op. cit.* Liv. II, Cap. XII, *ed. cit.* Vol. I, p. 235; João de Barros, *op. cit.* Déc. I, Liv. VIII, Cap. 9, *ed. cit.* p. 342.

62 F. Lopes de Castanheda, *op. cit.* Liv. II, Caps LXXVII – LXXIX, *ed, cit.* Vol. I, pp. 390–99.

When Francisco de Almeida later defeated the Turkish fleet at Diu in 1509, he afterwards collected together no less than six hundred pieces of the enemy's artillery, some hundred and eleven pieces of which were '*falcões* with bronze chambers, *camelos* like those of the Portuguese and *meas esperas*.' Unable to load into a ship a *meo basilisco* (a heavy gun), Almeida ordered it to be taken ashore and broken up.[63]

The Portuguese by this time seem to have been using bronze guns exclusively, for Almeida at Diu (1509) and Lopo Vaz de Sampayo (1528), ordered captured iron guns to be thrown into the sea, because they were of no use in their ships, (*porque nom servia nos nossos navios*).[64] Their bronze guns appear to have been muzzle loaders, for at Chaul in 1521, Diogo Lopes de Sequeira was to find that the enemy's rapid firing of a breech-loading *roqueira* was of greater advantage to them than the slower muzzle loading of the Portuguese, (*milhor que dos nossos tiros, que tirando os avião de tornar a calhar avante*).[65]

In spite of the possession of heavy artillery of Turkish provenance and the tactical advice of a Sicilian renegade, the ruler of Calicut was unable in 1526 to defend successfully the fort of Calicut against the superior artillery of the Portuguese, since the aiming of his gunners was so bad, (... *os mouros vendo que os seus bombardeiros erão tão pouco certeiros*).[66]

In 1538 the Turks laid siege to the Portuguese fortress of Diu for twenty-five consecutive days with a fleet of eighty-five ships armed with four hundred heavy guns, of which fifteen were 'bastard-galleys', which each carried a *basilicus* (heavy gun), firing forward as well as two other heavy guns. Each had two *camelos* (medium gun) on the poop and on each side four iron breech-loading *roqueiras* and firing forward three *falcões* (light guns) on swivels. Other ships were similarly armed. Some eight hundred Christian slaves under the command of a Venetian captain of ten galleys under Turkish orders participated in the siege.[67]

The guns used for the actual siege consisted of nine enormous *basilisci*, each firing cast iron shots of from 90 to 100 *arrateis* [= 41kg31 to 45kg90], five *espalhafatos* firing stone shots of 35cms, 43cms and 49cms in diameter, fifteen *leões* and *águias*, four *colubrinas* and a few *canhões de bater* (battering cannon). There were also eighty other guns among which were *esperas, selvagens, meias esperas* and *falcões* and finally a *quartau*, 'a fearful weapon'.[68] In spite of the terrible pounding of their superior heavy guns, the Turks were unable to reduce the Portuguese

---

63 G. Correia, *op. cit. ed. cit.* Vol. I, Cap. IV, p. 951.

64 *Ibid.* Vol. III, Cap. I, p. 232.

65 *Ibid.* Vol. II, Cap. XXI, p. 668.

66 F. Lopes de Castanheda, *op. cit,* Liv. II, Cap. CIX, *ed. cit.* Vol. II, p. 322; João de Barros, *op. cit,* Déc. III, Liv. IX, Cap. VIII, *ed. cit.* f.243^{v}.

67 G. Correia, *op. cit.* Vol. III, Cap. CXVI, pp. 869–870.

68 João de Barros, *op. cit.* Déc. IV, Liv. X, Cap. X, ed. (Lisbon, 1946), p. 594.

fortress, (the most strongly fortified place held by them in India), and to bring the garrison to surrender. They abandoned the siege and withdrew, though they returned again in 1546, with no better success, leaving behind them some forty pieces of heavy artillery.[69]

The Portugese met the challenge of the heavy guns of the Turks by continually increasing the size and power of their own artillery, both on land and at sea. By 1531 at Diu they had heavy *basilisci* mounted on the prows of galleys,[70] and by 1542 the Governor Martim Afonso de Sousa was arming a galleon and four caravels each with a *basiliscus* on the prow and with six big guns on the sides.[71]

Throughout the 16th century and later, the whole Indian Ocean was to be the scene of an intense competition to bring heavier and heavier artillery to bear on fortifications that were continually being made stronger and stronger. A singular example is that of the repeated attacks on the Portuguese fortress of Malacca by the small Moslem state of Aceh. The unsuccessful attack of the Acehnese in 1537 led the Portuguese to re-enforce their fortifications in 1564 and 1568. When the Acehnese attacked again in that year with 200 bronze cannon obtained under Turkish or Portuguese renegade guidance, their guns did not prove powerful enough to batter the 'solid fortified walls of durable materials built by the Portuguese'.[72] The sequence of later failed attacks vividly shows, as P.Y. Manguin has written, ' ... the inability of the Acehnese to keep up with the rapid pace of evolution of European military technology and strategy'.[73]

All through the 16th century the Portuguese were to supply non-Moslem societies in the East with guns and gunners;[74] two *berços* and six half *berços* together with a hundred muskets were sent on gun carriages to Ethiopia in 1541;[75] in 1564 the Portuguese of Macau were called on by the Chinese to help repress the activity of pirates with their artillery,[76] and in 1568 the Daimyo of Bungo (Japan) wrote to the

---

69 Letter of D. João de Castro to the City of Goa, 15 Nov. 1546, in D. João de Castro, *Obras Completas*, ed. (Lisbon, 1981), Vol. IV, pp. 39–41; see also G. Correia, *op. cit.* Vol. IV, p. 569.

70 F. Lopes de Castanheda, *op. cit.* Liv. VIII, Cap. XXXI, p. 618.

71 G. Correia, *op. cit. ed. cit.* Vol. IV, Cap. III, p. 243.

72 P.Y. Manguin, 'Of fortresses and galleys. The 1568 Acehnese siege of Melaka after a contemporary bird's eye view' in *Modern Asian Studies*, Vol. 22, No 3, (1968), p. 625.

73 *Ibid. loc. cit.* and C.R. Boxer, 'The Achinese attack on Malacca in 1629', in *Malayan and Indonesian Studies . Essays presented to Sir Richard Winstedt*, ed. J. Bastin & R. Roolvink, (Oxford, 1964), pp. 105–121 – reprinted in C.R. Boxer, *Portuguese Conquest and Commerce in Southern Asia 1500–1750*, Variorum Reprints, (Aldershot, 1990).

74 C.R. Boxer,'Asian Potentates and European Artillery in the 16th –18th centuries', in *Journal of the Malayan Branch of the Royal Asiatic Society*, Vol. XXXVIII, (1966), pp. 156–172, reprinted in C.R. Boxer, *Portuguese Conquest and Commerce...*, pp. 156–172.

75 G. Correia, *op. cit. ed. cit.* Vol. IV, Cap. LI, p. 353.

76 B.Videira Pires, 'Os tres heróis do IV Centenário' in *Boletim Eclesiástico da Diocese de Macau*, Vol. 62, (1962), pp. 699–700.

Jesuit priest D. Belchior Carneiro referring to an *espera* (medium sized gun) which the Viceroy had promised him, but which 'had been lost in the [wreck] of the great ship from Malacca. He hoped that another would come and added that ' ... if I ask again for this *espera*, it is because I am on the sea-shore with enemies on the border.'[77]

By 1535 the Portuguese, in accordance with a treaty of friendship of 1516, were founding cannon and making powder for the King of Siam (Thailand) as well as training his subjects in these arts.[78]

Not enough is known about the exact circumstances and extent to which Portuguese artillery technology was taken over and copied by the societies around the Indian Ocean.

Although certain Chinese sources state that the Portuguese iron breech-loading culverin or *berço* was first introduced into China when taken from the captured ships of Martin Afonso de Melo Coutinho's ill-fated expedition of 1523, there is evidence that the *fo-lang-chi* (engine of the *farangi* or Franks) was known and copied as early as 1510.[79] It may perhaps have been introduced by Chinese merchants who could have been given one by Diogo Lopes de Sequeira in gratitude for the friendliness they showed him when he visited Malacca in 1509, warning him of the hostility being worked up against him by jealous Mourish traders.[80] A Chinese report of 1525-1530 written by Ku-Ying-Hsiang declares of the *fo-lang-chi* that ' ... one single shot, finding its target, will smash the hull and send the enemy to the bottom. With this arm one can sail about at will on the high seas and no other country's ships can match it.'[81] Of Chinese artillery João de Barros (1552) wrote less contemptuously than the earliest sources we have seen, that ' ... in firearms for naval warfare, from the experience we, the Portuguese, have, the Chinese have no reason to be envious of ours in Europe'.[82] Yet he noted the Chinese recognition of the superiority of European technology: ' ... when they saw how ours was made, they took up the method, for they are excellent founders [...] but since their iron is impure cast iron (*ferro pedrez*) it breaks like glass'.[83] Later in the early 17th century Chinese proficiency in founding cannon came to be recognized by the Portuguese to the point of contracting the services of Chinese founders to work at

---

77 Letter from Otomo Yoshishige, Daimyo of Bungo to D. Belchior Carneiro, 13/9/1568, in C.R. Boxer, *The Great Ship from Amacon,* (Lisbon, 1959), p. 318.

78 Fernando da Silva Amaro, 'Fundições e Fundidores Artilheiros Portugueses na Ásia e em África', *Boletim Eclesiástico da Diocese de Macau*, Vol. LVIII, (1960), p. 251.

79 J. Needham, *Science and Civilisation in China,* (Cambridge, 1986), Vol. 5, Pt 7, pp. 369–372.

80 F. Lopes de Castanheda, *op. cit.* Liv. II, Cap. CXIII, *ed. cit.* pp. 462–64. (*por estarem* [*os chins*] *afeyçoados aos nossos pela fama que tinhão deles [...] offrecendolhe sua amizade...*)

81 J. Needham, *op. cit.* pp. 375–76.

82 João de Barros, *op. cit.*, Déc. III, Liv. II, Cap. VII, *ed. cit.* f.47^r.

83 *Ibid. loc. cit.*

both their Macau and Goa foundries. The Chinese authorities on the other hand sought the services of Portuguese to instruct them in the art of gunnery.[84]

The story of rapid adoption by the Japanese of the European musket, introduced by the Portuguese, is well known,[85] yet the Japanese are not known to have founded artillery before the early 17th century, even then preferring European guns to their own.[86] When in the 16th century they were able to acquire European artillery, they do not seem to have been able to serve it themselves. Luís Fróis tells of a Japanese feudal lord who in 1584 possessed two *meios-falcões,* but lacking a gunner, it was an African black (*cafre*) who did the loading of the guns, a Japanese soldier who did the aiming and a Malabar Indian who fired them.[87] The loading and firing were at that time obviously skills transmitted by Portuguese mentors.[88]

The Portuguese began founding cannon in India at least as early as 1525, first at Cochin, later at Goa.[89] The Bocarro family, well known for their work as founders of cannon at Goa, were eventually to move their activities to Macau in 1623.[90] In mid-17th century, production was such at Macau that cannon were exported to Goa and to Portugal.[91] Both bronze and iron guns were produced at Macau, the copper being imported from Japan according to a document of 1643.[92]

To conclude, we have sought, in what has been said above, to outline the development of the use of artillery by the Portuguese first in North Africa and then later in the East. There still remains however much research to be done on the identification and dating of the crucial turning points in the development of founding technology and in the art of using cannon on ships at sea, as well as to destroy or to protect land fortifications. It is still unclear to what extent the technology and skills in the making and serving of artillery were imported from northern Europe or developed locally. Nor are the sources and methods of supply of the raw materials, especially copper and tin to found bronze cannon, adequately known. However incomplete our knowledge of the subject remains, it may yet be confidently said that if nautical astronomy was the key to the Portuguese conquest of the Atlantic, artillery was the key to their mastery of the Indian Ocean.

---

84 Fernando da Silva Amaro, *art. cit.* Vol. LIX, Nº 681, p. 75. and C.R. Boxer, 'Expedições militares portugueses em auxílio dos Mings contra os Manchus (1621–1647)' in *Boletim Eclesiástico da Diocese de Macau,* Março, (1940), pp. 6–7.

85 J. Needham, *op. cit.* pp. 429–430 and João Paulo Oliveira e Costa, 'A introdução da armas de fogo à luz da história do Japão de Luís Fróis', in *Estudos Orientais*, Vol. III, (Lisbon, 1992), pp. 113–129.

86 C.R. Boxer, 'Asian Potentates...', p. 170.

87 Luís Fróis, *op. cit.* Vol. IV, p. 59.

88 João Paulo Oliveira e Costa, *art. cit.* p. 127.

89 Fernando da Silva Amaro, *art. cit.* p. 1037.

90 *Ibid.* p. 1043 and Manuel Teixeira, 'Os Bocarros' in *Congresso Internacional de História dos Descobrimentos*, (1960), *Actas*, (Lisbon, 1961), Vol. V, Pt. II, pp. 359–386.

91 F. da Silva Amaro, *art. cit.* Vol. LIX, Nº 681, (1961), p. 78.

92 *Ibid. loc. cit.*

Appendix

Tentative List of Types of Portuguese Cannon by order of method of loading, of calibre, length and weight of shot

BERÇO	-Breech-loader. -Calibre: 5 cms. -Length ? -Weight of shot: 1 lb (=450g).
CÃO-	Breech-loader? -Calibre: 8 cms. -Length? -Weight of shot: 3.5–4 lbs. (1kg575–1kg800)
FALCÃO	-Muzzle loader. -Calibre: 9–9.5 cms. -Length: 3.06m.–3.23m. -Weight of shot: 5–6 lbs (=2kg250–2kg700)
MEIA ESPERA.	-Muzzle loader. -Calibre: 9.5 cms. -Length: 1m80–2m. -Weight of shot: 6 lbs (=2kg700).
ESPERA.	-Loading? -Calibre: 12 cms. -Length: 2m20–2m40. -Weight of shot: 12 lbs. (=5kgs400).
CAMELETE	-Loading? -Calibre: 17.3 cms–18 cms. -Length: 2m4 -Weight of shot: 12–14 *arrateis* (=5kg500–6kgs430).

CAMELO

-Muzzle loader?
-Calibre: 18 cms–34 cms.
-Length: 2m64–2m76.
(1m80–3m according to Col.Valdez dos Santos)
-Weight of shot: 18lbs (8kg100).

AGUIA

-Muzzle loader?
-Calibre: 16 cms–17.5 cms.
-Length: 3m–3.5 m.
-Weight of shot: 30–40 lbs (13kgs500–18kgs).

LEÃO

-Muzzle loader?
-Calibre: 20 cms.
-Length: 3m60–4m.
-Weight of shot: 50 *arrateis* (=23kgs).

SELVAGEM

-Muzzle loader?
-Calibre: 25cms–28 cms.
-Length: 3m36.
-Weight of shot: 50 *arrateis* (=23 kgs)
(But according to Col.Valdez dos Santos: 42–66 lbs (= 18kg900–29kgs700).

BASILISCO

-Muzzle loader?
-Calibre: 'more than 20 cms'
-Length: 30 palmos (=6m84),
-Weight of shot: 90–100 *arrateis* (41kgs–46kgs).

ESPALHAFATO

-Muzzle loader?
-Calibre: 35 cms–49 cms.
-Length: 2m40 3m40.
-Weight of shot: 45kgs90–128kgs52.

QUARTAU

-Muzzle loader?
-Calibre: 56 cms.
-Length: ?
-Weight of shot: 450 *arrateis* (=206kgs500).

References:

-General Pereira do Valle, 'Nomenclatura das Bocas do Fogo Portugueses do século XVI', in *Revista da Artilharia*, Ano LVIII, (1962), Nos 439–440, pp. 381–390.

-Col. Nuno Valdez dos Santos, *A Artilharia Naval e os Canhões do Galeão 'Santiago'*, (Lisboa, 1986).

-João de Barros, *Ásia*, Déc. III, Liv. I, Cap. IV, and Déc. IV, Liv. X, Cap. X.

-Gaspar Correa, *Lendas da Índia* ed. (Oporto, 1975), Vol. IV, pp. 474–556. and *passim*.

We have not found exact definitions of the ROQUEIRA and of the ESMERIL. COLUBRINA appears to be a generic term.

* * *

Grateful thanks are here expressed to Dr Inácio Guerreiro, Vice-Presidente of the Instituto de Investigação Científica Tropical in Lisbon for his most patient kindness in providing me with photocopies of important material unavailable outside Portugal. Copies of this material were circulated to my colleagues Professor Georges Boisvert, Professor of Portuguese at the Sorbonne and Richard Barker of Derby, England to whom my warmest thanks go for their invaluable comments and suggestions.

XVIII

"Peuples sauvages" et "états despotiques": la pertinence, au XVIe siècle, de la grille aristotélicienne pour classer les nouvelles sociétés révélées par les Découvertes au Brésil, en Afrique et en Asie

INTRODUCTION

LES grandes synthèses historiques et géographiques du XVIe siècle, portugaises, espagnoles et autres, classent en deux catégories les nouvelles sociétés extra-européennes révélées par les Découvertes.

D'abord les Indiens du Brésil, classés comme «sauvages»: on relèvera l'absence chez eux de toute forme de religion, de structure sociale et politique ou de propriété privée. On dira d'eux qu'ils sont des gens «sans Foi, sans Loi et sans Roi» (1). Sous des formes parfois plus développées, cette expression est employée par le chroniqueur portugais Damião de Góis (1556) (2), par l'humaniste Jerónimo Osório (1571) (3), par l'historien du Brésil Magalhães de Gândavo (1576) (4) et par les jésuites italien Pietro Maffei (1588) (5) et français Pierre Du Jarric (1610) (6).

En revanche, à propos des Ethiopiens et des Chinois, on soulignera le fait que leur monarque dispose d'un pouvoir absolu, que les corps intermédiaires, tels qu'on les trouve dans le féodalisme européen — Comtes, Ducs, Marquis —, n'existent point et que le droit à la propriété privée est inconnu.

Les deux concepts de «société sauvage» et d'«état despotique» ont leur origine lointaine dans la *Politique* d'Aristote.

«L'homme», dit-il à propos du sauvage, «est par nature destiné à vivre en cité; celui qui est sans cité est, par nature et non par hasard, un être ou dégradé ou supérieur à l'homme; il est comme celui à qui Homère reproche de n'avoir 'ni clan, ni loi, ni foyer'.» (7)

Autrement dit, il manque à l'homme sauvage système de parenté, institutions sociales et demeure fixe. La *Vetusta translatio* latine du texte du stagirite utilisée au Moyen Age emploie les termes: *insocialis*, *illegatus*, *sceleratus* (8),

(1) Nous ne traiterons pas ici le problème de la religion des Indiens brésiliens.

(2) Damião de Góis, *Chronica do felicissimo rei D. Manuel*, Parte I, Lisbonne, 1556, Cap. LVI, éd. mod. Coimbra, 1949, p. 130 («nem adoram, nẽ sabem ler, nem escrever, nẽ tem egrejas, nem usam imajẽs [...] nem tẽ lei, nem peso, nẽ medida, nẽ moeda, nẽ Rei, nem senhor...» — ils n'adorent rien, ne savent ni lire ni écrire, n'ont pas d'églises, ne connaissent pas d'images [...] ils n'ont pas de loi, ni poids ni mesures, ni monnaie, ni Roi, ni seigneur...)

(3) Jerónimo Osório, *De rebus Emmanuelis Regis Lusitaniae invictissimi virtute et auspicio gestis libri duodecim*, Lisbonne, 1571, p. 66. Autres éditions, Cologne, 1574, 1576, 1581, 1586 («Nullas literas norunt: nullam religionem colunt: nullis legibus alligantur: nullis ponderibus aut mensuris utuntur: nullius Regis imperio subjiciuntur»).

(4) Pero Magalhães de Gândavo, *História da Província da Santa Cruz*, Lisbonne, 1576, trad. franç. Paris, 1837 («... ils n'ont en effet ni Foi, ni Loi, ni Roi, et vivent sans ordre...»)

(5) Giovanni Pietro Maffei, *Historiarum Indicarum libri XVI*, Florence, 1588 (traduction française Lyon, 1603, p. 98: «... ils n'ont ni Foy, Loy, ne Roy»).

(6) Pierre Du Jarric, *Seconde Partie de l'Histoire des choses plus memorables advenues tant ez Indes Orientales q'autres pais de la descouverte des Portugais*, Bordeaux, 1610, 252-253 («... ils vivoient sans Foy, sans Loi, & sans Roi»). Il est évident que Du Jarric ne fait que copier Magalhães de Gândavo ou Maffei.

(7) Aristote, *Politique*, I, II, 6, 125b. La citation d'Homère est de l'*Iliade*, 9, 63.

(8) Giacomo dal Sasso, *La Politica di Aristotele riprodotta nel latino della vetusta translatio e recata*

termes qui avaient peut-être inspiré au Moyen Age l'expression «sans Foi, sans Loi et sans Roi». Dans son commentaire de la *Politique* d'Aristote, publié à Salamanque en 1502, l'Espagnol Fernando de Roa emploie l'expression *sine tribu, sine iure, sine domo* (9).

Des régimes despotiques, Aristote dit que:

«outre [la royauté chez les Grecs] il existe une autre forme de monarchie: telles sont des royautés barbares; elles ont toutes à peu près le même pouvoir que les tyrannies, mais elles sont fondées sur la loi et héréditaires: les barbares étant, par leur caractère, naturellement plus serviles que les Hellènes et les peuples d'Asie plus serviles que ceux d'Europe, ils supportent le pouvoir despotique sans aucune gêne» (10).

La *Vetusta translatio* Médiévale dit des barbares qu'ils *sufferunt despoticum principatum nihil contristati* (11), et la traduction latine de Rome de 1492 dit: *preferunt servile iugum aeque animo.*

A partir de ces termes évoluera la notion que les sujets du prince despotique ont le statut et le comportement d'esclaves.

A la Renaissance, dans son ouvrage *Le Prince* (1513), Machiavel (1469-1527), s'inspirant peut-être d'Aristote (12), opposera avantageusement le régime politique de la France à celui de la Turquie, et mettra en évidence, avec ces deux exemples, l'inexistence dans le modèle oriental de corps intermédiaires héréditaires (13) entre le monarque et le peuple.

«Toutes les principautés, dont il nous reste quelque trace dans l'histoire, sont gouvernées de deux manières différentes: ou par un prince absolu, et tous les autres sont des esclaves à qui, comme ministres et par grâce, il accorde la faculté de l'aider à gouverner le royaume; ou bien par un prince et des grands, ces derniers ne gouvernant pas par la faveur du prince, mais seulement par un droit inhérent à l'ancienneté de leur race; ceux-ci ont des états et des sujets particuliers, qui les reconnaissent pour leurs seigneurs, et qui ont pour eux une affection particulière [...] La Turquie et la France fournissent de notre temps des exemples de ces deux espèces de gouvernement.» (14)

Le régime politique de la Turquie, exemple du schéma aristotélicien du despotisme, allait-il donc servir de modèle à tous les peuples de l'Orient? C'est ce que nous allons voir.

Mais d'abord élargissons le champ de notre question. Quel fut le degré d'adéquation de ces deux schémas, empruntés à l'Antiquité par la Renaissance, pour rendre compte de la structure sociale et politique, d'abord des Indiens du Brésil, ensuite des Ethiopiens, enfin des Chinois, trois sociétés inconnues ou peu connues de l'Europe, et dont les Portugais venaient de rapporter des relations détaillées?

S'est-on servi de ces schémas, en les plaquant sur la réalité au moment des tout premiers contacts, pour ensuite les abandonner dès que fut reconnue la plus grande complexité de la société en question (15), ou au contraire, devant une complexité immédiatement reconnue et extrêmement difficile à appréhender, a-t-on eu recours aux catégories des Grecs pour communiquer une réalité que l'on ne parvenait pas à déchiffrer en termes familiers?

Nous soutiendrons ici le première de ces hypothèses en ce qui concerne le Brésil, la seconde pour ce qui a trait à la Chine.

LE BRÉSIL

C'est par la lecture d'une petite plaquette d'actualité, décrivant le voyage portugais de 1501, que l'Europe a pris connaissance pour la première fois de l'existence des Indiens brésiliens. Cet opuscule, intitulé *Mundus Novus,* publié probablement à Paris en 1501 et attribué à Améric Vespuce, a connu un immense succès: soixante éditions avant 1529, en six langues (16).

---

in italiano con note a compendio del commento di San Tommaso d'Aquino, Padoue, 1949, p. 5.

(9) Fernando de Roa, *Commentarii in politicorum libros*, Salamanque, 1502, f. vii (v.).

(10) Aristote, *op. cit.*, III, 14, 6, 128a.

(11) Giacomo dal Sasso, *op. cit.*, p. 146.

(12) R. G. Mulgan, *Aristotle's Political Theory,* Londres, 1957, p. 68 (royauté barbare), pp. 69-72 (Aristocratie).

(13) Nicolas Trigault n'a-t-il pas écrit, en 1616, de la Chine, que: «Le nom d'Aristocratie, Democratie ou Polycratie est inouy entre les Chinois»? Cf. Nicolas Trigault, *Histoire de l'expédition chrétienne au royaume de la Chine*, Lyon, 1616, Liv. I, Chap. VI, p. 65.

(14) Machiavel, *Le Prince*, éd. Garnier, Paris, 1987, Chap. 4, p. 281.

(15) Certains des compilateurs que nous avons cités au début, et qui répètent le cliché «sans foi, sans loi et sans roi», nuancent aussitôt ce jugement péremptoire.

(16) Rudolf Hirsch, «Printed Reports of the early Discoveries and their Reception», in Fredi Chiapelli (ed.), *First Images of America*, Los Angeles, 1975, Vol. II, p. 540.

A propos des moeurs des Indiens, voici ce que dit l'auteur, dans la traduction française de Mathurin Redouer (1515):

«Aussi ilz nõt point de biens propres mais toutes les choses sont cõmunes. ilz viuẽt ensemble sans loy, sans empereur & chascun de soy mesmes est seigneur. ilz ont autãt de femmes qu'ilz en veulẽt. Le filz se mesle avec la mere & le premier avec la premiere toutes & quanteffois q'ilz veullent [...] Outre cela ilz nont nulles eglises ny tiẽnẽt aucune foy.» (17)

Grâce aux recherches d'Alberto Magnaghi, on admet aujourd'hui que ce texte n'est pas de Vespuce lui-même, mais a été rédigé, avec des additions, à partir d'un manuscrit reconnu authentique et que l'on appelle la «lettre de Bartolozzi», nom de celui qui l'a éditée au XIXe siècle (18).

Confrontons le texte du *Mundus Novus* avec celui de la «lettre de Bartolozzi», dont l'auteur du premier s'est inspiré. Des Indiens du Brésil, Vespuce dit en effet que:

«Non tengono ne legge, ne fede nessuna, e vivono secondo natura. Non conoscono immortalità d'Anima non tengono fra loro beni proprie, perchè tutto è commune: [...] non anno Re: non obediscono a nessuno, ognuno è Signor di se.» (19)
(Ils n'ont ni loi, ni aucune foi et vivent selon la nature. Ils ignorent l'immortalité de l'Ame, ils n'ont pas entre eux de biens propres, car tout est commun [...] ils n'ont pas de Roi: ils n'obéissent à personne, chacun est son propre maître).

Si l'on compare ces deux textes, il apparaît, tout de suite que, si le thème de l'absence de religion, de lois, de chefs, de propriété privée, est commun aux deux, celui du non respect de l'interdit de l'inceste a été gratuitement introduit par le rédacteur anonyme.

Une comparaison avec un autre récit portugais de la même époque confirmerait cette conviction. L'imprimeur allemand Valentim Fernandes, établi à Lisbonne, a dressé en 1503 un acte notarié concernant les voyages portugais vers ces terres australes.

«Les habitants de ce monde-là», écrivait-il, «n'ont ni foi, ni religion, ni idolâtrie et ne savent rien de leur Créateur; ils ne sont pas Soumis à des lois, ni à aucun gouvernement, mais uniquement aux conseils des anciens; ils ne possèdent rien en propre, mais tout leur est commun, sauf les femmes [...] Les hommes s'unissent aux femmes, mais non en public et à l'exception de ces deux degrés de parenté: le fils avec sa mère, le père avec sa fille, le frère avec la soeur.» (20)

Le non-respect de l'interdit de l'inceste, attribué à toute société sauvage, est un vieux thème qui remonte à l'Antiquité (Diodore de Sicile (21), Saint Jérôme) (22) et qui traverse tout le Moyen Age (Brunetto Latini (1260/1266) (23), Antonio Malfante (1447) (24). Il semble avoir été tout simplement projeté sur le Brésil par le rédacteur du texte attribué à Vespuce.

Au fur et à mesure qu'on avance dans le XVIe siècle, les observateurs nuancent les jugements sévères des premières années.

Magalhães de Gândavo écrit en 1576:

«Ils ont l'habitude de se marier avec leurs nièces, filles de leurs frères ou de leurs soeurs; il les regardent comme leurs femmes légitimes.» (25)

Le Français Jean de Lery rapportera en 1578 des propos semblables:

«Touchant le mariage de nos Ameriquains, ils observent seulement ces degrez de consanguinité: que nul ne prend

(17) Améric Vespuce [attribué à], *Sensuyt le Nouveau Mõde & navigations faites par Emeric Vespuce Florentin*, Paris (1515), éd. facsimilé Princeton University Press, 1916, p. 154.

(18) Alberto Magnaghi, *Amerigo Vespucci*, 2e édition, Rome 1926, Chap. II.

(19) Améric Vespuce, «Lettre à Laurent di Pier Francesco de Medicis», (Lisbonne, 1502), Recension de Bartolozzi, in H. Vignaud, *Améric Vespuce, (1451--1512)*, Paris, 1917, p. 410.

(20) António Alberto de Andrade, «O Auto notárial de Valentim Fernandes (1503) e o seu significado como fonte histórica», in *Arquivos do Centro Cultural Português*, Fundação Gulbenkian, Paris, Vol. V, 1972, p. 538.

(21) Diodore de Sicile (60 av. J.C.), *Bibliothèque historique*, trad. F. Hoefer, Paris, 1846, Vol. I, Livre III, Chap. 14 («Les Ichthiopagi ont en commun leurs troupeaux ainsi que leurs femmes et enfants»).

(22) St. Jérôme, «Adversus Jovianum», II, Cap. VII, in Migne, *Pat. Lat.*, Tome XXIII, cols. 307--309 («Les perses, mèdes, indiens et ethiopiens [...] copulent avec leurs mères, leurs grandmères et aussi avec leurs filles et leurs nièces»).

(23) Brunetto Latini, *Li livres dou Tresor*, éd. P. Chabaille, Paris, 1863, p. 171 («... la gent de Ethiopie et de Gartemanz ne savent que est mariage, ainz ont entr'eux femmes communes à tous, et par ce avaient que nul ne conoist pere, se mere non»).

(24) Antonio Malfante (1447), in Charles de la Roncière, *La Découverte de l'Afrique au Moyen Age*, Le Caire, 1925, Vol. I, p. 155. (ces peuplades [du Sahara] ont des moeurs bestiales, le père connaissant sa fille, le frere sa soeur»).

(25) Pero Magalhães de Gândavo, *op. cit., trad. cit.*, p. 115.

sa mère, ni sa soeur, ni sa fille à femme: mais quant à l'oncle il prend sa nièce & autrement en tous les autres degrez ils n'y regardent rien.» (26)

C'est sur la base de ces constatations qu' Alfred Métraux a rédigé son chapitre sur les Tupinamba dans le *Handbook of South American Indians* (27).

Sur les institutions politiques et sur l'existence de la propriété privée, les témoignages de la seconde moitié du XVIe siècle apportent des jugements qui ne s'écartent guère des tout premiers.

En 1554, le jésuite José de Anchieta écrit à propos des Indiens du Brésil qu'«ils ne vivent sous l'autorité d'aucun roi ou chef et ils n'ont d'estime que pour ceux qui ont accompli quelque prouesse digne d'un homme fort.» (28) Anchieta, parlant des Indiens du Pérou, dont il avait connaissance grâce au récit du voyage du Frère António Rodrigues dans le bassin du Rio de la Plata jusqu'à la frontière de ce pays, ajoutait cependant que ceux-ci «vivaient tous sous l'autorité d'un seul chef» (29). On voit alors poindre le début d'une appréciation d'une diversité géographique dans les structures sociales des Indiens d'Amérique du Sud.

De l'organisation politique des Indiens en général, Damião de Góis écrivait, à l'instar d'Anchieta: «... nem tem rei, nem senhor; obedecem somente àquelles que nas guerras que tem huns com os outros sam mais valentes & destes fazem cabeça enquanto não cometem couardia» (30), (ils n'ont ni roi, ni seigneur; ils n'obéissent qu'à ceux qui se montrent les plus braves dans les guerres qu'ils se livrent entre eux, et ils les mettent à leur tête, tant qu'ils ne commettent aucune lâcheté), et Magalhães de Gândavo (1576) reprend ces mêmes propos en termes presque identiques: «Ces gens n'ont ni roi ni souverain; cependant il y a un chef dans chaque village à qui ils obéissent volontairement, mais non parce qu'ils s'y croient obligés.» (31)

Quant à l'absence de propriété privée, les auteurs de la seconde moitié du XVIe siècle confirment pleinement les jugements des premiers témoins.

En 1552, Manuel de Nóbrega écrit que les Indiens «... non possegiano niuna cosa propria, ma tutte sono in commune, & di quelle che uno ha, deue far parte alli altri [...] ne si curano di accumulare richezze.» (32) (... ne possèdent rien en propre, mais tout est en commun, et ce qu'a l'un d'eux il doit le partager avec les autres [...] ils ne se soucient pas d'accumuler des richesses). Magalhães de Gândavo (1576) soulignait lui aussi que les Indiens «... ne possèdent pas de terres et ne se soucient pas d'en posséder, de sorte qu'ils vivent sans cette avarice qu'on trouve chez toutes les autres nations.» (33)

Pour le Brésil, ce n'est donc qu'à propos de l'existence d'un système de parenté que l'on voit une modification notable; en ce qui concerne les autres thèmes, les témoignages des premiers observateurs restent valables, mais avec quelques nuances.

LE CONCEPT DE DESPOTISME

Examinons maintenant le concept de «despotisme» et l'histoire du terme depuis le Moyen Age.

Dans sa traduction latine de la *Politique* d'Aristote (1260), Guillaume de Moerbeke employa l'expression *despoticus principatus* (34). Il fut suivi par Nicole Oresme qui, dans sa traduction en français, *Le Liure de politiques d'Aristote* (1371-4) imprimé en 1489, se servait du terme «Princey despotique». «Princey des-

---

(26) Jean de Lery, *Histoire d'un voyage fait en la terre de Bresil*, La Rochelle, 1578, Chap. XVII (p. 262 de l'éd. de Genève, 1580). 4 autres éditions à Genève au XVIe siècle et une en latin, Genève, 1586.

(27) Alfred Métraux, «The Tupinamba», in *Handbook of South American Indians*, New York, 1963, Vol. III, p. 111.

(28) Fr. José de Anchieta, Lettre à Ignace de Loyola, 1er Déc. 1554, in Serafim Leite S.J., *Monumenta Brasiliae*, Vol. II (1553-1558), Rome, 1957, p. 96. Cette lettre fut publiée, en résumé, dans *Copia de unas cartas de las Indias, Iapon, y Bresil*, Lisbonne, 1555, Barcelone, 1556.

(29) Ibid. p. 99.

(30) Damião de Góis, *Chronica do felicissimo Rei D. Manuel*, Parte I, Lisbonne, 1556, éd. Coimbra 1949, Parte I, p. 131.

(31) Pero Magalhães de Gândavo, *op. cit., trad. cit.*, p. 111.

(32) Manuel da Nobrega, «Informatione delle parti del Brasil» (1552), in *Diversi avisi particolari dall' Indie di Portogallo*, Venise, 1558, f. 40 (v.).

(33) Pero Magalhães de Gândavo, *op. cit., trad. cit.*, p. 118.

(34) R. Koebner, «Despot and Despotism: vicissitudes of a political term», in *Journal of the Warburg and Courtauld Institute*, Vol. XIV, (1951), pp. 278-279.

potique», écrit-il, «est princey sur serfs et ilz souffrent pource qu'ilz sont de serville nature [...] Et sont à ce accoustumez et des enfance virent leurs pères en telle servitude. Et per aventure nont pas memoire que leurs pays feust oncques en liberté.» (35)

A la Renaissance, les humanistes bannirent les termes «despotisme» et «despotique». L'humaniste italien Leonardo Bruni Aretino, dans sa traduction de 1438, imprimée en 1478, leur substitua *dominatio* et *servile iugum* pour «pouvoir despotique» (36). L'humaniste espagnol Juan Gines de Sepúlveda, dans sa traduction latine de la *Politique* d'Aristote, parue à Paris en 1548, inventa le terme *herile imperium*. Sepúlveda fut le premier, à notre connaissance, à appliquer, dans le commentaire de sa traduction, le concept aristotélicien de despotisme au régime politique des Turcs (37).

Dans sa traduction française de la *Politique* d'Aristote, éditée à Paris en 1568, le Français Loys Le Roy employa les expressions «monarchie seigneuriale» et «empire seigneurial», donnant comme exemples: «l'estat du Turc, du Moscovite et du Prêtre Jean [= Negus d'Ethiopie]» (38).

Le philosophe français Jean Bodin, dans ses *Six Livres de la République*, publiés à Paris en 1576 (huit éditions au XVIe siècle, avec une traduction en latin et une en espagnol), reprendra les termes de Loys Le Roy et définira trois types de gouvernement: la «Monarchie royale», la «Monarchie tyrannique» et la «Monarchie seigneuriale». De cette dernière, il citera en exemples la Tartarie, la Moscovie, les Turcs, l'Ethiopie et l'empire de Charles V au Pérou (39).

L'ETHIOPIE

Pour ce qui a trait à ce pays, Bodin indiquera sa source: la *Verdadeira Informaçam das Terras do Preste Joam* de Francisco Álvares, parue à Lisbonne en 1540. L'ouvrage d'Álvares eut un succès considérable en Europe: trois traductions en espagnol, une, réduite, en italien, dans la compilation de Ramusio (1550), deux en français publiées à Anvers en 1558, deux en latin (1556 et 1558), une en allemand en 1566 et une en anglais en 1600.

«Vray est», écrit Bodin, «que François Alvares écrit qu'il a veu en Ethiopie fesser tout nud le Grand Chancelier, & autres grands seigneurs comme vrais esclaves du prince, & tiennent cela à grand honneur. Et par le discours de son histoire on peut aisement recueillir, que le grand seigneur d'Ethiopie est Monarque seigneurial: mais les peuples d'Europe plus hautains & guerriers que les peuples d'Asie & d'Afrique, n'ont jamais peu souffrir de Monarques seigneuriaux.» (40)

Voyons maintenant le passage où Álvares décrit le châtiment du «Justice majeur» [= Grand Chancelier]:

«Neantmoins nous le voulumes accompagner (le Justice majeur) devant la tente de Prete Ian, où étant entré, nous veîmes sortir un peu après deux pages qui l'accompagnèrent jusque au lieu ou l'on punit par batures les delinquants et feirent la venir deux ministres de justice qui feirent depouiller la Justice majeur; puis l'ayant jettée par terre, lui atacherent les mains à deux pieux, et les pieds bien serrés d'une couroye de cuyr, que deux hommes tenoyent. Ces bourreaux étoyent d'un coté et d'autre, lesquels le frapperent par plusieurs fois, mais la plupart des coups portoient à terre et quand le Prete Ian disoit qu'on le touchât, alors le coup penetroit jusques à l'os, et n'eut que trois de telles singlades cette justice majeur, laquelle j'ay veu étriller trois autrefois de cette sorte et au bout de trois jours exercer son office, tout ainsi qu'auparavant. Car cela ne leur tourne à honte ny deshonneur mais disent que c'est faveur, que le Prete-Ian leur porte, d'autant qu'il a encore bonne souvenance d'eux: puis dela à peu de temps il leur fait grace, et les remet en leur premiere dignité et etat.» (41)

Ce qui a frappé Bodin, ce n'est pas seulement l'humiliation d'un haut fonctionnaire aussitôt restauré dans ses fonctions, mais surtout le fait que la punition fut perçue non pas comme un «déshonneur», mais comme une «faveur». Ces sentiments lui suffisent

---

(35) Nicole Oresme, *Le liure de politique d'Aristote*, éd. Paris 1489, Fol. CI (v.), col. 2.

(36) R. Koebner, *op. cit.*, p. 282.

(37) Juan Gines de Sepúlveda, *De Republica Libri VIII*, Paris, 1548, f. 98 (v.) *(Quattuor hic regni genera memorantur [...] Alterum, quo regna Barbarorum, quae cum lege nitantur, & sint hereditaria, herile tamen imperium exercent cuiusmodi nunc est regnum Turcorum)*.

(38) Loys le Roy, *Les Politiques d'Aristote*, Paris, 1568, pp. 371-373.

(39) Jean Bodin, *Six Livres de la République*, Paris 1576 (éd. de Paris, 1579, Livre II, Chap. II, p. 190).

(40) Jean Bodin, *op. cit., loc. cit.*

(41) Francisco Álvares, *Historiale description de l'Ethiopie*, Anvers, 1558, Chap. CXVI, ff. 272 (r.-v.). Texte portugais: Francisco Álvares, *A verdadeira Informação das Terras do Preste João das Índias*, Lisbonne, 1540, Chap. CXVII (éd. Lisbonne, 1989, p. 233).

pour classer le régime politique de l'Ethiopie comme un despotisme barbare.

Pour sa part, Paolo Jovio, dans son *Historiarum sui temporis*, publié à Florence en 1551-52 et qui a connu treize éditions au XVIe siècle, nous dresse un tout autre portrait, presque idyllique celui-là, des moeurs politiques des Ethiopiens:

«... ayant leur republique très prudément establie; [ils] gouvernent les peuples en telle équité de loix, que, combien qu'ils ayent fort affectueux égard à l'autorité du Prince, & qu'ils portent honneur à son nom en tout jugement & action, [ils] règnent non autrement que les Gentilshommes Venitiens: environ lesquels la majesté du Prince est seulemét en vigueur pour contenance de puissance, qui est deguisée ou ostée.» (42)

Jovio cite bien Álvares à d'autres propos dans son texte, mais la source de la citation que nous venons de donner ne se trouve pas dans le récit du Portugais; on ignore d'où elle vient, à moins qu'il ne s'agisse du témoignage d'un moine éthiopien rencontré à Rome.

Dans ses *Relationi Universali*, Venise 1592 (onze éditions entre 1592 et 1612), le compilateur Giovanni Botero insiste fermement sur le caractère despotique du régime politique des Ethiopiens, sur le fait que les sujets du monarque sont ses esclaves et sur l'arbitraire avec lequel il octroie ou supprime charges et bénéfices à son gré:

«Il governo del Preste Gianni hà del despotico assai, perche tiene i suoi vassalli bassissimo; e non meno i grandi, che i piccoli: e li tratta piu presto come schiaui, che come sudditi [...] Il Re dà, e toglie qual si voglia grande Signoria a cui gli piace: nè è lecito à colui, che ne è private, mostrarsi pur di mala voglia [...] Nissuno hà vassalli, se non il Re...» (43)

(Le gouvernement du Prêtre Jean tient beaucoup du despotisme, parce qu'il garde ses vassaux très bas; et tout autant les grands que les petits; et il les traite plutôt comme des esclaves que comme des sujets [...] Le Roi donne ou retire quand il veut des grandes Seigneuries à qui il lui plaît: et il n'est pas permis à celui qui en est privé de se montrer seulement mécontent [...] Nul n'a de vassaux, sauf le Roi...)

La source de Botero est vraisemblablement toujours Francisco Álvares.

Le Français Pierre d'Avity, autre compilateur, copiera mot à mot Botero, dans sa somme sur les états, empires et royaumes du monde, publiée à Paris en 1614:

«Ce monarque tient ses sujets bas le plus qu'il luy est possible, & non moins les grands que les petits, & les traicte plustost comme esclaves que comme sujets [...] Le Roy oste et donne les Seigneuries à qui bon lui semble, & il n'est permis à celuy qui en est privé de montrer qu'il en est mal content [...] Il n'y a aucun qu'ait vassaux que le Roy...» (44)

Bernard de la Roche Flavin, dans ses *Treze Livres des Parliaments de France* (1617) va, de son côté, opposer les institutions du royaume de France à l'arbitraire du monarque d'Ethiopie et à l'absence de corps intermédiaires héréditaires. Le Prêtre Jean, écrit-il,

«... constitue Roys & grands seigneurs tels personnages de son obéissance que bien luy semble, les emparant de Royaumes & Seigneuries, les demettant aussi d'icelles, toutes & quantes au droict sans qu'ils en osent montrer aucun semblant de tristesse...» (45)

Dans le récit d'Álvares, source de tous ces exposés de Botero, de d'Avity et de la Roche Flavin, il s'agit très précisément du cas de deux hauts personnages: le Betudete et le Tigremahon, tous deux destitués de leurs fonctions par le Prêtre Jean et condamnés à mort (46). Telle est la précarité du statut des grands seigneurs devant le caprice du monarque, dit Álvares, que lorsqu'ils se rendent à la cour du Prêtre Jean, aucun des seigneurs détenant des terres

«não pode sair nem abalar para a corte em nenhuma maneira, sem ser chamado do Preste, e, sendo chamado, não deixerá de vir por nenhuma cousa, e quando da terra que senhoreia abala, não deixa em ela mulher, nem filhos, nem fazenda nenhuma, porque vai em esperança de nunca tornar, porque, como acima dito é, o Preste dá quando quer e toma quando quer...» (47)

(ne peut s'en éloigner ni se rendre à la cour en aucune manière sans être convoqué par le Prêtre et, s'il est

---

(42) Paolo Jovio, *Histoire... de Paolo Jovio* [trad. franç.], Lyon, 1552, p. 194.

(43) Giovanni Botero, *Delle relationi universali*, Venise, éd. de 1595, Libro Terzo, pp. 103-104. Première édition Venise 1592, puis Venise 1595, 1599, 1602, 1607-1608, 1612 et Turin 1601; une traduction latine Cologne 1598, une traduction allemande Cologne 1596 et deux traductions espagnoles: Barcelone 1603 et 1605.

(44) Pierre d'Avity, *Les Estats, empires, royaumes et principautés du Monde*, Paris, 1614; Paris, 1619, 1625 (éd. Paris, 1625, p. 368). Une traduction anglaise parut en 1615.

(45) Bernard de la Roche Flavin, *Treze Livres des Parliaments de France*, Bordeaux, 1617, p. 704.

(46) Francisco Álvares, texte portugais, Cap. CXIX et Cap. CXX, p. 237-239.

(47) Idem, Cap. CXXVI, éd. cit. p. 245.

convoqué il ne manquera pour rien au monde d'y aller; quand il quitte le territoire qu'il gouverne, il n'y laisse ni sa femme, ni ses enfants, ni aucun bien, car il s'attend à n'y pas revenir parce que, comme il est dit plus haut, le Prêtre donne quand il veut et prend quand il veut...)

On peut donc conclure que, à l'exception du seul Jovio, tous les auteurs reflètent assez fidèlement le témoignage d'Álvares qui, par la grande diffusion de son ouvrage, a contribué à modeler, au XVIe siècle, l'idée que se faisait l'Europe de l'Ethiopie.

LA CHINE

Le premier auteur de l'un des nombreux compendiums géographiques du XVIe siècle à juger le gouvernement des Chinois est l'Espagnol Jerónimo Román, dont l'ouvrage, publié en 1575, a connu deux éditions au XVIe siècle.

«La potencia del Rey de China necessariamente ha de ser grande, porque no ay hombre que en el reyno sea señor de una almena, ni vassallo, ni imperio, ni juridicion de un palmo de tierra. No ay Condes, ni Duques, ni otro titulo mayor que de Loyta (48) que es cavallero.» (49)

(La puissance du Roi de Chine doit forcément être grande, car il n'y a pas un homme dans le royaume qui possède quoi que ce soit, ni vassaux, ni pouvoir, ni juridiction sur un pied de terre. Il n'y a ni comtes, ni ducs, ni aucun autre titre que celui de Loyta, qui veut dire chevalier.)

Un jugement analogue sur l'absence de propriété privée et de corps intermédiaires apparaît dans l'oeuvre de Juan González de Mendoza, moine augustinien à qui le Pape Grégoire XIII avait demandé de rédiger une histoire de la Chine, bien qu'il n'y fût jamais allé. Dans son *Historia de las cosas mas notables* [...] *de la China*, publié à Rome en 1585, il écrivait ceci:

«Il n'y a point en tout le Royaume aucun Seigneur de vassaux, non plus qu'en Turquie, & n'ont entre eux rien de propre, soit en patrimoine, ou en biens meubles: & ce que le Roy donne de grace soit en faict de gouvernement, ou à raison de services, ou pour autre respect particulier tout cela se pert avec la personne à laquelle aura esté fait le don, & retourne au Roy comme devant. Que s'il luy plait de continuer au fils les biens ou l'estat que tenoit le père, il le faict plus de grace que d'obligation ou de devoir, dõnant à entendre que ce qu'il fait, ce n'est point pour avarice ou autre profit particulier mais que c'est seulement pour obvier aux inconvénients & occasions de trahison que pourroyent venir, s'il y avait de grands & riches Seigneurs en son royaume.» (50)

Les quarante-six éditions, en sept langues, de l'ouvrage de Mendoza ont solidement ancré dans la conscience européenne sa vision de la Chine et l'assimilation qu'il en fait à la Turquie (51).

L'historien jésuite Giovanni Pietro Maffei comparait, lui aussi, dans son *Historiarum Indicarum* (Florence 1588), les Chinois aux Turcs. Son ouvrage a connu trois éditions, dont une traduction en français.

«Il y a encore une autre chose qui importe fort pour une grande richesse, c'est qu'à la manière des Turcs & de toute la secte de Mahomet, en tout l'Empire des Sines, personne entierement ne peut imposer tribut ou gages, sinon le Roy seul. Il n'y a point, comme entre nous, aucuns Comtes, aucuns Ducs, Marquis, Barons ou autres Potentats. Le Roy donne les Magistrats & Preuostes, non tant pour le bien-faict ou recompense, que pour labeur & don: & souvent ces célébres noms d'honneur ne suyuent pas l'héritier qui dégénère. Tout chacun s'acquiert ses richesses & dignities par étude & diligence. Et la paresse ou faineantise oyseuse n'occupe point les places des vaillans hommes & vigilans soubs un fard de noblesse. Voyla a peu pres les biens & louãges des Sinois, lesquels sans doute sont surmontés par des maux & vices plus griefs.» (52).

Le jugement de Maffei sur les Chinois est relativement favorable, ce qui n'est guère le cas des autres auteurs que nous avons cités.

Giovanni Botero, auteur de la compilation *Relationi Universali* (1592), fait encore partie du premier groupe.

«Il gouerno della China hà del despotico assai, conciosa che, non è in tutta la China altro Signore, che il Rè; ne sanno che cosa sia Conte, Marchese, ò Duca;

---

(48) Lao-t'ai = officier de l'état. Cf. D. Lach, *Asia in the making of Europe,* Vol. I, Bk II, Chicago, 1965, p. 754. Le terme *loutea* est employé par Galiote Pereira; voir plus loin.

(49) Jerónimo Roman, *Republicas del Mundo*, Medina del Campo, 1575. (éd. de Salamanque, 1595, Vol. III, f. 227 (v.).

(50) Juan Gonzalez de Mendoza, *Histoire du Grand Royaume de la Chine*, trad. franç., Paris, 1589, Pt I, Liv. III, Chap. II, f. 47 (v.). (1ere trad. franç. Paris, 1588).

(51) Mendoza n'a pas inspiré que des commentaires critiques à l'égard de la Chine. Dans ses *Essais*, Montaigne fit l'éloge du *censorate*, tel que Mendoza l'avait présenté. Cf. Michel de Montaigne, *Essais*, Livre III, Chap. XIII (éd. de La Pleiade, Paris, 1962, p. 1049).

(52) Giovanni Pietro Maffei, *Historiarum Indicarum*, Florence, 1588. (trad. franç. Lyon, 1603, p. 340).

nè vi è altro a cui si paghe tributo, ò gabella. Il Rè conferisce tutti i magistrati, e la nobilità istessa. Egli somministra loro il modo di mantenersi, e non fanno cosa nissuna d'importanza delle quale ne sia avisato. Onde egli è non solamente ubbidito come Rè, ma adorato come Dio. Conciosa che in ogni provincia vi é un ritratto d'esso Rè d'oro, che si tiene couerto cõ un velo, fuor che ne i nouilunij: perche allhora si scuopre, e tutti i magistrati lo visitano, e s'inginocchiano inanzi, come all'istesso Rè. e non solamente il Rè; ma i presidenti ancora, e i giudici sono riueriti di tal sorte, che non si parla loro, si non in ginocchio. Il che auilisce grandemente gli animi de i popoli, e li rende schiavi anzi che sudditi del loro Prencipe.» (53)

(Le gouvernement de la Chine tient beaucoup du despotisme, du fait qu'il n'y a dans toute la Chine d'autre Seigneur que le Roi; il n'y a rien de tel que Comte, Marquis ou Duc; il n'est personne d'autre à qui l'on paie tribut ou gabelle. Le roi nomme tous les magistrats ainsi que la noblesse. Il leur indique comment se comporter et ils ne font rien d'important, dont il ne soit informé. Aussi n'est-il pas seulement obéi comme un Roi, mais adoré comme un Dieu. Le fait est que, dans chaque province, il y a un portrait en or de ce roi, qu'on garde recouvert d'un voile, sauf à la nouvelle lune; parce qu'à ce moment-là, on le découvre, et tous les magistrats lui rendent visite et s'agenouillent devant lui, comme devant le Roi lui-même. Et non seulement le Roi, mais encore les présidents et les juges sont tellement vénérés, qu'on ne leur parle qu'à genoux. Ce qui avilit l'âme des peuples, et en fait des esclaves plutôt que des sujets de leur Prince).

Botero, à la différence de Maffei, n'est pas tendre pour les Chinois, en qui il reconnaît toutes les caractéristiques du despotisme, tel qu'Aristote l'avait défini.

Si l'on considère que, dans toutes ces compilations, les appréciations concernant la société chinoise se ressemblent et que la première en date est de 1575, on peut leur attribuer une source commune, plutôt portugaise qu'espagnole: soit le chapitre sur la Chine de João de Barros, dans sa troisième Décade publiée en 1563, soit l'ouvrage de Gaspar da Cruz, paru à Lisbonne en 1569-70, soit le récit du soldat Galiote Pereira, composé en 1553-63 (54) et publié pour la première fois en italien en 1565 (55).

Etant donné que Cruz s'est directement inspiré du récit de Galiote Pereira, et que les propos de Barros ne comportent de jugements ni sur les droits de la propriété privée, ni sur le pouvoir absolu du Roi, mais seulement sur la faculté du monarque de nommer ou de déplacer ses administrateurs à son gré (56), il est vraisemblable que la source première, sur laquelle les compilateurs s'étaient mis à broder, est le récit de Pereira.

Galiote Pereira, troisième fils de l'Alcaide--Mor d'Arraiolos, est parti pour l'Inde en 1534. Boxer, qui loue le sérieux de son témoignage et la perspicacité de son regard, note qu'il faisait bien la distinction entre ce qu'il avait vu et ce qu'on lui avait raconté (57). Voici un extrait de son texte, traduit du portugais original, resté inédit jusqu'au XXe siècle:

«E tem mais este Rei [de la Chine] pera segurança de seus reinos a não haver nunqua alevantamento en todos elles, não há nenhum homem que se possa chamar senhor, mais de sua casa somente, [il y a] muitos e grandes regedores, os quais emquanto o são o seu serviço e estado he tamanho como o dum muito grande principe. E estes, ainda postos e tirados tantas vezes, que não há nunqua tempo pera poderem criar nunqua malicia. Os quais tem grandes ordenados emquanto servem, e depois d'apousentados tambem tem em suas vidas hum certo tanto, mas nas cidades onde estão hão-no de receber mês entrado, mês saido, dos que já são ordenados pera aquillo, de maneira que [le Roi] he senhor de tudo, sem o ninguem ser, como tenho dito, somente de sua casa» (58)

(De plus, ce roi [de Chine], pour la sécurité de son royaume et qu'il n'y ait nulle part de soulèvements, fait en sorte qu'il n'y ait pas un homme qui puisse se dire maître, d'autre chose que de sa maison. [Il y a] beaucoup de grands gouverneurs qui, tant qu'ils sont en poste ont un service et un train de vie de très grand prince. Et ceux-là sont nommés et démis si souvent

---

(53) Giovanni Botero, *Relationi Universali*, Vicenza, 1596, Lib. II, Cap. II, p. 65.

(54) Galiote Pereira, «Algûmas cousas sabidas da China por Portugueses», in C. R. Boxer, «A Portuguese account of South China in 1549-52», in *Archivum Historicum Societatis Iesu*, Vol. XXII (1953), pp. 57-92.

(55) In Nuovi *Avisi delle Indie di portogallo venuti nuovamente dalli Reverendi Padri della Compagnia de Gesu*, Pt IV, Venise, 1565, ffs. 63 (r.) — 68 (r.).

(56) João de Barros, *Terceira Década da Ásia*, Lisbonne, 1563, liv. II, Cap. VII, f. 45 (v.). (Os quais cargos [...] nam lhe dura mais de tres annos, & ajnda muytas vezes no meyo tempo sem o elles saberem, sam sobressaltados, com que os tiram dos taes cargos & os mudam pera outre parte... / Ces postes [...] ne sont pas attribués pour plus de trois ans, il arrive même souvent qu'ils soient destitués avant ce délai sans en être avertis, on leur fait quitter leur poste et on les envoie ailleurs...).

(57) C.R. Boxer, «A Portuguese account of South China», in *Archivum Historicum Societatis Iesu*, Vol. XXII, Rome, 1959, p. 57 et p. 61.

(58) Galiote Pereira, «Algûas cousas sabidas da China por Portugueses», (1563), in *art. cit.*, p. 90.

qu'ils n'ont jamais le temps de comploter. Ces hommes sont très bien payés tant qu'ils servent et, une fois retraités, ils ne vivent pas mal, mais dans les villes où ils sont ils reçoivent leur pension au début ou à la fin du mois de ceux qui ont des ordres pour cela, de sorte que [le Roi] est maître de tout et que personne, je l'ai dit, ne l'est que de sa maison.)

Si ce texte affirme que personne ne possède de propriété privée autre que sa propre maison, que les administrateurs sont très souvent autoritairement nommés ou démis de leurs fonctions, et que le monarque dispose sur tout d'un pouvoir absolu, on n'y trouve l'esquisse d'aucune comparaison avec les structures de la société européenne (absence de Ducs, de Comtes, etc.) et aucun rapprochement avec celles de la société turque. Il est douteux que Pereira ait jamais entendu parler de la définition aristotélicienne du despotisme. Ce n'est que le hasard qui a fait que sa description de la Chine correspondait à l'attente des compilateurs.

Machiavel est-il la source où se sont abreuvés les compilateurs, dans leur constante référence à la Turquie? C'est possible, mais non pas certain. La Turquie constitue, durant tout le XVIe siècle, une menace sérieuse pour l'Europe et, comme l'a noté Atkinson, le nombre d'ouvrages publiés alors sur la Turquie est plus du double de ceux publiés sur l'Amérique (59). Elle est donc au coeur des préoccupations européennes plus que tout autre pays extra-européen.

Que les essais de comparaison soient essentiellement l'oeuvre des compilateurs plutôt que des témoins portugais, on pouvait s'y attendre: ceux-ci se bornent à constater ce qu'ils voient, sans se soucier de le juger; ceux-là s'attachent à faire comprendre à leurs lecteurs une réalité «autre».

Pour donner un sens à la réalité inédite des nouvelles sociétés extra-européennes, les compilateurs ont eu recours à des grilles, naturellement issues de la culture commune aux lecteurs européens. Parmi ces grilles on trouve, outre la grille aristotélicienne, la grille judéo-chrétienne. Celle-ci premettra surtout de savoir si les gens de la société en question croient ou non en un dieu unique, ou du moins en révèlent des traces dans leurs croyances; s'ils descendent d'Adam et par quel chemin leurs ancêtres sont parvenus au Nouveau Monde; si la prédication des Apôtres est arrivée jusqu'à eux et si le souvenir, aussi vague soit-il, en est décelable.

De ces grilles, c'est la judéo-chrétienne qui sera le plus volontiers appliquée par les protagonistes de la Contre-Réforme dans les pays ibériques; l'aristotélicienne aura la préférence des philosophes laïques en Italie et en France.

Pour autant que nous sachions, la grille aristotélicienne n'est évoquée par aucun des auteurs portugais du XVIe siècle. L'absence, au départ, de toute théorie du politique informant au préalable leur vision, rend d'autant plus précieux leur témoignage à l'état brut, mais pour le reste de l'Europe, le recours à un modèle pour organiser le foisonnemente de faits inédits paraissait non seulement légitime, mais indispensable. Bien que d'une subtilité assez rapidement épuisée, la grille aristotélicienne a donné aux Européens, durant ce premier siècle de contact avec les sociétés extra-européennes, sous forme d'«idéal-type» au sens donné par Max Weber, le meilleur instrument d'analyse dont disposait l'Antiquité.

(59) G. Atkinson, *Les nouveaux horizons de la Renaissance française*, Paris, 1935, p. 10.

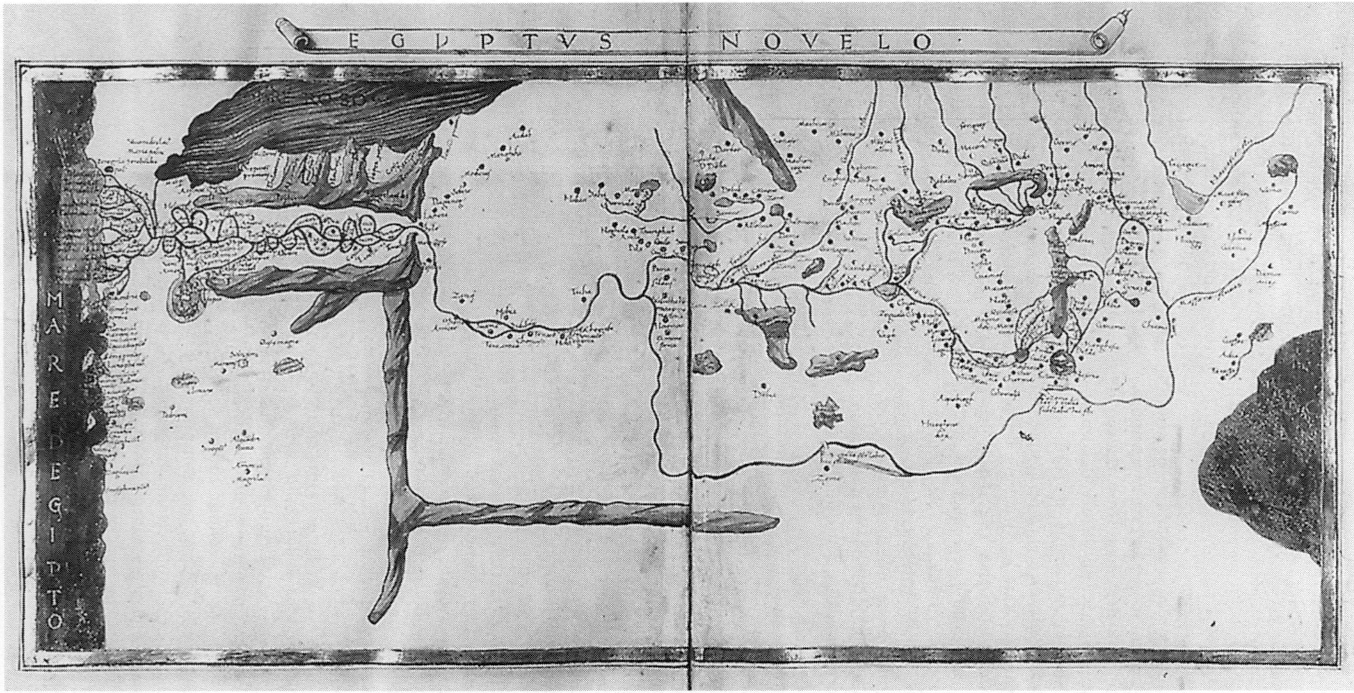

Fig. 1. "Egyptus Novelo" by Jacopo Angiolo in his Translation of Ptolemy's "Geography", 1441–1456
(Bibliothèque Nationale, Paris)

XIX

South-east Africa as Shown on Selected Printed Maps of the Sixteenth Century*

One of the results of improvement in the technique of printing in the 16th century was the large scale production of printed maps. These maps were principally made for the amusement and instruction of the ordinary man avidly curious for knowledge of the strange new distant lands brought to light by the discoveries. Because of the public for whom they were destined, they were unlike the nautical charts or portolans from which they were constructed. These maps are often still easy to come across, and for this reason are much sought after by collectors.

What often puzzles collectors of maps of Africa is the plethora of names of towns and rivers in regions which were only discovered to Europe late in the 19th century. To ascribe these names to the imagination of mapmakers anxious to fill up the empty spaces is just not possible. No man ever gratuitously invents. He may twist and develop certain facts, but there is always some source not his own behind them.

It is our intention, therefore, to trace down the origins of the wholly fanciful cartographical picture of the interior of Africa which was formed in the 16th century and transmitted through the 17th till it began to be revised in the 18th century.

The printed map of the 16th century grew out of two main traditions: first, the Ptolemaic cartography which took on a great vogue at the close of the 15th century; and, second, the nautical charts embodying the results of the discoveries, themselves a development of the Mediterranean portolan. The Ptolemaic cartography later rapidly fell out of favour as soon as it was found impossible to reconcile with the experience produced by the discoveries.

The importance of the nautical charts lay in the fact that, right from the very voyages of Dias, Gama and Cabral, they established with remarkable accuracy the definitive outline of the African continent.[1] Since these charts were designed to a technical end, i.e., the facilitation of navigation, they rarely carried any indications concerning the interior of the continent. At the most, they show the Nile and the Mountains of the Moon as conventional decorations.

The Juan de la Cosa Map, 1500-1508?[2]

The first map to show the African continent approximately as the discoveries had shown it to be is the world chart of Juan de la Cosa. Juan de la Cosa was a Biscayan pilot who had served under Columbus. Though this is a manuscript map, we shall include it in our study, for it is essential to the understanding of later printed maps. It represents the African coastline correctly as far as the Rio do Infante, the point reached by Bartolomeu Dias. The trace of the rest of the coastline shows that the map's author knew of Vasco da Gama's voyage, but had seen no chart resulting from it. The names on the coast from the Rio do Infante northwards are concocted and repetitive. This map is interesting for two reasons. It is the first to show the Congo rising in a lake, which connects, incidentally, with the Nile. It also shows two rivers coming from a single source debouching on the east coast. We shall return to the matter of these rivers later.

---

* The term "South-east Africa" is intended to refer to that part of the African coast and its hinterland from the Cape of Good Hope up to but not including the Island of Mozambique.

[1] The position of the Cape of Good Hope on the Caverio map is out by only 22′. Cf. E. G. Ravenstein, *Vasco da Gama's First Voyage* (London, 1898), p. 212.

[2] Preserved in the Madrid Naval Museum. Reproduced in Santarém's Atlas (Paris, 1849), P. xxi, also in Lelewel's Atlas, Pl. xlii. George E. Nunn, in *The Mappemonde of Juan de la Cosa* (Jenkintown, Pennsylvania, 1934), dates the map about 1508, but his opinion is not universally accepted. It is probable that different parts of the map were completed at different times.

The Juan de la Cosa map is not, strictly speaking, a nautical chart. It is in fact untypically so, for the very reason that it contains indications on the interior of the continent.

Of the nautical charts still in existence dating from the period immediately following the discoveries, there are four which are of interest for the representation of the African continent. They are the so-called Hamy chart, 1502,[3] the so-called Anonymous Munich Chart, 1502 or 1503,[4] the so-called Cantino chart, 1502,[5] and the Caverio chart, 1502.[6] None of these charts carry any indications of the interior of the continent beyond the conventional ones of the Nile and the Mountains of the Moon. Although they mark the names of rivers on the coast, they do not show their courses in the interior. It is on these charts or on copies of them that the first printed maps were based.

The Contarini Map[7]

The first printed map to show the African continent as the voyages of Dias, Gama, and Cabral had revealed it is the Contarini map of 1506. It was designed by Giovanni Matteo Contarini, engraved by F. Roselli, and published in Venice. This map, which is considered as the oldest known printed map showing the results of the discoveries, is not copied directly from a portolan; it is clearly inspired by Ptolemy and only diverges from him on the outline of the African continent as revealed by the discoveries.[8] In the interior, we find the Nile and the Mountains of the Moon conventionally represented. But we find also four rivers having their sources fairly deep in the interior, one debouching on the west coast, north of the Cape of Good Hope, and the other three along the south east African coast. These rivers cannot be of Ptolemaic inspiration, for no "pure" edition of Ptolemy shows this part of the African coast, and, as we shall see later, it is improbable that knowledge of them came by way of the discoveries. Turning to the accounts of Vasco da Gama's voyage, we find that the first land his fleet touched in the neighbourhood of the Cape of Good Hope was St. Helena Bay. The Roteiro of Vasco da Gama's voyage says that the River Samtiagua (S. Tiago—St. James) enters the bay four leagues to the south east of the anchorage and is a "stone's shot across at the mouth and from two to three fathoms in depth at all states of the tide".[9]

Of the other three main rivers we shall take the most northerly first. From the curve of the coast at this point we can fairly safely identify it with the Rio dos Bons Sinais, so named by Vasco da Gama. Which modern river the Rio dos Bons Sinais represents we cannot be sure. Ravenstein[10] identifies it with the Kilimani river. The remaining two rivers we notice have their source in mountains deep in the interior. At this period (1506), the Portuguese could not possibly have had any knowledge of the interior beyond what the local Arabs might have told them, and, though possibly of Arab origin, knowledge of these rivers could not have been derived from questioning local inhabitants. These two rivers appear again on the next map we shall study. Between the two rivers, we find a place called *Vigict Magna*, which appears on numerous later maps as *Vigiti Magna*. *Vigict* may be a corruption of the river Wanshit in Abyssinia.[11]

The Waldseemüller 1507 Map (Fig. 2)

Waldseemüller used as the basis for his 1507 map the nautical charts of Caverio and Cantino. He was not, however, disposed, even in the face of the new facts they revealed, to jettison his faith in Ptolemy. His outline of the African continent is crude. We find the Ptolemaic *Prassum Promontor* shown near Cape Delgado. We find also the two rivers already shown on the Contarini map rising in the Mountains of the Moon in the interior. They are named in turn as the *Rio de S. Vincenco* and the *Rio de Lago*. The *Rio de S. Vincenso* (Rio de São Vicente) is identified by Ravenstein as the Pungwe river, the *Rio de Lago* as

[3] Reproduced in A. Kammerer, *La Mer Rouge, l'Abyssinie et l'Arabie depuis l'Antiquité*, II, part 2 (Cairo, 1935), pl. CXLV.
[4] *Ibid.*, pl. CLIII.
[5] Reproduced in E. G. Ravenstein, *l.c.*, map VI.
[6] Reproduced in A. Kammerer, *l.c.*, II, part 2, pl. CLI.
[7] In British Museum, B. M. Maps. C. 2. cc.
[8] A. Kammerer, *l.c.*, II, part 2, pl. 73.
[9] Ravenstein, *l.c.*, p. 6.
[10] Ravenstein, *l.c.*, p. 19, note 1.
[11] Cf. Crawford, "Some Medieval Theories of the Nile", *Geogr. Jour.*, Sept. 1949, p. 13.

Fig. 2. Part of Walsdeemüller's 1507 World Map (Wolfegg-Waldsee Schloss-Bibliothek).

the Umbelasi river.[12] There is the conventional representation of the Mountains of the Moon and the two lakes from which the Nile takes its source, having in turn their sources in the Mountains of the Moon. We also find a third feature—a lake labelled *Sacaff Lacus* and containing an island, situated just west of the Mountains of the Moon and linked by a river to another lake in the middle of the Sahara. This feature of an inland river linking two lakes does not appear in some "pure" editions of Ptolemy,[13] but it is to be found on many medieval and even earlier maps.[14] Sometimes there are variations. The river, instead of linking up with the lake in the Sahara, is joined to the Nile just near the point where its two arms branch off to lead to the two lakes over the Mountains of the Moon.[15] In some maps, it just ends in the Sahara.[16] The *Sacaff Lacus* is sometimes called *Sachaf Lacus* or *Saphat Lacus*. Sometimes, as on the Contarini map, in which the third lake also appears, it is duplicated. We shall discuss the problem of the lake and the river further on.

The Ruysch Map of 1508

The printed map of Johannes Ruysch Germanus appears as the *Tabula Nova* in the 1508 edition of Ptolemy. The outline of the African continent is very crude. In the interior we find the Mountains of the Moon and the third lake which is doubled on the Contarini map. We find also the two rivers debouching on the east coast. One is called the *Infante Fl.* (Rio do Infante), and the other as *Fl. Dagva* (Rio de Lago?). The Rio do Infante is identified by Axelson as the Great Fish.[17]

The phenomenon of the two rivers debouching on to the east coast is common to many other contemporary maps of this period. It only disappears as a recognizable feature in Waldseemüller's 1516 map which we shall study presently. The rivers are present on the maps of Jan Seversz, Leyden, 1514,[18] Madrignano of 1508,[19] Louis Boulenger, 1514,[20] Apianus, 1520,[21] Orontius Finaeus, 1531,[22] on the globe of Johann Schöner, 1515,[23] and on the so-called "Green Globe" or "Quirini Globe" in the Bibliothèque Nationale, Paris.[24]

Study of the Two Rivers

We have thus been able to see that the two rivers rising in the interior and debouching on to the south east coast of Africa are represented as a separate feature unrelated to any of the rivers named on the nautical charts. The problem is, then, how and whence this feature came to be introduced into European cartography. Before proceeding further, let us examine a passage from João de Barros' First Decade. He writes of the south-east African coast as follows: "The country which we treat of as the Kingdom of Sofala is a vast region ruled by a heathen prince called Benomatapa. It is surrounded like an island by two arms of a river which proceeds from the most notable lake in the whole continent of Africa. The river which flows towards Sofala, after leaving this lake and running some distance, divides into two arms, one of which reaches the Ocean on this side[25] of Cape Correntes, and is that which our people formerly called

12 Ravenstein, *l.c.*, 217. Today also called the Umbelusi.

13 Cf. map in Gregorius Reisch, *Margarita Philosophica* (Fribourg, 1503). Reproduced in Kammerer, *l.c.*, II, part 2, fig. 80. The third lake is still present, but it is placed symmetrically beside the other two *above* the Mountains of the Moon.

14 E.g., the Hereford world map, ca. 1280; the Ebstorf map (after 1270); the Anglo-Saxon world map, ca. A.D. 1000 (B. M. Cotton M.S. Tiberius B. V., p. 56 G); and the Beatus world map, ca. A.D. 776.

15 Cf. Contarini map, 1506.

16 Cf. *Hydrographia sive Charta Marina*, Ptolemeus (Argentinae, 1513), repr. in Kammerer, *l.c.*, II, part 2, fig. 86.

17 In a personal communication. This is his revised opinion on the identification given in his book *South East Africa, 1488-1530* (London, 1940), p. 19.

18 Reproduced in A. Kammerer, *l.c.*, II, part 2, pl. CXLVIII.

19 Madrignano, *Itinerarium Portugalensium e Lusitania in Indiam et inde in Occidentem et demum ad Aquilonem* (Milan, 1508) (frontispiece map). Also reproduced in L. Bagrow, *Geschichte der Kartographie* (Berlin, 1951), p. 108.

20 Reproduced in Nordenskiöld, *Facsimile Atlas* (Stockholm, 1889), pl. XXXVII.

21 Repr. *ibid.*, pl. XXXVIII.

22 Repr. *ibid.*, pl. XLI.

23 Repr. in E. L. Stevenson, *Terrestrial and Celestial Globes* (New Haven, 1921), p. 83, fig. 43.

24 Bibliothèque Nationale, Paris, Dépt. des Cartes et Plans, Ge. A 335, in glass case at left of Sebastian Cabot map. Carries no date or place. Estimated by Bibl. Nat. catalogue to be between 1513-1515.

25 I.e., south of Cape Correntes.

Rio da Lagoa, and now Espírito Santo, which name was given by Lourenço Marques, who went to explore it in the year 1545.

The other arm enters the sea twenty-five leagues beyond Sofala and is called *Cuama*, although in the interior other people call it Zambesi".[26]

We see here the same two rivers indentified by Barros as the Espírito Santo and the Zambesi, but here representing in reality the Incomati-Limpopo and the Zambesi. The Limpopo was confused with the Incomati. It is clear, therefore that the idea of the two rivers having their sources far in the interior and debouching on the east coast is a regular feature of African cartography of the early 16th century. The very fact that they do not consistently coincide with the names of the rivers on the nautical charts leads one to believe that knowledge of them goes back before the period of the Discoveries. This knowledge could have reached Europe from Arab sources via Pero de Covilhã, who is considered to have visited Sofala (1489), or from Arab sources via Italy.[27]

The Waldseemüller 1516 Map[28] (Fig. 3)

This map was published in a series of large sheets from woodcuts; a thousand copies were printed. It carries a much greater elaboration of detail than his 1507 one. The outline of the continent is more accurate and closely follows the Caverio chart. The interior of the continent is filled with names. We find the Nile, the Mountains of the Moon, and the "third lake". The river leading from the third lake just reaches the Nile, but is not joined to it, as if Waldseemüller had remained in doubt and had resolved to show it. On the east coast, we find five rivers which have their sources in the interior, and one of these (the *Gomormager Fl.*) is linked to the now elaborate river system round the third lake. Below this river we find the *Rio dagoada*, which inland is called the *Huilaste Fl.* Nearby we notice the *Rio de Lago*, which, in the 1507 map, was treated as one of the "two rivers" and which here tails off a short way from the coast. There are three remaining rivers (*Rio dos medões*),[29] the *Rio de S. Thomé*,[30] and the *Rio do Infante*. The first two have their sources in the *Zaet Mons. Zaet* or *Zethe* is a province in southern Abyssinia.[31]

Returning to the first-mentioned river, the *Gomormager Fl.*, it is obvious that this is not a Portuguese name, and, further, it does not appear on any of the nautical charts. As it debouches at Sofala, we can safely identify it as the Sofala river. What is true of the *Gomormager Fl.* is also true of the *Huilaste Fl.*, for neither is this a Portuguese name. We can see, then, that Waldseemüller is here taking the step of correlating the river systems of his cartography of the interior with the rivers marked on the nautical charts. Where did he get his information on the cartography of the interior? The answer to this problem seems to lie in a map of Egypt and Abyssinia entitled *Egyptus Novelo*, to be found in Jacopo Angiolo's translation of Ptolemy's *Geography* (Fig. 1).[32] Roberto Almagià dates this map before 1454.[33] Kammerer dates it after 1441, but before 1456.[34] If we compare this map with Waldseemüller's there is certainly a remarkable similarity in the arrangement of the lakes and rivers, but not, as we shall see later, in the names. It is for this reason that it seems likely that Waldseemüller copied, not from this map, but from one like it.

The most important fact about the *Egyptus Novelo* map is that it is constructed without scale, while the nautical charts were constructed on a very accurate scale. The people who probably gave the author the

---

26 João de Barros, *Décadas da Ásia* (Lisbon, 1552), Dec. I, Book X, Chap. I. English translation from George McCall Theal, *Records of South East Africa*, VI (Cape Town, 1898).

27 The infusion of Arab culture into Western Europe via Sicily was not inconsequential during the Middle Ages. Cf. J. K. Wright, *Geographical Lore of the Time of the Crusades* (New York, 1925), *Amer. Geogr. Soc. Research Series*, no. 15, chap. III.

28 Reproduced in J. Fischer and F. von Wieser, *The World Maps of Waldseemüller (Ilacomilus) 1507 and 1516* (Innsbruck, 1903) (in English and German). See Fig. 3.

29 Today the Kosi river. Cf. Ravenstein, *l.c.*, p. 217.

30 Today the Umtata river, Cf. *ibid.*, p. 216.

31 Cf. R. Biasutti, "La Carta dell' Africa di G. Gastaldi e lo Sviluppo della Cartografia Africana nei Sec. 16 e 17", *Bol. d. R. Soc. Geogr. Ital.*, Serie V, IX (1920), p. 402.

32 In Bibliothèque Nationale, Paris. Fonds Latin No 4802. Cf. also H. Omont, *Géographie de Ptolémée*, trad. Latine de Jacopo Angiolo, Florence, reproduction réduite des Cartes et Plans du ms. 4802 (Paris, s.d. (1924)).

33 Quoted by Crawford, *l.c.*, p. 7.

34 A. Kammerer, *l c.*, III, part 3, p. 11.

information for his map were Abyssinian monks who came to Florence during the Council of 1441.[35] Europe's contacts with Abyssinia during the Middle Ages were much more frequent than might be expected.[36] But it would be unwise to estimate that the *Egyptus Novelo* map was solely based on the reports of Abyssinian monks. The two symmetrically opposed lakes suggest a Ptolemaic framework behind a local colouring of place-names. The attempted fusion of this Abyssinian cartography with the nautical charts which was to put the whole of African cartography on a false basis until the 18th century has its origin largely in the exaggerated notion of the southern extent of Abyssinia which at this time was identified with the legendary kingdom of Prester John. Such was the conviction of the immensity of Prester John's kingdom that we find Afonso de Albuquerque writing to the King of Portugal as late as 1513 as follows:

"The land of Prester John is very big. It extends along the borders of the *Mogadishu* hinterland towards *Sofala*, from the Red Sea to *Suaquem* (Suakin on the Red Sea coast north of Massawa), and in the interior it is said to extend as far as Nubia, which we call Ethiopia, and as far as the land of certain Moors called *Ajaje*, from whence gold comes to *Suaquem* in cubes like dice. And thus the land of Prester John extends towards *Manicongo* (the Congo) and the coastal lands of that region and towards the coast which runs to the Cape of Good Hope. There are gold mines in the land of Prester John, and in my opinion the gold which goes to Sofala, Mogadishu and Mombasa comes from a land which owes allegiance to Prester John."[37]

The notion of a vast African continent inhabited by primitive politically unorganised negro tribes was something Europe had still to grasp. If the vast length of the African coastline, in comparison with that of Europe, was being slowly appreciated, there was no comparable appreciation of the immense unknown in the interior. Thus Abyssinia was stretched down to fill the vacuum, and the consequences of this are to be seen in maps of Africa well into the 18th century.[38]

There were, however, certain grounds for the exaggeration of the extent of Abyssinia. The Kingdom of Prester John was reputed, as we have seen, to be rich in gold, and had not the Portuguese discovered at *Sofala* as rich a source as had ever been heard of? Chá Masser (ca. 1505) says of *Sofala:* "The King is a Christian and under Prester John."[39] The author of *Calcoen* (1504) calls the negroes of *Sofala Paepians*,[40] a Dutch corruption of Pape Jan (Prester John). *Sofala* was also thought to be the *Ophir* of the Bible,[41] and more than one reference can be found to this among 16th century writers.[42]

The origin of the "Ophir theory" can be traced to the account of Tomé Lopes of Vasco da Gama's second voyage to India in 1502. This, as far as is known, was first published in 1550 by Ramusio in the first edition of his *Navigationi e Viaggi*,[43] but it is probable that it may have been printed elsewhere previously. Tomé Lopes says that the Moors of Sofala told him that "they had ancient books and writings from which it might be stated that the mine from which King Solomon drew so much gold every three years was this same one".[44]

We can see now the grounds on which Waldseemüller might have felt himself justified in identifying the *Gomormager Fl.* with the Sofala river. But it was Stabius, a contemporary of Waldseemüller, and not

---

[35] A. Kammerer, *l.c.*, III, part 3, p. 12.

[36] Crawford, *l.c.*, p. 8, and also Pe. Marcellin de Civezza, *Saggio di Bibliografia Geografica Storica Sanfrancescana* (Prato, 1879), pp. 217-222.

[37] Letter from Afonso de Albuquerque to the King, 4 December 1513, in *Algunos Documentos da Torre do Tombo* (Lisbon, 1892), p. 329.

[38] The *Egyptus Novelo* map gives proof of the illusion of an extended Abyssinia by the portrayal of the Gulf of Guinea in the south-western corner.

[39] *Relazioni di Lunardo de Chá Masser* (ca. 1505). A document published in *Mem. Acad. R. Ciencias de Lisboa*, Nova Série, Tomo IV da Colecçao, Classe de Ciencias Morais, Políticas e Belas Letras, p. 67. Also in Appendice del *Archivo Storico Italiano* (Firenze, 1846), pp. 13-48.

[40] *Calcoen* (Antwerp, 1504). Duplicated edition and English translation by J. Ph. Berjeau (London, 1874).

[41] Cf. I Kings IX. 28, X. 11, 21-22; II Chron. IX. 10, 21, VIII. 18; Gen. X. 29, II. 11; Job XXII. 24, XXVIII. 16; Psalms XIV. 9; Isaiah XIII. 12; I Chron. XXIX, 4.

[42] *E.g.*, Raphael Volaterranus, *Commentariorum Urbanorum* (Basilae, in Officina Frobeniana, Anno MDXXX), f. 138 v. Of Sofala he says: "Hanc esse putant ex argumentis quae Solomonis libris Ophir appellatur, ob auri venas magna copia repertus." Milton says of Sofala: "Sofala, thought Ophir" (*Paradise Lost*, XL, 400).

[43] Giovanni Battista Ramusio, *Navigationi e Viaggi*, I, 1st ed. (Venice, 1550).

[44] *L.c.*, I, 3rd ed. (Venice, 1563), f. 134 r.

Waldseemüller himself, who, in his map of 1516, engraved by Albrecht Dürer,[45] first took the bold step of trying to correlate the Abyssinian cartography with the nautical charts. Stabius' map, unlike Waldseemüller's, shows rivers only and no place-names. It is on this map that we first find the *Gomormager Fl.* and the *Huilaste Fl.*

The *Gomormager Fl.* probably represents the Jamma river[46] in Abyssinia, and the *Huilaste Fl.* really the Webi Shabelle; this latter on account of the direction of its course and the fact of the name "Bali" being shown south of it. Bali was a province of Abyssinia which stretched from the Webi Shabelle in the east to the Omo river in the west.[47]

In the east, we find on Waldseemüller's map a lake fed by a river and with the legend below it: *Vabi lacus dulcis*. Near it are found the place-names *Tasso*, *Gafat*, and *Adia*. The lake and the place-names are all to be found on the *Egyptus Novelo* map, with the exception that Waldseemüller's *Tasso* is *Tarassa*. This lake is, I think, lake Zway, the largest of the south Ethiopian chain of lakes. The river running into it is the Machy or Maki.

Such comparisons as we have made should be sufficient to prove the relation between Waldseemüller's cartography of the interior of the continent and the cartography of the *Egyptus Novelo* map.

It should not be thought that the *Egyptus Novelo* map can easily be paralleled with a modern map of Abyssinia. Identification of its names, rivers, mountains, and lakes has been attempted by Crawford.[48] The principal feature of Crawford's article is his identification of what become on Waldseemüller's map the Mountains of the Moon, and which on the *Egyptus Novelo* map appear as lying between the two symmetrically opposed lakes and named *Zehecuder* and *Zabacuber Mons*, as the Lasta range of mountains in Abyssinia which divides the Tsellari river from the Tacaze or Takaze river, the highest point of which is Mt. Abuna Josef (4,190 meters).[49]

Crawford believes that the three lakes all reduce to one—lake Tana. "Sachaf" is not a topological name but the title of the ruler of the province of Shoa—*Sahafa Lam*.[50] The conclusions of his article have been contested[51] by those who are unprepared to abandon the belief that the term Mountains of the Moon refers to the snow-capped mountains of the Nile basin on the Equator.[52] The Arabic for "Mountains of the Moon" is "Djabal al-Qoumr". "Djabal" means "mountain" and "Qoumr" can mean "moon" or anything which shines or glistens.[53]

Through the *Egyptus Novelo* type of map, the Mountains of the Moon became integrated into the cartography of Abyssinia *before* the discoveries and, after the appearance of the nautical charts at the beginning of the 16th century, they (together with the three lakes and the connecting river system and the Abyssinian place-names) were all projected southward when the first attempts were made to integrate the cartography of the interior with the orographic cartography of the nautical charts.

Evidence that the representation of the Moutains of the Moon and the two symmetrically opposed lakes existed independently and anterior to the *Egyptus Novelo* type of cartography of Abyssinia, in so far as the latter became localized by the addition of recognizable place-names, is provided by the following text from the Arab geographer Dimisqi (died 1326): "The Nile", he says, "comes from the Mountains of the Moon, which separate the habitable part of the earth from the part scorched by the sun of which we have no knowledge. Its sources are ten rivers. The most eastern is fifteen days' journey from the most western

---

[45] Reproduced in Nordenskiöld, *Periplus* (Stockholm, 1897), Pl. XLVII. Reduced reproduction in E. L. Stevenson, *l.c.*, p. 88, fig. 45.

[46] Cf. map on p. XX in C. F. Beckingham and G. W. B. Huntingford, *Some Records of Ethiopia* (London, 1954).

[47] *Op. cit.*, p. 230.

[48] Crawford, *l.c.*, pp. 8-29.

[49] *Ibid.*, pp. 9 and 18.

[50] Cf. F. M. Esteves Pereira, *Chronica de Susenyos, Rei de Ethiopia*, II (Lisbon, 1900), p. 211, quoted by Crawford, *l.c.*, p. 21.

[51] See discussion at the end of Crawford's article, *l.c.*, pp. 24-29, and later correspondence in the *Geogr. Journ.*, June, 1950, p. 268, and July 1950, p. 123. See also Max Kratochwill, "Das Mondgebirge des Claudius Ptolemäus", *Mitt. Geogr. Ges. Wien* XXII (1950), pp. 222-224.

[52] For a thorough defence in favour of this theory see T. Langenmaier, "Alte Kenntnis und Kartographie der Zentralafrikanischen Seeregion", *Mitt. Geogr. Ges. München*, XI (1916), pp. 1-114.

[53] Cf. Th. Simar, "La Géographie de l'Afrique centrale", *Revue Congolaise*, 1912, p. 94.

They all flow into two great lakes, separated one from the other by a distance of four day's journey. The circumference of the eastern lake with its islands and mountains is three days' journey, and that of the western nearly five days. The lands round about these lakes are peopled by savage tribes of negroes whose ways resemble those of fierce animals. They eat those who fall into their power. The two lakes are situated between 50° and 56° longitude and between 6° and 7° latitude on the other side of the equator. The eastern lake is called *Lake Koutou* or *Tamin* of the Sudan, and the western lake the lake of *Demadem, Qaldour* or *Hedjami.* From each of these lakes flow four rivers, which flow into a third lake of considerable size, 7 degrees south of the equator and called the lake of *Djaus*, the lake of "Confluence" or *Lake Kouri* of the Sudan, and having a circumference of six days' journey. From this lake flow three great rivers: one flows towards the west called the *Ghanah* (Niger); the other towards the east and then southward, called the *Demadem* or *Mogadishu* river (Webi Shabelle) and the third northward called the Nubia river or Nile."[54]

We can discover certain points of similarity between Dimisqi's hydrography and that of the *Egyptus Novelo.* On the *Egyptus Novelo* map we recognize the two lakes, *Koutou* and *Demadem*, but there is no lake *Kouri* into which the two former flow. The chief point on which the two maps differ is that the three lakes are placed by Dimisqi *below* the Equator. The notion that all the great rivers of Africa link with the Nile and have their sources in one vast central lake is a common Arab idea and it appears to exist as such, independent of that of the Mountains of the Moon and of the two symmetrically opposed lakes. We shall return to the origins of this idea further on.

The feature of the Mountains of the Moon and the two symmetrically opposed lakes is commonly supposed to appear first in Ptolemy's geography (ca. A.D.150). Sometime ago it was argued on the basis of a close study of the text that these were a later interpolation.[55] Quite recently it has been put forward that Ptolemy's *Geography* in the form in which we have it is not the work of one man, but that the text was compiled from various sources by an unknown Byzantine scholar of the 10th or 11th century, and from it the 26 maps were drawn by a Greek monk, Maximos Planudes, about 1300.[56] Only the general geographical theories derive from Ptolemy. We find, however, traces of the idea of the Mountains of the Moon already in classical literature, for Aristotle speaks of the Nile rising in the "Silver Mountains".[57]

We can see the connection between "silver", something shining, and the Arab word *Kamar* or *Qoumr. Qoumr* has been noted above as meaning "moon" or anything glistening. It seems likely also that the Mountains of the Moon had a cosmological significance such as the Atlas Mountains were supposed to have by the writers of Antiquity. The latter were held to be the support upon which the vault of heaven rested and which kept heaven and earth asunder.[58] On the Catalan world map of 1450,[59] there is a legend under the Atlas Mountains as follows: *Aquesta montana es appellada carana* p *sarains e* p *cristians mõtis claris* (= this mountain is called *carana* by the Saracens, and "bright mountains" by Christians).

On the map of Portugal in the Atlas containing the *Egyptus Novelo* map, the *Montes Claros* hills outside Lisbon are labelled *Lune Mõtis.*[60] Here one sees again the connection between *Kamar*, *claris*, or *Claros* (bright), and the moon. The cosmological significance of high mountains is not limited to Greek and Arab literature, but it is to be found also in Chinese literature, where one finds mention of the *K'un-lun* or *K'ouen-louen* mountains in the west. The word originally appears to have referred to the round vault of the sky, of which the gigantic Tibetan mountains seemed to be the prolongation or support. Etymologically, the word is connected with the binom *hun-lun*, meaning chaos.[61] The Chinese have applied the term to the native races of the countries round the Indian ocean, including the natives of South East Africa.[62]

[54] Ch. de la Roncière, *Découverte de l'Afrique au Moyen Age*, II (Cairo, 1925), p. 108, quoting A. F. Mehren, *Manuel de la Cosmographie du Moyen Age* (Copenhagen, 1874).

[55] Cf. William Desborough Cooley, *Claudius Ptolemy and the Nile* (London, 1854), p. 3.

[56] Cf. L. Bagrow, "The Origin of Ptolemy's Geography", *Geografiska Annaler* (Stockholm, 1945), Heft 3-4, pp. 318-387. See also suppl. notices in *Imago Mundi*, IV (1947), pp. 71-72.

[57] Aristotle, *Meteorology*, I, pp. 13 and 21.

[58] Cf. Homer, *Odyssey*, I, 54.

[59] Reproduced De la Roncière, *l.c.*, I, pl. X.

[60] Cf. Reproduction in H. Omont, *l.c.*, pl. VII.

[61] Cf. L. de Saussure, "L'Etymologie du Nom des Montagnes K'ouen-louen", *Toung Pao*, XX, p. 370.

[62] See J. J. L. Duyvendak, *China's Discovery of Africa* (London, 1949), p. 2.

FIG. 3. PART OF WALDSEEMÜLLER'S 1516 WORLD MAP (Wolfegg-Waldsee Schloss-Bibliothek)

De Saussure says it is probable that this word has a peculiarly Chinese meaning like that which associated the Atlas mountains with the foundations of the universe. The expression "K'ouen-louen", he adds, seems to mean "the heights at the end of the world".[63] The Chinese also placed their Paradise on the K'ouen-louen Mountains, from which they believed several rivers flowed.[64]

There further seems to be a complex association of the moon and mountains with cosmological and religious ideas suggesting something of the character of an archetypal myth in the human collective unconscious whose origin goes back far into early man's history. In Christian literature of the Middle Ages we find the Terrestrial Paradise traditionally situated on a mountain. Gervase of Tilbury, a writer of the Middle Ages, describes the Terrestrial Paradise as separated from our inhabited world by a long tract of land and sea elevated so high that it reaches the sphere of the moon, so high that the waters of the Deluge failed to disturb it.[65] Dante describes the Terrestrial Paradise as situated on the Mount of Purgatory in the southern hemisphere.[66] Numerous writers of the Middle Ages placed the Terrestrial Paradise in Africa. Their reason for this was founded on the Biblical tradition[67] of the four rivers that flowed from it: the Phison, the Euphrates, the Tigris and the Geon. The Phison was usually identified as the Ganges, and the Geon as the Nile.[68] Many geographers, impressed by the impossibility of representing these rivers as flowing from a single source, abandoned the task and arbitrarily placed the Terrestrial Paradise near the head waters of the Nile.[69]

It is interesting to compare Ibn Said's description of the Mountain Djabal al-Nadāma (Cape Correntes, the limit of Arab navigation) as being three days' journey high.[70] The *Libro del Conoscimiento de todos los Reynos* (Book of Knowledge of all Kingdoms), an anonymous work by a Spanish monk of the 14th century, places the Terrestrial Paradise in the "lofty mountains of the Antarctic pole",[71] but places the Mountains of the Moon in Senegambia.[72] Duarte Pacheco, in his *Esmeraldo in Situ Orbis* (ca. 1508), mistakenly taking Ptolemy for his authority, identifies the Mountains of the Moon with "the rocky mountains of the Cape of Good Hope".[73]

The Mountains of the Moon, it would then appear, have their origin in an archetype based more on man's need to find a symbolic expression to answer the cosmological problems of the universe, than on the mistaken interpretation of a geographical feature. Like the Terrestrial Paradise, the Mountains of the Moon were placed on the borders of the known world, and, as, with the progress of discovery, geographical horizons were pushed back, they retreated until, as we see in Duarte Pacheco, they could go no further than the Cape of Good Hope. After that, they retreated into the interior of the continent, where finally, in the 19th century, they became identified with the Ruwenzori range.

Giacomo Gastaldi's[74] Map of 1564[75] (Fig. 4)

This map is important, for it is a type map which later became the inspiration of other cartographers. Gastaldi published an earlier world map in 1546,[76] but it is a very crude one. In his 1546 map, the "two rivers" are shown, and so is the "third feature", but they are not in any way integrated with one another.

---

[63] L. de Saussure, *l.c.*

[64] A. Graf, *Miti, Legende e Superstizioni del Medio Evo* (Torino, 1892-93), p. xii.

[65] See J. K. Wright, *Geographical Lore in the Time of the Crusades* (New York, 1925), p. 261.

[66] Dante, *Purgatory*, Canto XXVIII.

[67] Genesis II, 10.

[68] In particular by Zaccaria Lilio, *Breviarum Orbis* (1493), quoted by Gustavo Uzielli, *Il Prete Gianni* (Florence, 1893), p. 9.

[69] *E.g.*, the Catalan world map of 1450, reproduced in G. H. T. Kimble, "The Catalan World Map of the R. Biblioteca Estense", *R. Geog. Soc. Reproductions of Early Maps*, II (London, 1934).

[70] In Youssouf Kamal, *Monumenta Cartographica Africae et Aegypti*, IV, fasc. 1 (Cairo, 1936), f. 1081.

[71] *The Book of Knowledge of All Kingdoms*, ed. by Sir Clements Markham (Hakluyt Society, 1912), p. 33.

[72] *Ibid.*, p. 34.

[73] Duarte Pacheco Pereira, *Esmeraldo in Situ Orbis*, English translation by G. H. T. Kimble (Hakluyt Society, 1937), p. 16.

[74] Sometimes called Jacopo Castaldi.

[75] One of the original copies of this map exists in the Paris Bibliothèque Nationale, Dépt. des Cartes et Plans, under Ge D D 5077. See Fig. 4.

[76] Reproduced in Youssouf Kamal, *loc. cit.*, V, fac. 1 (Cairo, 1951), f. 1522.

Fig. 4. Giacomo Gastaldi's Map of 1564 (Universitäts-Bibliothek, Basle)

Gastaldi is known to have produced a world map in 1561, but it has been lost.[77] He began to design his 1564 map about 1545.[78] He was obliged to revise it in accordance with the new facts which appeared concerning the geography of the interior of Africa in João de Barros' first *Década da Ásia*, published at Lisbon in 1552. The relevant chapter from Barros was extracted and appeared in Italian in the second edition of Ramusio's *Navigationi e Viaggi* published in Venice in 1554. Ramusio and Gastaldi were friends and co-workers.[79] It has been shown by Biasutti that Paolo Forlani, in his map of 1562 closely similar to Gastaldi's map of 1564 which appears to precede it, did no more than copy the latter, for Gastaldi's map, although unpublished, lay already completed in manuscript in 1562.

In his map of 1564, Gastaldi reveals scarcely any influence of the Abyssinian cartography of the *Egyptus Novelo* type; rather he is considerably influenced by João de Barros' chapter describing the Kingdom of Sofala, otherwise known as the Empire of *Monomotapa*. An extract from this chapter has been quoted above, but we will now quote it at greater length.

"The country which we treat of as the Kingdom of Sofala is a vast region ruled by a heathen prince called Benomotapa. It is surrounded like an island by two arms of a river which proceeds from the most notable lake in the whole continent of Africa, the knowledge of which was greatly desired by ancient writers, as it is the hidden source of the famous Nile and from it also issues our *Zaire* (river Congo), which flows through the Kingdom of Congo. On this side we may say that this great lake is nearer to the western than to the eastern ocean according to the situation given it by Ptolemy. On the side of the said Kingdom of Congo the following six rivers flow into it: *Bancare*, *Vamba*, *Cuylu*, *Bibi*, *Maria-Maria* and *Zanculo*,[80] all of which have abundant waters, besides others that have no name, which make it almost a sea, navigable by many vessels and containing an island which can send forth more than thirty thousand men to fight with those of the mainland. Of the three notable rivers which are at present known to proceed from this lake and which flow into the sea at a great distance from each other, that which has the longest course is the Nile, which the Abyssinians of the land of Prester John call *Tacui*,[81] into which two other notable rivers flow, which Ptolemy calls *Astabora* and *Astapus* and the natives *Tacazi*[82] and *Abanhi*.[83]

Though the *Abanhi*—which according to them means "father of waters", because of its great volume—proceeds from another lake called *Barcena*[84] and by Ptolemy *Coloa*, which also contains islands, in which there are monasteries of religious (as may be seen from our geography)[85] this lake is not to be compared with the other just mentioned, for, according to information obtained from Congo and Sofala, the latter appears to be more than a hundred leagues in length.

The river which flows towards Sofala, after leaving this lake and running some distance, divides into two arms, one of which reaches the Ocean on this side[86] of Cape Correntes, and is that which our people formerly called *Rio da Lagoa*,[87] and now Espírito Santo, which name was given to it by Lourenço Marques, who went to explore it in the year 1545.

The other arm enters the sea twenty-five leagues beyond Sofala and is called *Cuama*, although in the interior other people call it Zembere.[88] This arm has greater volume than that of Espírito Santo, being

[77] Cf. R. Almagià, "In torno ad un grande mappemondo perduto di Giacomo Gastaldi, 1561", *La Bibliofilia*, XLI (Florence, 1939), pp. 259-266.

[78] Cf. R. Biasutti, "La Carta dell' Africa di G. Gastaldi (1545-1564) e lo sviluppo della cartografia africana nei sec. 16 e 17", *Boll. Soc. Geogr. Ital.*, Serie V, IX (Rome, 1920), p. 417.

[79] Biasutti, *l.c.*, pp. 414-419.

[80] The Vamba is today the Wamba, which rises in the Lunda, district of Angola and joins the Kasai seventy miles above its confluence with the Congo. Cf. Crawford, *l.c.*, 23. The other rivers are unidentifiable.

[81] *Tacui* may be a misapplied name deriving from the place-name in Abyssinia, Dagosa or Taqusa (cf. C. F. Beckingham & G. W. B. Huntingford, *l.c.*, p. 243).

[82] *Tacazi* is the river Takaze.

[83] The *Abanhi* is the Blue Nile. One recognizes the above two rivers as the feeders of the Nile both rising from the two lakes situated above the Mountains of the Moon on the *Egyptus Novelo* map.

[84] *Barcena*—Arabic Bahr-Tsana—Lake Tana in Abyssinia.

[85] Never published, and lost.

[86] *I.e.*, south of Cape Correntes.

[87] This is now the Umbelasi river.

[88] *I.e.*, the Zambesi river.

navigable for more than two hundred and fifty leagues; and into it flow these six navigable rivers: *Panhames, Luanguoa, Arruya, Manjovo, Inadire,* and *Ruenia,* all of which drain the land of *Benomotapa,* and most of them carry down much gold which is formed there.

Thus with these two arms, and the sea on the other side, the great Kingdom of Sofala is an island, with a circumference of more than seven hundred and fifty leagues".[89]

In Barros' description the *Sachaf lacus* of Waldseemüller's maps appears to become one enormous central lake from which come all the great African rivers, the Nile, the Congo, the Zambesi and the Incomati-Limpopo. He makes no mention of the two symmetrically opposed lakes situated above the Mountains of the Moon, but introduces *Lake Barcena* (lake Tana). This lake is shown on Waldseemüller's 1507 map, where it is called *Colona,* a corruption of *Coloa.* Lake Tana contains many islands with churches or monasteries on them, but, in relation to the largest, that of Dek, which is three miles in diameter, they are none of them of comparable size.[90]

We have seen earlier in the Juan de la Cosa map the first cartographic evidence of the belief that the Congo rose in a central lake. João de Barros and Juan de la Cosa were not the only ones persuaded that the Congo flowed from a lake. In 1520, Gregorio de Quadra was sent by the king of Portugal to try to penetrate up the Congo to reach this lake with the intention of opening up a route to Abyssinia. He failed because of opposition from the king of Congo.[91] In 1536, Manuel Pacheco wrote to D. João II that the king of Congo had provided wood and given him permission to build two brigantines above "those falls which are in the river" so that he might reach the lake, and that he hoped to set out that year.[92] Nothing is known of the results of the expediton. Martin Fernandes Enciso wrote in 1519: "The people of *Manicongo* say that the river of *Manicongo* rises in the great and high Mountains of the Moon, in a great lake, and that from this same lake rises another river which is greater and which flows in another direction, and that this is the Nile".[93]

The idea of one immense central lake can be seen in numerous other writers of the 16th century. Alphonse de Saintonge, in his Cosmography of 1544, writes as follows: "The people of *Manicongue* say that the river of *Manicongue* comes from the Mountains of the Moon, as I have said, and from a great lake which lies at the foot of them and which is more than thirty leagues across. Most of the rainfall which falls on the southern slopes of these mountains runs into this lake. That which falls on the northern slopes runs into the river Nile. From the aforementioned lake another river flows in the other direction to the ocean in southern Ethiopia near Melinde. This river first runs into another lake from which at least two or three rivers flow to the southern Ethiopian coast." (coast of East Africa)[94]

In this text we can see the *Egyptus Novelo* cartography disguised by elements of the Arab theory of a central lake. The lake thirty leagues across situated *below* the Mountains of the Moon is really Waldseemüller's *Sachaf Lacus* (Lake Tana). The second lake is probably a misinterpretation of *Lake Zwai* in Abyssinia, and the river flowing into it the *Maki,* or *Machy.*[95] *Lake Zwai* with the *Maki* flowing into it is shown on the *Egyptus Novelo* map in the south western corner where it is labelled *Vabi lacus dulcis.* The "two or three other rivers" that flow to the coast from this lake probably reflect the three rivers shown on the Sanuto-Vesconte map of 1320,[96] as debouching on the east African coast and rising in a mountain in the interior.

Further empirical evidence of the existence of a central African lake in the 16th century is to be found in João dos Santos. He says of the Zambesi that the negroes "have a tradition from their ancestors that the

---

[89] João de Barros, *Décadas da Ásia,* Decade I, Book X, Chap. I (Lisbon, 1552), English transl. from G. McCall Theal, *Records of South East Africa,* VI (Cape Town, 1898) (amended by the author).
[90] Cf. Beckingham & Huntingford, *l.c.,* p. 243.
[91] Cf. Damião de Góis, *Crónica do Rei D. Manuel,* Pt. IV, Ch. LIV (Lisbon, 1556-7); and Jerónimo Osório, *De Rebus Emmanuelis,* Book XII, III (Lisbon, 1571) (p. 300, in Filinto's Port. transl. (Lisbon, 1804)).
[92] Cf. Paiva Manso, *Documentos para a História do Congo* (Lisbon, 1877), doc. XXXVIII, p. 66.
[93] Martin Fernandes Enciso, *Suma de Geographia* (Seville, 1519), p. 110.
[94] Alphonse de Saintonge, *Cosmographie* (1544), ed. by G. Musset, Collection Carl Schefer (Paris, 1904), p. 341.
[95] Cf. the sketch map in Beckingham and Huntingford, *l.c.,* p. xcii.
[96] Reproduced in Kammerer, *Mer Rouge,* II, part 2, pl. cxxxv.

river rises in a great lake in the middle of Ethiopia (Africa) from which rise other great rivers which flow in all directions, each having a different name. In the middle of this lake are many islands, peopled by negroes and which are rich with animals and food."[97] On his evidence, the myth of a central lake would appear to be common to the primitive tribes of Africa as well as to the Arabs.

It would seem possible that the Arab conception of a central lake from which all the rivers of Africa flow finds confirmation not only in the myths of African tribes, but also a primitive expression in classical literature as *Lake Tritonis*. The lake, which is associated by Herodotus with the Atlas Mountains, is at times associated with Lake Tana in Abyssinia, and at times becomes a vast central African lake which is later doubled, tripled, or even quadrupled, in the cartography of the 16th century.

We find yet further confirmation of the myth of a vast central African lake on a Chinese world map of ca. 1300 A.D.[98] This map, which, curiously enough, gives a triangular shape to the African continent, shows a central lake with an island in it so large that the continent is scarcely more than a ring round it. A river leads into it from the south just as the Abay leads into Lake Tana. To the east is a long river running nearly the whole length of the continent, dividing in the south into two forks as its sources.[99] This can only be the Nile. A similar map to this also showing a large central lake is another Chinese world map of 1402.[100] It seems that we are to be led to the conclusion that Lake Tana in Abyssinia is related to an archetypal myth present in Arab, Greek, Chinese, and Negro civilisation, of a vast lake in the African interior. How much is due to Lake Tana for the setting in motion of the life of this myth, and how much it owes its origin to other sources, is something we cannot yet be sure about.

We have now seen the chain of influences behind Gastaldi's cartography, the principal emphasis of which is on a vast central African lake which is placed slightly nearer the west coast than the eastern in accordance with Barros' note that it is "nearer the western than to the eastern ocean". Gastaldi, however, shows a second smaller lake symmetrically opposed to the main large one, and which is obviously the source for Barros' river, the Tacazi, although Barros does not mention this lake. Gastaldi calls this lake *Zaflam*, the name of which, as we have seen, comes from the title of the ruler of the Abyssinian province of Shoa—Sahafa Lam.[101] This representation of a second smaller lake, symmetrically opposed to the large one, shows that Gastaldi remained still faithful to Ptolemy. We now have Ptolemy's two symmetrically opposed lakes situated above the Mountains of the Moon. It becomes apparent what a completely different direction the development of central African cartography now takes with Gastaldi from that taken with Waldseemüller influenced by the *Egyptus Novelo* map. With Waldseemüller we have two symmetrically opposed Ptolemaic lakes with a third lake situated to the west and south of the Mountains of the Moon, which feeds the Nile, and one river (*the Gomormager*—Zambesi) debouching on to the east coast. With Gastaldi we have the same two symmetrically opposed Ptolemaic lakes with the more western enlarged in accordance with the tradition of a vast lake in the African interior. He makes rise from it the Nile, the Congo, the Zambesi, and the Zavora river[102] (and here, in this latter, Gastaldi parts company with Barros).

There is yet another lake to be considered on Gastaldi's map. This is the small lake well to the south of Barros' central lake. From it flows a river debouching on to the coast, where it is called *Infante fl.* (Rio do Infante). This lake is really Lake Zway in Abyssinia. In actual fact, no river flows from Lake Zway to the sea. Only the river Maki, or Machi, flows into it. Gastaldi's lake does indeed have a river flowing into it rising in the mountains north of it. How, then, did Lake Zway come to be linked to the sea? If we look at Fra Mauro's map of 1460,[103] we find a river called the *Xebe* running along the southern edge of the map.

[97] João dos Santos, *Ethiopia Oriental* (Évora, 1609), Book II, Chap. II (I, p. 165 in ed. 1891).

[98] Walter Fuchs, *The Mongol Atlas* (*Monumenta Serica*, Monograph VIII), map 44. Bibl. Nat., Paris, Dept. Mss. (8° Imp-Or 1303-viii).

[99] Cf. The Nile forking to the two lakes from which it rises at the base of the Mountains of the Moon in Ptolemy.

[100] Walter Fuchs, "Was South Africa known in the 13th century?", *Imago Mundi*, X (1953), p. 50, reprod. of map.

[101] Esteves Pereira, *Chronica de Susenyos*, II (Lisbon, 1900), p. 211.

[102] That the river shown by Gastaldi is the Zavora can be proved from the legend at its mouth *Aguada de boa paz*. The *Aguada de boa paz* is identified by E. G. Ravenstein with the Zavora river.

[103] Reproduced in De la Roncière, *l.c.*, pl. xxxvi.

It rises in a lake marked *Sacnaelt* and debouches into the canal which divides the island of Diab from the African continent. In its lower course it is called the *Galla*. On Gastaldi's map, we find the name *Gale* near where it leaves the lake. The word itself refers to the Galla tribe. This river is the *Mogadishu* river of Dimisqi, and probably also Ptolemy's *Raptus fl.* It is difficult to say which modern river it corresponds to. It may represent a complex of both the Jub, the Webi Shabelle, and the Ghibbe, all confused as one river. There is no evidence to show how it came to be thought that it had its source in Lake Zway in particular, except the probable application of the Arab "a priori" belief in African rivers rising in a central lake. We find the presence of this lake and river, albeit somewhat blurred, in the text quoted from Alphonse de Saintonge's Cosmography. He, however, speaks of a lake from which "at least two or three rivers" flow to the east African coast; but we have shown that the increase of the number of these rivers to three can be traced back to the Sanuto-Vesconte map of 1321. Lake Zway with the Maki flowing into it is shown on the *Egyptus Novelo* map, where it is called *Vabi lacus dulcis*,[104] and on Waldseemüller's map of 1516, where it has the same name. In neither of these maps, however, is it linked by a river to the sea.

Now, on Gastaldi's map we find Lake Zway linked by the Jub-Ghibbe-Webi Shabelle river to the coast where the latter is equated with the Rio do Infante in accordance with the attempt practised by all cartographers of the 16th century (and already seen in Waldseemüller's map of 1516) to correlate the nautical charts with the cartography of Abyssinia. Below this lake we find the place-name Cumissa, which is declared by Biasutti as a non-Ethiopian name and which is unidentifiable.[105] Along the south east African coast we find a town "Cortada". This means "Inlet" in Portuguese and refers only to a landmark on the coast.[106] Next along the coast we find a town *Alagoa*. This refers to Delagoa Bay. [107] Up the *Alagoa* river (Umbelasi river) we find *Belugaras*, and northwest of it *Bauagul*. The latter sometimes appears on later maps as *Davagul*.[108] Both names are unidentifiable and are not Portuguese or negro words. They are probably Abyssinian.

Continuing our examination of Gastaldi's map, we shall now work south-east, following the northern "watershed" of the Cuama-Zambesi "basin". South of the *Regno de Amazone* we find the *Regno de Armeta* and a place called *Armeta*. This is unidentifiable, but probably a place in Abyssinia.

South-east of this we find *Regno de Maitachazi* and a place called *Maitachazi*. The word *Maitachazi* is compounded of the Abyssinian word *Mai*, meaning water, and *tachazi* the river Takaze in Abyssinia. North-east of *Maitachazi* we find *Regno de Camur* and a place *Camur*. This may refer to the Khamir tribe, who live in the Lasta mountains of Abyssinia.[109] East of Camur appears *Regno de Agag* and a place *Agag*. Agag refers to a collective name for certain Abyssinian tribes, the Agau.[110]

East of *Agag* we are brought once more into proportion with *Regno de Monzabique*—Kingdom of Mozambique. South of this appear the unidentifiable rivers, the *Viado*, the *Vindo*, and the *Basso*. The second may be a duplication of the first. Coming down the coast, we find *Angos*, the modern Angoshe; *opadram*, the ***padrão*** of Saint Raphael erected by Vasco da Gama on his second voyage. There are two places which are unidentifiable, *Marsuzo* and *Zuaroa*. Neither of these two places, nor the three rivers, *Viado*, *Vindo*, and *Basso*, appear on any nautical chart of the period before 1530.[111] At the mouth of the Zambesi-Cuama we find *Mongallo*. This place is first mentioned by Duarte Barbosa as *Mangaló*.[112]

The place-names in the region circumscribed by the *Cuama*-Zambesi and Spirito Santo rivers are practically entirely taken from Barros' chapter on *Monomotapa*.[113] This region Barros describes as being "like an island", having a circumference of seven hundred and fifty leagues. The Portuguese league equals 3,2

---

[104] Cf. Vabi and Webi.
[105] Biasutti, *l.c.*, p. 403.
[106] Cf. João de Lisboa, *Livro de Marinharia*, ca. 1514, first published in Lisbon (1903), p. 162: "esta cortada".
[107] Cf. *ibid.*: "Hũa grãde alagoa co hũ rio para o sertão".
[108] *E.g.*, Petrus Plancius' map of Africa of 1594.
[109] Cf. Beckingham and Huntingford, *l.c.*, p. 56, note 5.
[110] Cf. *ibid.*, and also Th. Simar, *Le Congo au 16e siècle* (Brussels, 1919), p. 67.
[111] Cf. Kammerer, *l.c.*, II, part 2, app. I.
[112] Cf. *The Book of Duarte Barbosa*, transl. by Mansel Longworth Dames, I (London, 1918), p. 14.
[113] João de Barros, *Décadas da Ásia* (Lisbon, 1552), Decade I, Book X, Chapt. I.

miles.[114] This gives an average circumference of approximately 2400 miles, and, therefore, an average "diameter" of approximately 760 miles. How Barros arrived at his figure we have no means of knowing, nor can we guess what persuaded him to use the simile of an island. However, comparison of our calculated diameter with a modern map shows that Barros' seven hundred and fifty leagues reflects at least a fairly accurate appreciation of the southward bend of the Zambesi in its upper reaches. Barros declares the Zambesi to be navigable for two hundred and fifty leagues, i.e., 940 miles.[115]

Gastaldi's cartography formed the basis of many other cartographers' maps of Africa in the 16th century. Among those influenced by him, particularly those who followed his hydrographical system, are Abraham Ortelius in his world map of 1564[116] and in his map of Africa of 1570 (a popular map with collectors as so many were printed), Livio Sanuto in his map of Africa in his *Geografia*, published in Venice in 1588,[117] Berteli in his very crude map of Africa of 1565[118] (this map is often found in editions of Ramusio's *Navigationi e Viaggi*), and Fausto Rughesi in his map of 1597.[119]

Gerard Mercator's World Map of 1569[120] (Fig. 5)

Gerard Mercator published four cartographic works in the 16th century, of which only his world map of 1569 is of significant interest for the cartography of the African interior. His first work was a world map published in 1538.[121] Central Africa is blank except for the Congo, shown rising in a lake. There is no connection shown with the Nile. In 1541 he produced a globe.[122] On it the central African cartography, considerably developed on his 1538 map, is manifestly copied from Waldseemüller's 1516 map. There is one exception, however. The interior river, which on Waldseemüller's map is joined to the *Sachaf lacus*, is broken a short way north of the *Sachaf lacus*, giving the impression that the Congo, the interior river, and the *Sachaf lacus* all have their sources in this region north of the *Sachaf lacus*. Next among Mercator's works comes his world map of 1569.

Mercator's principal contribution to the development of the cartography of central and south-east Africa was in attempting to harmonise the Abyssinian cartography of the Waldseemüller (1516) tradition with that of Gastaldi (1564). The result was that the already complicated network of rivers and lakes (the legacy of Waldseemüller rather than Gastaldi) was made ever more complicated and tortuous. The cartography of Africa such as Mercator left it in 1569 was repeated by some mapmakers with little modification, even late in the 18th century, and was dutifully copied by most of the great Dutch school of mapmakers in the 17th century.

Though this map is an attempt to harmonise Waldseemüller with Gastaldi, the Waldseemüller tradition remained the strongest influence, and knowledge of Gastaldi's 1564 map only results in the introduction of modifications on a thoroughly Waldseemüller framework.

Mercator accepts the Barros-Gastaldi notion of the Cuama-Zambesi, the Congo, and the interior river (in Barros the Nile) as rising in the central lake. He does not accept Gastaldi's identification of the Espírito Santo with the Zavora river. Instead, he follows Barros' identification of it with the Incomati-Limpopo river. He preserves Waldseemüller's *Sachaf lacus* as Barros' central lake instead of, like Gastaldi, making the latter one of Ptolemy's symmetrically opposed lakes. The *Rio do Infante* is shown, as in Gastaldi, linked to the *Vabi lacus* in the interior, though the latter was not so named. Another small lake is shown in the course of the Espírito Santo which is not mentioned by Barros or shown by Gastaldi. This lake

114 Cf. Beckingham and Huntingford, *l.c.*, p. xcvi.

115 João de Barros, *l.c.*

116 Only known copy preserved in University Library of Basle. Fascimile reproduction in Bibliothèque Nationale, Paris Dépt. des Cartes et Plans, under (Ge. DD. 3018).

117 Sanuto does not show the Rio do Infante rising in a lake.

118 Reproduced in Kammerer, *l.c.*, III, part 3, fig. 103, p. 198.

119 Reproduced in R. Almagià, *Monumenta Cartographica Vaticana*, II (Rome, 1948), tab. XXI.

120 Original copy in Bibl. Nat., Paris, Dépt. des Cartes et Plans under (Rés. Ge A 1064). The map has been reproduced and published by the Service Hydrographique de Monaco, 1932, also in *Imago Mundi*, XII. See Fig. 5.

121 Reproduced in Nordenskiöld, *Facsimile Atlas*, Pl. xliii.

122 Reproduced in Youssouf Kamal, *l.c.*, V, fasc. 1, f. 1521.

appears on the *Egyptus Novelo* map where it is fed by the *Abia fl.* It reappears also on Waldseemüller's 1516 map. The *Abia fl.* probably reflects the Abay or the little Abay in Abyssinia, and the lake is just another repetition of Lake Tana. Mercator appears to have joined the Espírito Santo to it quite arbitrarily.

On the whole, Mercator's cartography of South-east Africa is a regression on that of Gastaldi. The Abyssinian cartography is projected far further south than in Gastaldi, and Abyssinian place-names are heedlessly intermixed with those from Barros' chapter on Monomotapa.[123] The Cuama-Zambesi is confused with the Sofala river and made to debouch at Sofala. The real Cuama is labelled Zuama and shown debouching north of Mongalo. The course of the false Cuama seems to be dictated by the map of Abyssinia used by Waldseemüller rather than by any study of Barros' chapter on Monomotapa.

The following is a list of place-names on Mercator's 1569 map with identifications where possible. The origins of names first found on Gastaldi's 1564 map are not considered, as these have been already dealt with above.

"Adia" — on Mercator's 1541 globe, on Waldseemüller's 1516 map, on *Egyptus Novelo* map = province of Hadya in Abyssinia.[124]

"Agrasele" — on Mercator's 1541 globe, on Waldseemüller's 1516 map, as "Acnase-gue"; on *Egyptus Novelo* map as "Achaia Thegie"?

"Alagoa" — cf. Gastaldi's 1564 map.

"Amara" — on Mercator's 1541 globe, on Waldseemüller's 1516 map, on *Egyptus Novelo* map = Amhara in Abyssinia.

"Armeta" — unidentifiable. Cf. Gastaldi's 1564 map.

"Augesa" — on Mercator's 1541 globe, on Waldseemüller's 1516 map as "Argesa" = Achefer, a district immediately south of Lake Tana.[125]

"Bafat" — *not* on Mercator's 1541 globe, on Waldseemüller's 1516 map as "Gafat", on *Egyptus Novelo* map as "Gaffat".

"Baía Ferrada" — This may mean "Horseshoe Bay" in Portuguese.

"Beif" — on Waldseemüller's 1516 map as "Beit", a corruption of "Zaet", repeated elsewhere on the same map? Cf. C. F. Beckingham & G. W. B. Huntingford, *Some Records of Ethiopia*, (Hakluyt Society, London, 1954), p. 10.

"Belugaras" — cf. Gastaldi's 1564 map.

"Benamataxa" — on Gastaldi's 1564 map as "Benamatapa".

"Bera" — "Dera" on Waldseemüller's 1516 map and on *Egyptus Novelo* map.

"Buda" — *not* on Mercator's 1569 map, on Mercator's 1541 globe, on Waldseemüller's 1516 map and on *Egyptus Novelo.*

"Buro mina de oro ubi est aurifodina" — Cf. Gastaldi's 1564 map.

"Butua" — on Gastaldi's 1564 map.

"Calburas" — Cf. Gastaldi's 1564 map.

"Camur" — Cf. Gastaldi's 1564 map.

"Cast. Portugal" — Origin untraceable. O. G. S. Crawford speaks of a 17th-century Portuguese castle at Gondar, but this would be, anyway, north of Lake Tana. The legend may refer to some fort built during the expedition of Cristóvão da Gama in 1542.

"Cible" — On Mercator's 1541 globe, a corruption of the Arabic "Djabal" (= "mountain").

"Cortada" — cf. Gastaldi's 1564 map.

"Cumissa" — cf. Gastaldi's 1564 map.

"Dauagul" — cf. Gastaldi's 1564 map.

"Dege" — On Mercator's 1541 globe, on Waldseemüller's 1516 map, on *Egyptus Novelo* map as "Dego".

"Deger" — On Mercator's 1541 globe, on Waldseemüller's 1516 map as "Degheri", on *Egyptus Novelo* map as "Degreti".

---

123 João de Barros, *Décadas da Ásia*, Dec. I, Book X, Chapt. I.

124 Cf. Beckingham and Huntingford, *Some Records of Ethiopia* (London, 1954), map on page xx.

125 Cf. *ibid.*, p. 226.

Fig. 5. From Mercator's 1569 Map
(Universitäts-Bibliothek, Basle)

"Degme" — On Mercator's 1541 globe, on Waldseemüller's 1516 map as "degino"? — unidentifiable.
"Derazara" — On Mercator's 1541 globe, on Waldseemüller's 1516 map as two separate places, "Dera" and "Zara". "Zara" = "Gara" on *Egyptus Novelo* map. "Dera" = modern Dara.[126]
"Doldel" — On Mercator's 1541 globe, on Waldseemüller's 1516 map, on *Egyptus Novelo* map as "Delda". Today Dildil, near Lake Tana.[127]
"Dosa" — On Waldseemüller's 1516 map as "Daun"?
"Euggi" — not on Mercator's 1541 globe, on Waldseemüller's 1516 map as "Heuggi", on *Egyptus Novelo* map as "Huoggi". Today Wagē, a province in Abyssinia.
"Gale" — Cf. Gastaldi's 1564 map.
"Gamba" — On Mercator's 1541 globe as "Amba", on Waldseemüller's 1516 map as "Amba" = Amba Geshen ("Ghisen" on *Egyptus Novelo* map) in Amhara, in Abyssinia.[128]
"Garma" — not on Mercator's 1541 globe, on Waldseemüller's 1516 map, on *Egyptus Novelo* map.
"Gebage" — on Mercator's 1541 globe, on Waldseemüller's 1516 map, "Geberge" on *Egyptus Novelo* map.
"Giero" — On Mercator's 1541 globe as "Ghiero", on Waldseemüller's 1516 map as "Ghiera", on *Egyptus Novelo* map as "Gara".
"Hiera" — On Mercator's 1541 globe, on Waldseemüller's 1516 map as "Hiere". Same as "Giero"? See above.
"Ionros" — On Mercator's 1541 globe, on Waldseemüller's 1516 map as "Huorob", on *Egyptus Novelo* map as "Huorab". Today Guerāgē, a province in Abyssinia.
"Kagala" — On Mercator's 1541 globe as "Hagala", on Waldseemüller's 1516 map as "Hagala", on *Egyptus Novelo* map as "Huadla".
"Kobal" — On Mercator's 1541 globe as "Gobal", on Waldseemüller's 1516 map as "Gobal", unidentifiable on *Egyptus Novelo* map, a corruption of the Arabic "Djabal" = mountain?
"Maitagasi" — On Gastaldi's 1564 map.
"Mara" — On Mercator's 1541 globe as "Mara", on Waldseemüller's 1516 map as "Mara", unidentifiable on *Egyptus Novelo* map.
"Meacho" — On Mercator's 1541 globe, on Waldseemüller's 1516 map, on *Egyptus Novelo* map as "Meath".
"Mogar" — on Mercator's 1541 globe, on Waldseemüller's 1516 map = modern "Muger". [129]
"Mongalo" — Cf. Gastaldi's 1564 map.
"Mossara" — On Mercator's 1541 globe, on Waldseemüller's 1516 map as "Mossata", on *Egyptus Novelo* map as "Massara Fluvius".
"Muhamabana" — This legend does not appear on any previous maps. It refers to the negro kingdom of Inhambane, in the hinterland of Cape Correntes, and the legend as such is considerably displaced.
"Quiticui" — Cf. Gastaldi's 1564 map.
"Ruenia" — Cf. Gastaldi's 1564 map.
"Samot" — On Mercator's 1541 globe, on Waldseemüller's 1516 map as "damot", on *Egyptus Novelo* map as "Damot".
"Tialso" — On Mercator's 1541 globe, on Waldseemüller's 1516 map as "Tasso", on *Egyptus Novelo* map as "Tarassa".
"Totama" — On Mercator's 1541 globe, on Waldseemüller's 1516 map, on *Egyptus Novelo* map as "Totumaa".
"Valonta" — On Mercator's 1541 globe, on Waldseemüller's 1516 map as "Vallama", on *Egyptus Novelo* map as "Vamo"?
"Vigiti Magna" — First appears on Contarini's 1506 map as "Vigict Magna" (cf. supra, p. 70).

[126] Cf. Beckingham & Huntingford, *Some Records of Ethiopia* (London, 1954), p. 233.
[127] Cf. O. G. S. Crawford, "Some Medieval Theories about the Nile", *Geographical Journal* (London), Sept. 1949, p. 12.
[128] Cf. Beckingham & Huntingford, *l.c.*, p. 227.
[129] Cf. Crawford, "Some Medieval Theories about the Nile", in *Geographical Journal*, Sept. 1949, p. 12.

"Zedaia" — On Mercator's 1541 globe, on Waldseemüller's 1516 map, on *Egyptus Novelo* map as "Sedaicion".
"Zephal" — On Mercator's 1541 globe as "Zephala", on Waldseemüller's 1516 map as "Zefalo" = modern Sofala.
"Zet" = Zethe, a province in Abyssinia.
"Zimbaos" — On Gastaldi's 1564 map as "Simbaoe".
"Zimbro" — Cf. Gastaldi's 1564 map.

Petrus Plancius' Planisphere of 1592[130]

After Mercator, the only significant modification in the 16th century of the tradition he had followed of the cartography of south-east Africa was introduced by Petrus Plancius in his planisphere of 1592. Plancius' contribution was to restore the features of Gastaldi's cartography at the expense of the Waldseemüller tradition. The more easterly of the two Ptolemaic symmetrically opposed lakes becomes once more the vast central lake. The *Sachaf lacus* he separated from the Congo-Nile system and reduced to insignificant proportions. As a result, the Cuama-Zambesi and the Espírito Santo are no longer linked to the Nile system. Thus, the notion that all these four rivers rise in the central lake, a basic feature of Barros' and Gastaldi's African hydrography, was abolished. Plancius otherwise retains the "macédoine" of place-names all over South-East Africa, left by Mercator.

Plancius' cartography was followed by the great Dutch map-makers Blaeu and Jansson.

Filippo Pigafetta's Map of Africa of 1591

Filippo Pigafetta's map, published in his book *Relatione del Reame di Congo*,[131] is quite unrelated to the traditions of either Waldseemüller or Gastaldi. Pigafetta based his book on information received in person from a Portuguese, Duarte Lopes, to whom he refers in his text and who had visited the Congo. Though he reports Duarte Lopes' views on the hydrography of central Africa, he does not follow them. Lopes' hydrography followed the theory of the single vast central African lake. Pigafetta tells us that "... in this region (in the latitude of Ptolemy's lakes and of the Kingdom of Congo and the Empire of Monomotapa), Lopes asserted that there is but one lake, which lies on the borders of Angola and Moçambique, and is 195 miles in diameter. Of this lake the people of Angola give information respecting its western side, and those of Sofala and Monomotapa of its eastern".[132] Lopes, as can be seen, was influenced by the same sources as João de Barros, though it is hard to explain the figure of 195 miles. Pigafetta then continues to give his view of the hydrography of central Africa. "It is true", he says, "that there are two lakes, but situated in quite a contrary direction to that of which Ptolemy writes; for he, as has been said, places his evidently from west to east, and those we now see lie almost in a direct line north and south, and nearly 400 miles apart. ... The Nile certainly flows from the first lake, which lies in latitude 12 degrees south, and is like a shell, and surrounded by very lofty mountains, and the largest of which, called Cafates, are on the east, and on both sides are mountains producing saltpetre and silver. The Nile flows thence 400 miles due north, and enters another very large lake, which the natives call a sea. It is larger than the first, for it is 220 miles across, and it lies under the equinoctial line. Regarding this second lake very certain information is given by the Anzichi, near the Congo, who trade in these parts, and say there are people who sail on the lake in large ships, and who write, and use weights and measures, such as they have not in Congo, and that their houses were built of stone and lime and their customs resembled those of the Portuguese. Hence it might be suggested the Empire of Prester John was not far off. From this second lake the Nile flows to the Island of Meroe, a distance of 700 miles, other rivers running into it, the principal of which is the river Colves, so called because it issues from a lake of that name on the borders of Melinde. When the Nile reaches Meroe it divides into two branches, and encompasses a high land, called *Meroe*, to the

[130] Reproduced in F. C. Wieder, *Monumenta Cartographica*, II.
[131] Filippo Pigafetta, *Relatione del Reame di Congo* (Rome, 1591).
[132] Cf. *ibid.*, English translation by Margarite Hutchinson (London, 1881), p. 129.

right of which on the east, is a river named *Abagni*, which rises in the lake *Bracina* and traverses the Empire of Prester John, till it reaches the aforesaid island."[133]

Reading between the lines of this text, Pigafetta's southern lake would seem to derive from Waldseemüller's *Sachaf Lacus*. The reasons for this are that the Mountains of the Moon are shown below it, and it is bounded on the east by the *Cafates* Mountains. The name *Cafates* probably derives from "Gafat", a province in Abyssinia south of the Abay adjoining Damot.[134] The name *Cafates* is mentioned by Francisco Alvares.[135] Pigafetta's northern lake would seem to derive from the vast central lake of João de Barros and Gastaldi.

If we now take Pigafetta's map and compare it with the anonymous portolan ascribed to Pedro de Lemos,[136] we can see that the system of Pigafetta's hydrography was not constructed from separate data taken from diverse sources. Pedro de Lemos' portolan is a curious work, for none of the rivers leading from the lakes are produced to the coast. Did he perhaps consider his central African hydrography as no more than speculative ornamentation and resist the temptation of linking imagination with reality by joining the rivers to the coast as did Pigafetta with a certain rashness later? We know very little about Pedro de Lemos except that there is a signed map of his dated 1594 in the British Museum[137] representing the coasts of Western Europe and of the Mediterranean as far as Sicily and of Africa as far as the Gulf of Guinea.[138] The portolan from which Pigafetta may have drawn carries no date and is dated by Kammerer ca. 1590, though he is only certain that it must have been completed after 1580. Such a dating would give ample possibility for Pigafetta to have benefited from it. It is however more likely that Pedro de Lemos' central African hydrography was not his private contribution but may have existed already as a tradition on other Portuguese manuscript maps, from which Pigafetta may equally well have drawn. That Pigafetta copied from a more elaborated Portuguese map than Pedro de Lemos' portolan is suggested by the fact that the language of the place-names in the African interior on Pigafetta's map is Portuguese and not Italian. Since Pedro de Lemos' portolan is the latest known manuscript map of a Portuguese map-maker which carries any elaborated system of central African hydrography, we must accept it as the highest advance made by the Portuguese in their speculative theories on the nature of the river systems of central Africa before the period of maps of the Duchesse de Berry type.[139] That there existed a tradition of a central African hydrography of the type represented in Pedro de Lemos' portolan is confirmed by examination of the world map of Bartolomeu Velho of 1561.[140] Besides an almost identical arrangment of the lakes and river systems, a number of place-names on Bartolomeu Velho's world map reappear on Pigafetta's map in their original Portuguese forms.

As on Pigafetta's map, Bartolomeu Velho shows two central lakes one above the other, both linked together by an interior river. Only the Cuama-Zambesi (and not the Espírito-Santo-Limpopo) is linked to the lower lake, thus breaking with the Barros tradition otherwise here followed by Pigafetta. Instead the *Rio Cuauo* links the lower lake to the east coast. This river would be today either the Rufiji or the Matandu. Just west of the Rufiji, modern maps show a very small lake. The Congo is shown joined to both lakes by two arms bearing the same name—"Rio de Manicomguo". Pigafetta's two small lakes in the Congo and in Angola, Lake Aqualuna and Quinbebe are not shown by Bartolomeu Velho.[141]

Pigafetta scatters the Abyssinian place-names of Waldseemüller and Gastaldi with as little attention to their southward displacement as Waldseemüller had already shown himself guilty of.

---

133 Cf. *op. cit.*, pp. 129-130.

134 Cf. Beckingham & Huntingford, *l.c.*, p. 235.

135 Francisco Alvares, *A Verdadeira Informaçam das Terras de Prester João*, 1st ed. (Lisbon, 1540), Chapt. XXXIIII.

136 Original in Bibliothèque Nationale, Paris, Dépt. des Cartes et Plans, Depôt des Cartes de la Marine, Portef. 1, Div. O, no. 4. Reproduced in A. Kammerer, *l.c.*, pl. LXXXV.

137 British Museum (Cotton Ms. Aug. I.i. 15).

138 Cf. A. Kammerer, *l.c.*, III, Part 2, p. 216.

139 Reproduced in A. Cortesão, *Cartografia e Cartógrafos Portugueses dos Séculos XV e XVI* (Lisbon, 1935), pl. LVI.

140 Original in the Istituto di Belle Arti, Florence. Two separately made reproductions (poor) exist in the Biblioteca da Marinha, Lisbon.

141 Th. Simar identifies Lake Aqualuna (or Achelunda as it is spelt in Pigafetta's text) with Lake Yanga Kulu, a small lake from which flows the Kugo, an affluent of the Kwango (cf. Th. Simar, *Le Congo au XVIe siècle* (Brussels, 1919), p. 51).

That Pigafetta had no direct knowledge of Bartolomeu Velho's world map is revealed by the fact that a number of place-names in ordinary script (not in block capitals) appear along the Zambesi on the latter's world map and none of these are to be found on Pigafetta's map. The presence of these names on Bartolomeu Velho's world map makes it the first map to carry any accurate information on the interior of the African continent. The Zambesi basin is the only part of the map on which place-names in the interior are shown.

It is interesting to note that nowhere on either Pigafetta's map or on Bartolomeu Velho's world map do the names of Tete and Sena appear. There is no known document which states when these settlements were discovered. Their names are not mentioned by João de Barros in his chapter on Monomotapa. The Portuguese settled at Sena and Tete ca. 1531.[142]

There is one intriguing name on Bartolomeu Velho's world map. This is the *R. dimbili* leading into the lower central lake. This is an African and not an Abyssinian name. It suggests that the origins of the representation of the lower central lake are not altogether mythical nor derive from distorted Abyssinian sources. However, from its positioning the name would appear to have a Congo rather than a South-east African origin. We cannot be certain to what it refers.

Conclusion

In the preceding pages we have seen how, at the beginning of the 16th century, the idea of the full and precise extent of the African continent installed itself in the European mind, and how the map-makers of the early 16th century (Waldseemüller and Stabius) took the rash step of trying to integrate the relatively scaleless cartography of Abyssinia of the 15th century with the Portuguese nautical charts. The resultant distortion was to persist in some cases until late into the 18th century.

The first to realize that something was wrong was Isaac Vossius, who in his book (with accompanying map) *De Nili et Aliorum Fluviorum Origine*[143] cleared away the whole overgrown grapevine of rivers, lakes, and place-names and made the African continent once more blank, mysterious, and "incognita". Of the maps of his time he wrote: "Plerae que itaque mappae Geographicae hodienae quae partim ex hoc (Leo Africanus), partim ex Francisco Alvares, admixtis etiam nominibus antiquis Ptolemaei, sunt compilatae adeo sunt ineptae et nullo judicio digestae ut ne unum quidem vocabulum recte locatum in iis invenies".[144]

After Vossius, De l'Isle (1708)[145] and D'Anville (1727)[146] both published maps of Africa. Their cartography of South-east Africa was based on João dos Santos' book *Ethiopia Oriental*,[147] the first volume of which was translated into French by Gaetan Charpy in 1684 and which appeared again in 1688.

In 1749 D'Anville published his critical map of Africa which while still further reducing the supposed revelation of the interior on both the west and the east coasts to more accurate proportions, really went to prove that the African continent was much more vast and extensive than had hitherto been admitted. In 1752 he wrote as follows: "... the sources of the Nile are still hidden to us, in a part of the interior of Africa further removed from that of which we have recently acquired some knowledge".[148]

After D'Anville's map of 1749 came the 19th-century's attack on the unexplored mysteries of the African continent.

---

142 Cf. George McCall Theal, *Records of South-east Africa*, VII (Cape Town, 1898), p. 486.

143 Cf. Isaac Vossius, *De Nili et Aliorum Fluviorum Origine* (The Hague, 1666). His map of Africa is reproduced in part (as far south as the Congo) in Youssouf Kamal, *Monumenta Cartographica Africana*, V., fasc. 1 (Cairo, 1951), f. 1555.

144 Cf. Isaac Vossius, *op. cit.*, p. 70, quoted by Th. Simar in *Le Congo au XVIe siècle* (Brussels, 1919), p. 45 (Simar considers this judgement excessive).

145 Reproduced in Youssouf Kamal, *l.c.*, V, fasc. 1, f. 1568.

146 Reproduced in *l.c.*, V, fasc. 2, f. 1580.

147 Cf. João dos Santos, *Ethiopia Oriental* (Evora, 1609).

148 D'Anville, "Dissertation sur les sources du Nil pour prouver qu'on ne les a pas encore découvertes", *Mém. Acad. Inscript. et Bell. Lettres*, XXVI (Paris, 1752), p. 46.

INDEX